HOLLEY

Rebuilding And Modifying

By
Jeff Williams

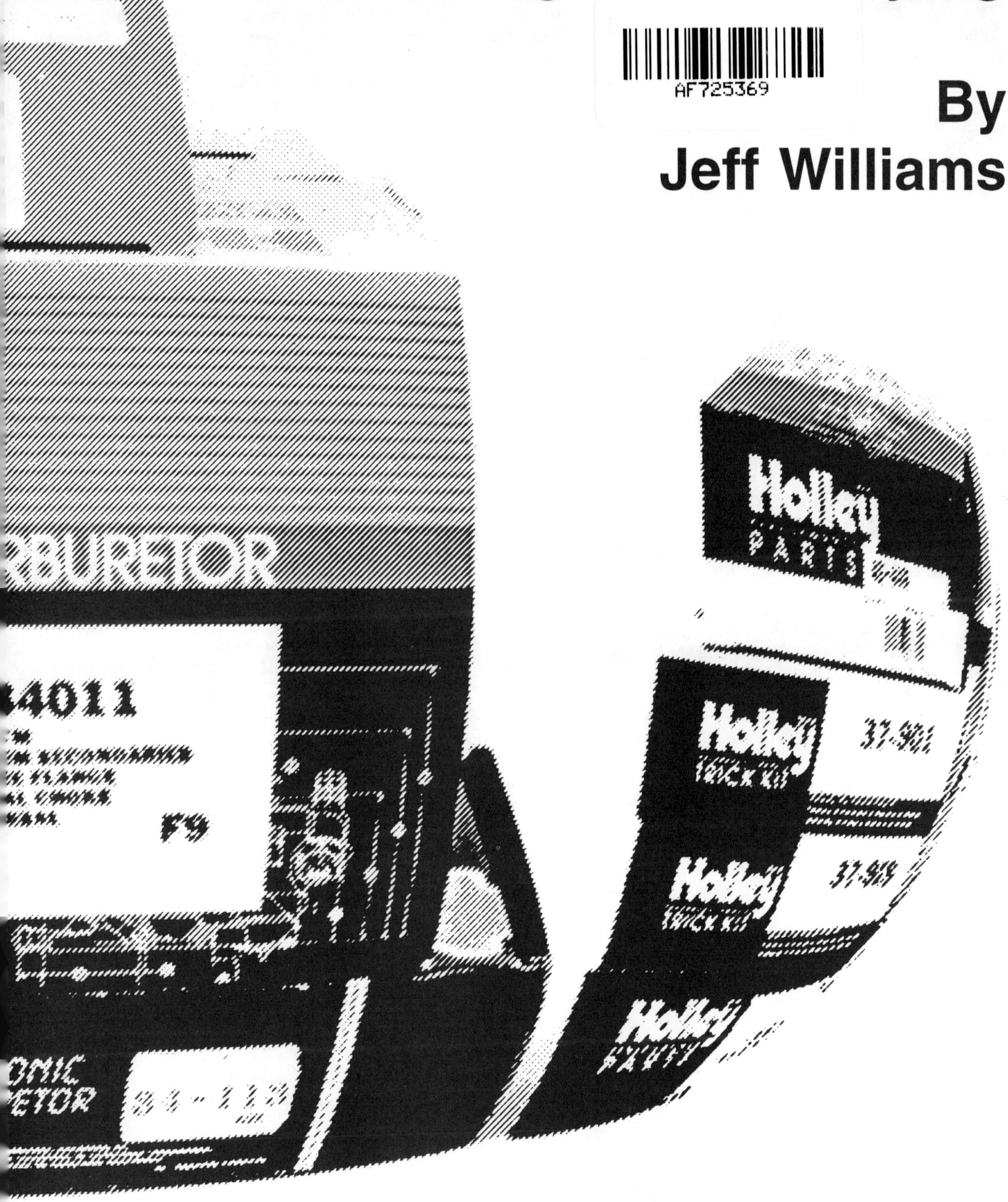

Holley
Rebuilding & Modifying

CARTECH®, Inc.,
39966 Grand Avenue
North Branch, MN 55056
651-277-1200 or 800-551-4754 • FAX: 651-277-1203
www.cartechbooks.com

OVERSEAS DISTRIBUTION BY:

Brooklands Books Ltd.
P.O. BOX 146, Cobham, Surrey, KT11 1LG, England
Telephone 01932 865051 • FAX 01932 868803
www.brooklands-books.com

Brooklands Books Ltd.
1/81 Darley Street, P.O. Box 199, Mona Vale, NSW 2103, Australia
Telephone 2 9997 8428 • FAX 2 9997 5799

GRAPHIC CAMERA BY
Brian Cale

EDITED BY
Larry Atherton

PRODUCED BY
Larry Atherton

ISBN 0-931472-27-X
Part No. SA27

Table Of Contents

TABLE OF SIDEBARS

ACKNOWLEDGMENTS

This book would not have been possible without the generous assistance from manufacturers and the dedication of many individuals that ensured this initial *Workbench*® Book from S-A Design would be up-to-date, complete, and accurate. The publisher, editors, and writers would like to extend their special appreciation and thanks to Christopher Kuczynski at Holley Replacement Parts for his relentless effort to answer all of our questions, respond to all of our faxes, and provide us the support that made this book a reality. We would also like to thank Brad Urban of The Carburetor Shop and Bryant Seller of Mike Jones Carburetion for their technical support. And for a most interesting interview that shed light on some of the idiosyncrasies of Holley carburetors, we would like to thank Murray Jensen, a carburetor professional par excellence.

A heart-felt thanks is given to all of those who helped us with this publication:

Coltec Industries Inc, Holley Replacement Parts:
- Christopher Kuczynski, Advertising Manager
- Robert Miller, Vice President Engineering
- Marvin Benoit, Manager, Specialized Markets
- Brain Salvati, Supervisor, Specialized Markets
- Lenny Golembiewski, Senior Product Merchandiser
- Jack Nelson, Holley Custom Shop
- Terry Carter, Holley Rebuilding Facility

Gale Banks Engineering, Gale Banks, Jim Rowlee

The Carburetor Shop, Brad Urban
The Eastwood Company
Edelbrock Equipment Company, Camee Edelbrock
Ken Mattern And His Super-Fast Corvette
Mike Jones Carburetion, Bryant Seller
Autotronic Controls, MSD Ignition
Snap-on Tools, Walt Commans, Gary Busha
Vericom, Rick Jobe
Weiand Automotive Industries, Joan Weiand

Introduction—What Is A *Workbench®* Book?

This is an S-A Design *Workbench®* book. It does not present a hasty look at Holley carburetor rebuilding and modifications, nor does it give vague advice or only cover the high points. Furthermore, this book does not require that you've had previous experience with Holley carburetors or a degree in fluid dynamics to use it. *This book is a complete reference that shows you how to work on modular Holley carburetors; step-by-step, every step, with nothing left out!* Hundreds of photos guide you through the easy and "tricky" procedures, and photo captions explain exactly where to look for possible problems and how to fix them. This book shows you how to rebuild your Holley and ensure that it's working perfectly.

But this book is more than a "repair" manual.

The author, Jeff Williams, and the editors at S-A Design Books have packed this book with information about components and accessories, including a wide range of performance tips and modification procedures to improve power and economy. And we didn't stop there. We wanted you to get the same professional results on your first rebuild that the pros get after years of experience. So we've detailed the special components, tools, chemicals, and other accessories you'll need to get the job done right, *the first time.* You'll even find tips that will save you money without compromising "top-notch" results. To top it off, we've included chapters on cleaning procedures and custom finishing, and an appendix of part numbers and other reference information, plus a *Work-A-Long* sheet to help you record vital statistics while you work on your Holley.

This book is not meant to replace other Holley carburetor books on your bookshelf. (In fact, if you are looking for in-depth information on Holley carburetor *function* and special application tips, refer to the best-selling S-A Design book *Holley Carburetors*, see inside back cover in this book.) *Holley Rebuilding & Modifying* is meant to detail, like never before, the "workbench" procedures required to troubleshoot and rebuild your carburetor, install any one of the hundreds of accessories from Holley and other manufacturers, and tune your Holley for optimum street or strip performance. If any one of these is your goal, we believe you are holding the right book in your hands.

PERFORMANCE STEP

39 Bellcrank Linkage

39—To remove the throttle plate on carburetors equipped with bellcrank secondary linkage, just lift the main body slightly and slide it sideways while pulling the intermediate arm off its stud. Unless the throttle plate will be disassembled, there is no need to remove the secondary actuation link (arrow) from the throttle linkage.

This is an example of the hundreds of step-by-step photos in this book. The photos will guide you through the easy and "tricky" procedures, while the captions explain exactly what to do, what not to do, and how to avoid possible problems.

HOLLEY MODELS COVERED IN THIS BOOK

This book covers the maintenance and modification of Holley modular-style carburetors. The Holley "modular" line includes those carburetors that have separate and removeable primary and secondary float bowls and metering systems, removeable throttle-plate assemblies (except for the Dominator®), and one-piece main bodies. This original design, developed in the 1950s, became one of the most widely used carburetors by both auto manufacturers and performance enthusiasts. In it's peak production, Holley produced over one million carburetors each month. Even today, Holley is universally the carburetor-of-choice for enthusiasts worldwide.

Holley modular (and non-modular) carburetors include many distinct models that are commonly referred to by various "model numbers." These numbers are different from Holley part numbers, since model numbers only indicate the overall design of the carburetor and not any one specific carburetor (further information on identifying Holley modular carburetors, their individual components, and the Holley part numbering system can be found in Chapter 3).

The Holley *models* covered in this book are the 4150, 4160, 4180, and 4190 four-barrels, the 3150 and 3160 three-barrels, the 4165 and 4175 spread-bore four-barrels, the 4500 Dominator four-barrel, and 2300 two-barrel carburetors. Within this broad range of Holley models can be found carburetors with two, three, and four venturis that flow anywhere from 390 to 1150cfm; equipped with single or dual accelerator pumps; mechanical or vacuum secondary actuation; idle adjustments on both the primaries and secondaries or just the primaries; and many other minor variations.

This book is designed to help you repair, modify, and tune any one of the hundreds of Holley carburetors based on these models. Here's how it accomplishes this.

WHY THIS BOOK IS DIFFERENT

While virtually every other performance book is simply a series of typical-looking chapters, this S-A Design *Workbench*

This book covers the maintenance and modification of Holley modular-style carburetors. These carburetors can be identified by their separate and removeable primary and secondary float bowls and metering systems, removeable throttle plate assemblies (except for the Dominator), and one-piece main bodies. In it's peak production, Holley produced over one million carburetors each month for both auto manufacturers and performance enthusiasts.

Details of special components, tools, chemicals, and other accessories you'll need to get the job done right can be found in the pages of this book. There's even a chapter on custom painting and finishing techniques that shows you how to turn your rebuilt Holley into a glistening show piece.

book is quite different. But why did we make it different? The answer to that is simple: *We had to!* This book not only helps you become familiar with Holley carburetors and the tools and materials you need to work on them, *it also shows you how to rebuild a Holley carburetor on your workbench, step-by-step.* This means that it must look different from other books. It must have photo sections that guide you through disassembly, cleaning, and reassembly. But this book also illustrates many of the hundreds of performance accessories, modifications, and tuning tips for power or economy that have been developed for Holley carburetors. This additional information is easy to find and clearly separated from the stock rebuild sequence. Through the use of *sidebars*, *icon-labeled* photos, special tables of contents, and even a separate tuning chapter, this book presents both stock and performance information and keeps each separate and accessible.

TEXT AND PHOTO CHAPTERS

The *text and photo sections* that make up the introduction that you are now reading, Chapters 2, 3, 4, part of 7 and 8, and all of Chapter 11 are the "familiar" part of this book. They consist of columns of text combined with photos, drawings, and charts. These familiar-looking pages introduce the basic information you'll need to identify carburetor components, determine if the carburetor is really at fault before you begin your rebuild, or even help select the best carburetor rebuilding kit for your needs. Also, hundreds of practical tips on tools, chemicals, and even custom painting and refinishing techniques are included in the text and photo chapters in this book. Finally, troubleshooting and tuning tips are discussed in Chapter 4 (*Before You Begin*), Chapter 10 (*Reinstallation And On-Car Adjustments*), and Chapter 11 (*Performance And Economy Tuning*).

Text and photo sections are the best—and most familiar—way to present information that covers a series of related subjects. However, the best way to detail each of the steps involved in rebuilding a carburetor is to *show how with step-by-step photos.*

STEP-BY-STEP PHOTOS AND ICONS

Removing the carburetor from the vehicle, disassembly, cleaning, reassembly, and on-car tuning are the subjects of the most important chapters in this book (Chapters 5, 6, 7, 9, and 10). Each of the *step-by-step photo sections* in these chapters include a sequence of numbered photos and captions that illustrate virtually every operation involved in rebuilding all Holley modular carburetors.

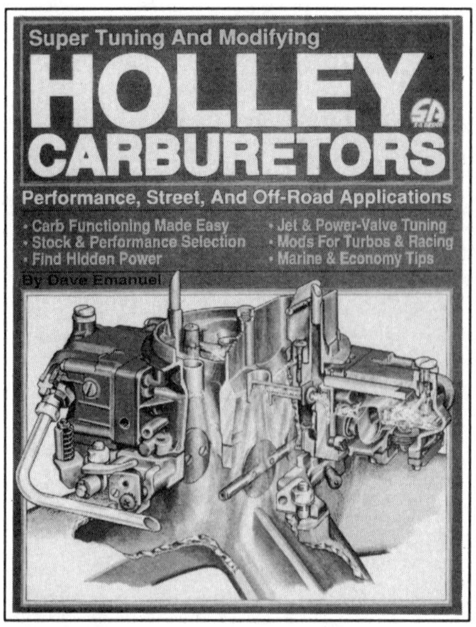

While this book provides step-by-step information on Holley rebuilding and customizing, if you're looking for in-depth information on Holley carburetor *function* and modifications for special applications (like racing or turbocharging), refer to S-A Design's best-selling book *Holley Carburetors* by Dave Emanuel.

As we have mentioned, the goal of this book is more than just rebuilding a single Holley in stock condition. The step-by-step chapters also include many additional photos that show how to install performance or economy accessories, how to modify stock components for special applications, or even call attention to assembly steps that are critical to proper operation or safety. To keep these photos separated from the main rebuild sequence, we have labeled them with unique icons. Icons are symbols that represent an idea, and photos marked with icons should be thought of as "special" information. An icon-labeled step may apply to your carburetor, in which case you should include the step in your rebuild. On the other hand, it may illustrate a modification or component that does not apply to your specific application; in that case, just skip it and continue with the next photo in the sequence.

Here are the icons that you will find in this book:

Other Holley Models—Photo illustrates how a step in the rebuild varies with different Holley models. For example, some models use different metering blocks, float bowls, linkage, or even fasteners. You will find *Other Holley Models* icons for 2300, 3-BARREL, 4150, 4165, 4175, 4180, 4190 & 4500 carburetors.

Torquing Fasteners—Photo illustrates a fastener that must be properly tightened with a torque wrench at this point in the rebuild.

Special Tool Used—Photo illustrates the use of a special tool that may be required or can make the job easier (caption with photo will explain further).

Precision Measurement—Photo illustrates a precision measurement or adjustment that is required at this point in the rebuild.

Important (!)—Photo indicates a step in the rebuild that is very important to the proper operation of the carburetor.

Safety Step—Photo indicates a step in the rebuild that is essential for the safe operation of the carburetor (to prevent fuel leaks, stuck linkage, etc.), or to guard the personal safety of the rebuilder.

Critical Inspection—Photo indicates that a component must be inspected to ensure the proper operation of the carburetor.

Performance Step—Photo indicates a procedure that applies to only high-performance or racing carburetors, or details a modification that can improve engine performance.

Economy Step—Photo indicates an optional step that details a modification or adjustment that will improve fuel economy.

Master Mechanic Tip—Photo illustrates a step in the rebuild that many non-professionals may be unaware of. It may illustrate a shortcut, a trick to improve reliability, prevent component damage, etc.

Notation Required—Photo illustrates a point in the rebuild sequence that the reader should note a particular measurement, size, part number, etc.

Custom Finishing—Photo shows a technique, procedure, or add-on equipment that can enhance the appearance of your Holley.

Save Money—Photo illustrates a method or alternate method of performing a carburetor-rebuild step that will save money but still give acceptable results.

The step-by-step photo chapters include many optional steps labeled with icons. This book also includes other optional information that, although it may not be essential for rebuilding a carburetor, the reader may find very useful. This "extra" information explains carburetor function, Holley part numbering, special tools and tuning equipment, and much more. These optional information "packages" are called *sidebars,* and they can be found throughout this book.

SIDEBARS

This book contains over thirty sidebars. They are easy to spot; look for a group of photos and some text enclosed in a grey box with a title along the side. The sidebars cover subjects that are generally too short for the text-and-photo chapters but require more detail than a single icon-labeled photo. The author and editors considered hundreds of possible subjects and selected those that they felt you would find the most interesting and valuable. Many contain facts about Holley carburetors that some professionals discover only after years of work. For example, there are sidebars on various ways to restore damaged threads, how to properly torque fasteners, identifying and repairing loose throttle shafts, understanding power valves, needle-and-seat tips, choosing an intake manifold, fuel bowl basics, details about Holley's Custom Shop, accelerator pump tricks, tuning with electronic dash-mounted "dynos," and much more.

Many sidebars are not presented in a specific order. However, you will find a complete subject list in the *Sidebar Table Of Contents* on page 4. You can read the sidebars one at a time as you progress through the book, you can look up specific sidebars at any time, or you can even skip everything else first and just read sidebars. However you choose to use sidebar information, we are sure that it will add to your understanding of Holley carburetors and help you get the most power, economy, and reliability from your vehicle.

In addition to the "workbench" procedures required to rebuild a Holley carburetor, this book introduces you to many of the hundreds of accessories available from Holley and other manufacturers. No other carburetor has as many "goodies" available to improve performance, economy, simplify modification, and even add a custom appearance.

WHERE TO BEGIN

Holley Rebuilding & Modifying is organized in a sequence that, theoretically, allows you to start at the first chapter and read straight through to the end. When you've finished, you should have a rebuilt carburetor installed on your vehicle. However, one reader may only need the troubleshooting information found in Chapter 4, while another may want help installing a performance accessory described in a sidebar, and others may be interested in the advice on the tools and chemical cleaners required for rebuilding found in Chapters 2 and 7. Because of these wide-ranging needs, here are some tips on how to use this book:

Want to rebuild your Holley and learn as much as possible about tools, chemicals, accessories, etc.: Proceed to Chapter 2 and continue in sequence to Chapter 11.

Have considerable experience and would like to diagnose and repair a problem that may or may not be related to the carburetor: Proceed to Chapter 4 and review troubleshooting information. If the problem lies in the carburetor, review disassembly, cleaning & parts inspection, and reassembly steps as needed in Chapters 5, 6, 7, 9 and 10. If carburetor problems persist after the rebuild, refer back to the troubleshooting information in Chapter 4.

Would like to custom finish your Holley for better appearance: Proceed first to Chapter 8 to find out what is possible and what is required, then review as much of the disassembly and reassembly information in the other chapters as needed.

For all readers: Review the tables of contents on pages 3 and 4 and study the information you are unfamiliar with. Then review the entire rebuild sequence before beginning work on your Holley. The more you are familiar with the material in this book before you begin your project, the more you will benefit from what this book has to offer.

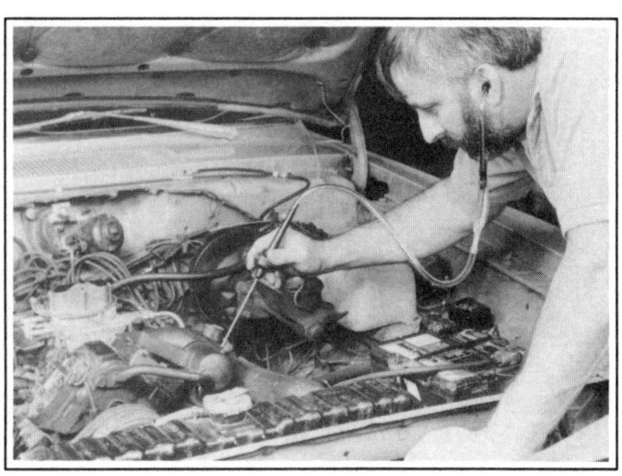

This book is organized in a sequence that allows you to start at the first chapter and read straight through to the end. When you've finished, you should have a rebuilt carburetor installed on your vehicle. However, if you just need troubleshooting information or assistance in installing a performance accessory, refer to the tables of contents to "tune in" to only those chapters or sidebars that provide the help you need.

HOLLEY Rebuilding & Modifying

2—Workspace, Tools, & Chemicals

As we will illustrate in this book, there are few difficult steps in rebuilding a carburetor. There are, however, *many simple steps*, each involving one or more of the numerous individual components that make up a Holley carburetor, including metal castings, screws, rubber and plastic parts, check balls, jets, tubes, springs, and many more. Each of these parts was designed to perform a specific function, and unless care is used in disassembly, cleaning, and reassembly, your Holley will not operate as it was intended.

In this chapter I will indicate what tools and accessories you'll need to successfully rebuild and modify Holley carburetors. You certainly don't need to buy a roll away full of tools or setup a professional shop, but you should take what you need from the information presented in this chapter and establish a comfortable, organized, properly-equipped work environment that will make it easy for you to follow the illustrated steps in this book and ensure a successful rebuild.

WORKSPACE CONSIDERATIONS

Rebuilding a carburetor involves several unique procedures, and some of these steps require special workspace considerations. As I've indicated, disassembly involves handling many component parts, some of which are quite small and easily lost, so adequate lighting, containers for parts organization, a comfortable work surface, and a keen eye for detail are all helpful. By the same token, a carburetor accumulates dirt and varnish that only can be removed with elbow grease and strong chemical cleaners, so a space should be made available for this "dirty" job, too. Finally, reassembly requires a clean work surface, precision tools, and plenty of patience to ensure that your rebuilt or modified Holley works right the first time you install it.

The work surface necessary for carburetor disassembly and reassembly can range from a slick, Formica- or metal-top workbench, to a humble "drip-pan" on a folding table. In fact, even if you have a quality workbench surface, disassembling a carburetor in a metal drip-pan can keep your work surface clean, and the raised edge of the pan will prevent small parts from rolling off into oblivion. Don't plan on disassembling your carburetor in a solvent tank. This is a sure-fire way to lose parts, and with many solvent tank setups, you'll be working at an uncomfortable, hunch-back position—not the best way to discover subtle carburetor problems, such as cracks, plugged up orifices, etc.

Carburetor Repair Stands

A Holley carburetor has no flat surfaces that can readily support the carbu-

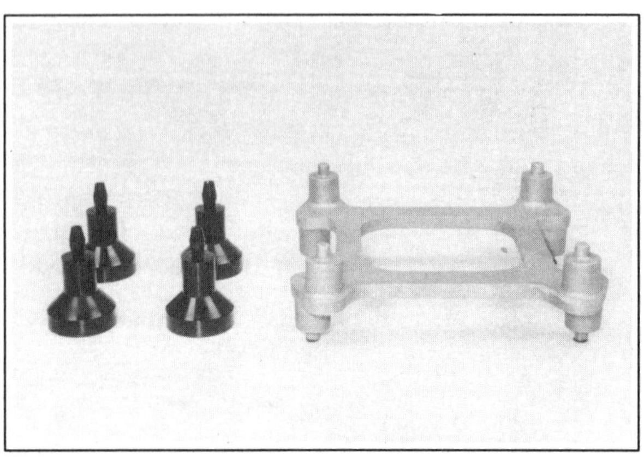

There are several "quick-change" carburetor stands that are both inexpensive and practical. *Thexton* makes a set of plastic, snap-in "feet" that install in a few seconds. Several companies manufacture aluminum stands that take no time at all to use (once you figure which set of pegs fit your carburetor base). This low-profile B&B stand is very handy for both square-bore and spread-bore carburetors.

A clever stand from Peters holds a Holley both upright and upside down by just flipping the stand over. Another unique stand is manufactured by MAC Tools. Fastened to the workbench, it can hold a Holley in any position.

retor during rebuilding. If the carburetor is set on the throttle-body base, the protruding throttle linkage prevents the carburetor from laying flat on the bench, and the throttle body and butterflies can easily be damaged, especially by metal-top workbenches.

Repair and rebuilding work on Holley carburetors is easier and more convenient when you use a carburetor stand, and there is no shortage of models from which you can choose. On one hand, some rebuilders use four, 2-inch long, 5/16-inch cap screws and eight nuts installed in the throttle plate mounting holes to support the carburetor during the rebuild. This setup is cheap and effective, but it does take a minute or so to install the bolts, and the carburetor is only supported in one position. I have used an adjustable bench-mounted stand, manufactured by MAC tools, that is very handy for some jobs. Like most stands, it attaches to the carburetor at the throttle plate, but unlike most stands, it allows the carburetor to be held in almost any position. This is particularly handy for custom work, modifications, or even some simple repairs. Although it doesn't take any longer to attach the carburetor to the MAC stand than to install two simple bolts, if you're just making a simple modification to the carburetor, it may take longer to mount the carburetor on the stand than it does to make the modification.

There are several "quick-change" carburetor stands that are both inexpensive and practical. Thexton® makes a set of plastic, snap-in "feet" that installs in a few seconds. They fit snugly into the throttle plate and provide good support. Several companies manufacture "drop-in" aluminum stands that take no time at all to use (once you figure which set of pegs fit your carburetor base). Both of these designs, or many of the variations found in auto part stores and speed shops, make rebuilding easier and are worth using.

Most carburetor stands, with the notable exception of the MAC bench-mount stand, do not support the carburetor in an inverted position. Although not essential, it is quite handy when performing service procedures on the "underside" of a Holley. A clever stand manufactured by Peters holds a Holley both upright and upside down by just flipping the stand over. Not too expensive, the Peters carburetor stand is a great Holley service aid for both the professional rebuilder or home enthusiast; I recommend it highly!

COMPRESSED AIR

There are few automotive maintenance or rebuilding jobs that do not require or cannot be done easier with compressed air. From blowing off accumulated dust and metal particles in an alternator, to operating pneumatic impact wrenches, compressed air is a uniquely effective tool. But is compressed air essential for rebuilding a Holley carburetor? Yes and no.

If you're planning on rebuilding a few carburetors and would like to do other, similar, home-repair tasks, a compressor rated at 3scfm @ 90psi, with a 7- to 10-gallon tank, will probably meet your requirements. Compressors of this type are available from many sources, including Grainger, Sears, and many tool stores for about $200.

If you are setting up a home shop to do a variety of automotive maintenance jobs and one of those jobs is rebuilding Holley carburetors, you are well advised to take the plunge and purchase an air compressor, the lines, fittings, and nozzles you'll need to complete the installation. However, if your goal is to rebuild one or two carburetors, it may not be necessary to make such a large investment. For many reasonably clean carburetors, you can perform a quality carburetor rebuild with just "canned-air."

Canned-Air

Compressed air is required to blow out the passages and dry the exterior of a carburetor during rebuilding. And that's exactly what you can get from canned air: compressed gas in an aerosol can. Although there are limits in what canned air can provide, by using a trick that I'll describe here, it can be a reasonable substitute for compressed air in most carburetor rebuilds; all for just a few dollars per rebuild (the cost of one or two 15-ounce aerosol cans).

Canned-air is regularly used to blow dust off of lenses and other optical and electronic equipment. At room temperature, the Freon pressure inside the can is about 60psi. But here's the catch: As gas is dispensed, the temperature of the can drops and so does the equalization pressure. So what was initially about 60psi, after a few blasts may drop to less than 30psi as the can gets cold. While 60psi will blow out carburetor passages and dry components removed from solvent or

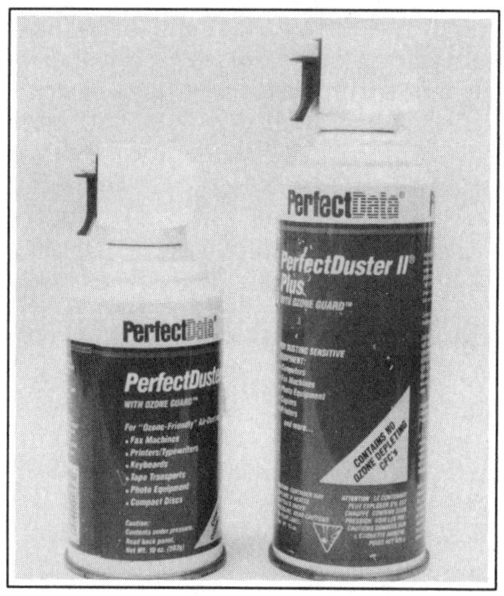

Compressed air is required to blow out the passages and dry the exterior of a carburetor during rebuilding. And that's exactly what you can get from "canned-air," a compressed gas in an aerosol can. Within limits, it can be a reasonable substitute for an air compressor when you are rebuilding a single carburetor or performing other small jobs.

A typical carburetor rebuild requires removing the carburetor from the vehicle, bench disassembly, thorough component cleaning, reassembly, initial bench adjustments, reinstallation on the vehicle, and final adjustment. Many of the tools you are likely to need are commonly found in most shops, like screwdrivers and pliers, but some are unique, such as clutch-head screwdrivers, inch-pound torque wrenches, or line wrenches.

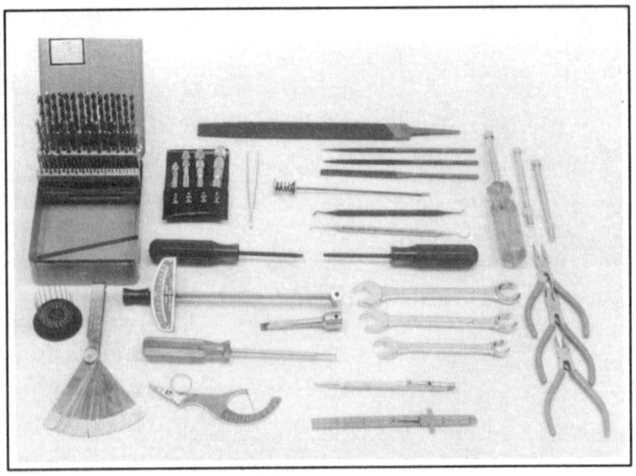

water, pressure much below 30psi will not. There is a simple solution to this problem. Keep canned-air in a bucket of warm water between blasts. This maintains a higher internal pressure and makes it quite suitable for small rebuilding projects. (WARNING: Don't use hot water; high pressures may cause the can to burst! Follow the specific warning instructions printed on the can and do not exceed recommended can temperatures).

For some home carburetor rebuilding projects, canned air will provide sufficient compressed air to perform a quality rebuild. But if you're really serious about carburetor and automotive repair work—or your Holley has hard deposits on internal surfaces or sandy particles have worked their way into interior passages—an air compressor should be at the top of your "must-have" list.

ESSENTIAL REBUILDING TOOLS

A typical carburetor rebuild requires removing the carburetor from the vehicle, initial cleaning, bench disassembly, thorough component inspection and cleaning, reassembly with new rubber parts and gaskets, initial bench adjustments, reinstallation on the vehicle, and final adjustment of linkage, idle, etc. I have made the assumption that all of these steps will take place, and I have included all the tools required to perform these steps in a list of essential tools. This list may include some tools you will not need, since individual Holley carburetors and rebuilding projects vary. But for the average vehicle, during the average Holley rebuild, the tools included in this list should provide you with nearly everything that's required to accomplish the job in a professional manner.

Although the essential tool list includes several items that may not be considered "essential" by all carburetor rebuilders, if you rebuild enough Holley carburetors, eventually you will find a need for every tool in the list. A good example of this is the straightedge, feeler gauge, and flat mill file that are used to measure the flatness and correct irregularities on the surfaces of the main body. Not every carburetor needs its sealing surfaces filed, but some do, and if you neglect this step, you may be faced with a carburetor that runs poorly despite new gaskets and an otherwise high-quality rebuild.

Another tool that some carburetor rebuilders may consider optional is an inch-pound torque wrench. The photos and sidebars in this book detail when and how to torque most of the fasteners on a Holley carburetor. Is torquing fasteners really important or just overkill? When you realize that *eventually every Holley carburetor will suffer from either external or internal leakage*, you begin to get an idea how important fastener tightness can be. When you further consider that many professionals routinely undertighten and overtighten critical fasteners and that this practice can easily result in a variety of difficult-to-locate problems from rough idle to hesitation or hard starting, I'm sure you'll realize that precision fastener tightening is not just important, it's absolutely essential! Take my word for it: An inch-pound torque wrench and its screwdriver- and socket-tip accessories are some of the most essential tools in my "essentials" list.

NON-ESSENTIAL TOOLS

Rebuilding a Holley carburetor can require—or at least can be easier with—some additional tools that I have classified as "non-essential." Examples of this type of tool include vacuum and pressure gauges and other test equipment. Although these items are not directly required for carburetor rebuilding, as I will explain in Chapter 4, *Before You Begin*, it is important to determine whether a performance problem is caused by the carburetor or some other defective component before you begin your rebuild. Nothing is more frustrating than spending a few dollars and a day's work rebuilding your carburetor only to find that your vehicle runs just as poorly afterwards. So the bottom line is: Are all the tools in the non-essential tools list really "non-essential?" The answer is maybe. The prob-

Essential Tool	Sample Usage
Screwdriver–1/8-inch flat tip	Idle adjustment
Screwdriver–3/16-inch flat tip	Pump cam screw
Screwdriver–1/4-inch flat tip	Bowl screws (early)
Screwdriver–#1 Phillips tip	Accelerator pump cover
Screwdriver—#2 Phillips tip	Throttle plate
Screwdriver–.170 Clutch tip	Secondary jet plate (early)
Screwdriver--T-20 Torx	Secondary jet plate (late) and some automatic choke covers
Nutdriver–3/8-inch hex	Bowl screws (late)
1/4- or 3/8-Drive Torque Wrench, Rated at 200 Lb. In.	All fasteners
1/4-inch flat screwdriver tip, 1/4- or 3/8-inch socket drive	Torque wrench accessory for float-bowl screws
#1 & #2 Phillips screwdriver tip, 1/4- or 3/8-inch socket drive	Torque wrench accessory for throttle plate, and a variety of other fasteners
Clutch tip or Torx screwdriver tip, 1/4- or 3/8-inch socket drive	Torque wrench accessory for secondary jet plate
3/8 Socket Set, 1/4- to 7/8-inch	Carburetor removal, fuel inlet fittings
3/8-Drive Socket, 1-inch (12-point)	Power valves
Open/Box Wrenches, 1/4- to 7/8-inch	Linkage, baseplate, etc.
Fuel-Line Wrenches, 1/2, 9/16, & 5/8	Metal fuel line to carburetor
5- to 7-inch Combination Pliers	Fuel line clamps, linkage
4- to 7-inch Needle-Nose Pliers	Linkage clips, E-clips
1-inch Blade Gasket Scraper	Removing old gaskets
Feeler Gauge Set	Checking flatness, clearances
Straightedge or Machinists Rule	Checking flatness
10-inch, Flat Mill File	Filing surfaces flat
Vaseline or Silicone Lubricant	O-rings
High-tack Adhesive	Holding some gaskets in place during assembly
Notepad & Work-A-Long Chart	Noting jet size, etc., during disassembly
Masking Tape	Marking vacuum lines during carburetor removal

Non-Essential Tool	Sample Usage
Carburetor Stand	Holding the carburetor during rebuilding
Float Scale	Determine buoyancy of composite floats
"Ice-Pick" Type Probe	Manipulating small parts, cleaning small passages, scribing identifying marks, etc.
Hemostats & Tweezers	Manipulating small parts
Spring-Clip Handlers	Installing and removing wire-type clips
Flexible Light Probe	Inspection of small passages and components, particularly in the venturis
Wire Drill Set And Pin Vise	Modifying air and fuel metering passages
Mity-Vac Handheld Vacuum Source	Various testing procedures
Vacuum Gauges	Various testing procedures
Fuel Pressure Gauge	Testing fuel pump, inlet pressure, etc.
Compression Tester	Determine condition of engine before carburetor rebuilding

lems you face in your repair work will ultimately determine what tools are required for troubleshooting and rebuilding a Holley carburetor.

GENERAL TOOL TIPS

If you're a casual mechanic, you can buy a tremendous amount of capability for very little money by shopping the swap meets for tools imported from Taiwan, China, Japan, and other off-shore sources. It is possible to nearly fill a modest rollaway with off-shore tools for the cost of one set of Snap-on combination wrenches. Although these tools are quite adequate for casual work, they often will not stand up to the hard daily use of a professional mechanic. But by keeping a keen eye when you shop for "cheap" tools, you can select the cream-of-the-crop that even professionals are able to use with some degree of success. For example, manufacturers of 3/8-inch drive socket sets are able to save money by using the same external forging die size for several sizes of sockets, say from 7/16-inch to 9/16-inch. This results in 7/16-inch sockets that have ridiculously thick walls making it difficult to get into tight spaces. And the 9/16-inch socket is so thin that the first time you encounter a stubborn fastener, the socket wall splits. To avoid the frustration of owning a box full of broken 9/16-inch sockets, look for socket sets that are more uniform in wall thickness. They are still inexpensive, but they are a higher quality tool that will provide longer service.

Most of the same advice about construction quality can be applied to other tools, including open, combination, and box-end wrenches. If the manufacturer uses the same external die over too wide a range of wrench sizes, some wrenches will be nearly unusable. Look for wrenches that have a smooth chrome finish, have nearly the same thickness in the box and open ends, and if possible, check them to make sure they fit the bolt head snugly. If you're buying screwdrivers, look for chrome-vanadium tips, smooth comfortable handles, and an overall look of quality. If they look good, they probably will work fine. Avoid screwdrivers with thin-plastic handles, cheap chrome finishes, or non-machined tips; they may twist more than the screws you're trying to remove.

Perhaps the greatest cost reduction in tools from off-shore sources can be found in air tools. Air impact wrenches, air ratchets, air drills, air sanders, air hammers, air chisels, and air screwdrivers are selling today for about 20% of cost of domestic air tools produced before the off-shore invasion began a few years ago. What about quality? Generally, most of these air tools are very good. They seem to be reliable, even with daily use. So is there any difference between a $35 1/2-inch impact from Taiwan and $175 1/2-inch Snap-on impact. Yes, there definitely is a difference. In more than one occasion, I've had to grab my Snap-on wrench to break loose a particularly stubborn bolt that my Taiwanese equivalent would just bounce around on. But except for the serious professional (and even for some of them), off-shore air tools are the best bargain in town!

Many automotive-related swap meets are held all over the country and are excellent places to purchase inexpensive tools. You can almost always find tools that will make your carburetor rebuilding job easier. For example, stainless-steel tweezer-like pliers, called hemostats, are super for removing spring clips and handling other small parts. Designed for medical uses and originally sold for nearly a hundred dollars each, hemostats are regularly found at swap meets for about a buck or two. Another terrific tool for carburetor work is a metal probe or pick. Designed for dental or other medical purposes, hundreds of these handy tools can be found at typical swap meets on sale for rock-bottom prices. Probes are useful for removing small pieces of gasket material, cleaning dirt from crevices, or scribing identifying marks. Swap meets are also common sources for tubing benders and cutters, flaring tools, air tools, various hand tools, and of course, new and used Holley carburetors.

Now that I have made a case for off-shore tools and other swap-meet bargains, let me muddy the waters a bit. Some enthusiasts and many professionals rarely invest in these inexpensive tools. They primarily use the highest quality tools, like Snap-on and MAC brands. Are they just for rich "techno-yuppies?"

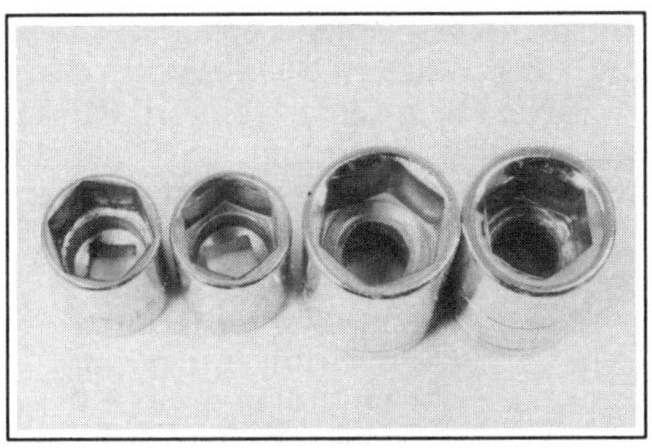

By keeping a keen eye when you shop for off-shore tools, you can select the cream of the crop. For example, avoid socket sets that use the same external forging die size for several sizes of sockets. When manufactured in this way, some sockets that have ridiculously thick walls making it difficult to get into tight spaces, while others are so thin that the first time you encounter a stubborn fastener, the socket wall splits.

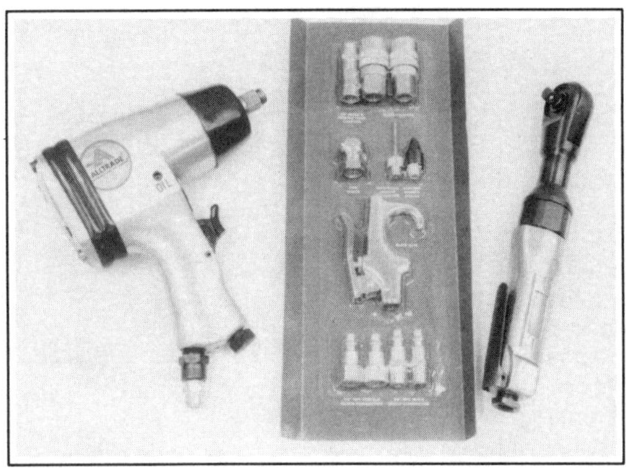

Perhaps the greatest cost reduction in tools from off-shore sources can be found in air tools and accessories. Air impact wrenches, air ratchets, air drills, air sanders, air hammers, air chisels, and air screwdrivers (particularly helpful for working on Holley carburetors), blow guns, and fittings are tremendous bargains in today's market.

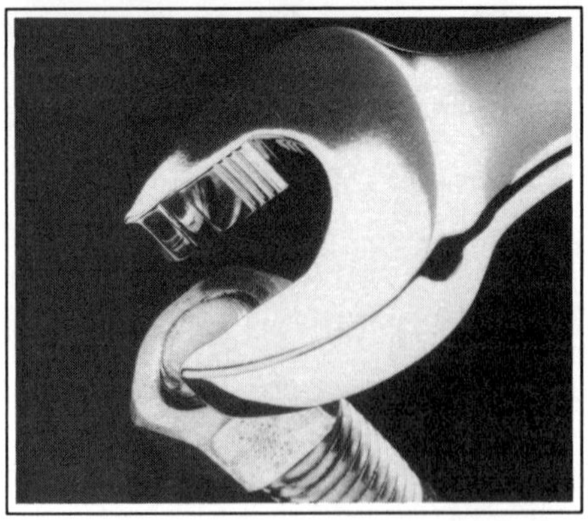

When you really "lean into" a stubborn bolt, the tips of a hex nut can round off, allowing the wrench to slip. To solve this problem Snap-on invented a wrench design called *Flank Drive®*. The interior ridges of the new wrenches improve the turning power an average of 40%. This is a remarkable achievement for a tool that has a design so simple that, at first, it seems impossible to improve.

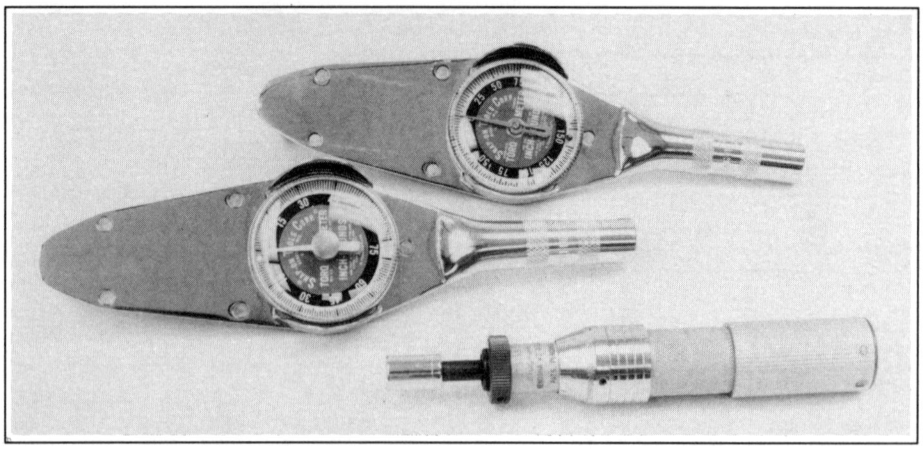

A Holley carburetor is a collection of hundreds of parts held together with fasteners. Torquing each fastener to its proper torque is an essential element in every quality rebuild. These Snap-on wrenches make applying the proper torque accurate and easy.

Maybe a few are, but despite the high cost of professional tools, the exceptional quality these tools offer draws a wide range of average-income enthusiasts to invest in them. For the reason why, let's take a close look at a few Snap-on tools.

If you examine most sockets, the gripping surfaces are simple ridges and valleys that contact the nuts or bolts near the tips of the hex. When the twisting force exceeds the torque capability of the nut, the tips of the hex "round off" and the wrench slips over the nut. It would seem that there is little that can be done about this phenomenon, but Snap-on re-engineered the common socket with its unique wrenching system called *Flank Drive®*. The visual appearance is only slightly different than other sockets—the ridges are separated by broader, smoother valleys—but the result is that 15-20% more torque can be applied to the same fastener before the hex nut rounds off. This advantage can make the critical difference when removing a stubborn fastener, or it can save your knuckles from painful bruises. Recently, Snap-on applied its Flank Drive® system to its open-end wrenches, resulting in an unusual-looking wrench that has serrated teeth in the "U"-channel that improve the turning power an average of 40%. This is a remarkable improvement in a tool that has a basic design so simple that, at first, it seems impossible to improve upon. That's just a small sample of what Snap-on offers; other professional tool companies, like MAC, Matco, and even the Sears brand Craftsman, produce tools with their own marque of quality. Professional tools cost more, but they really do deliver more quality and reliability.

CHEMICALS AND LUBRICANTS FOR CARBURETOR REBUILDING

Hand tools, compressed air, and adequate workbench space are all essential components to rebuilding a Holley carburetor. However, dirt is the most common cause of carburetor problems, so every rebuild must include procedures that will thoroughly clean both the exterior and, more importantly, the small interior passages and orifices in the main body and metering blocks. There is only one "tool" capable of performing this critical function: chemical solvents. Chemical cleaners range from common detergent- or petroleum-based solvents to the powerful chemical cleaners in spray- and immersion-type carburetor cleaners. First let's take a quick look at the state-of-the-art in solvent cleaning.

Solvent & Wash Tanks

When you need to remove dirt, oil, and grease from parts, the most frequent choice in both home and professional shops is to use solvent, a stiff-bristle brush, and a solvent tank. There are many types of solvent, but the most common type used in the automotive industry is a petroleum-based chemical made specifically for cleaning and degreasing. Solvent is not a very harsh chemical to the skin and nose (but always wear gloves and goggles when cleaning parts), and it is resistant to combustion unless heated (to typically over 120 degrees[F]). Never use gasoline instead of solvent. Gasoline vapors are carcinogenic and, needless to say, it is very flammable. Solvent is widely available, so do yourself a favor and use solvent for cleaning and leave gasoline in your fuel tank.

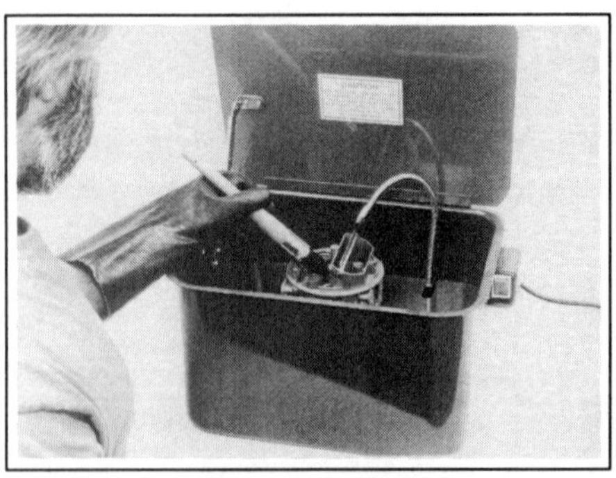

Small, benchtop parts washers are inexpensive and make parts cleaning a joy compared to using an old paint can and a brush. A typical benchtop cleaner holds from 2 to 5 gallons of cleaner and incorporates a small electric pump and filter to clean the re-circulating solvent.

When you need to remove dirt, oil, and grease from parts, the most frequent choice in both home and professional shops is to use solvent, a stiff-bristle brush, and a solvent tank. This is Snap-on's smallest, bench-top solvent tank. It has a pump, internal filter, a removeable work shelf, and a flame-proof lid.

Solvents can be used out of a bucket, in a simple wash tank, or in a professional-type parts washer with a solvent pump. Small, benchtop parts washers are inexpensive and make parts cleaning a joy compared to using an old paint can and a brush. A typical benchtop cleaner holds from 2 to 5 gallons of cleaner, incorporates a small electric pump that directs a stream of solvent from a hose, usually has a filter to clean the re-circulating solvent, and most have a safety cover that will drop shut if the solvent catches on fire. A solvent tank of this type sells for $75 to $200. If you are setting up even a modest home workshop, give serious consideration to a tabletop washer. Once you use one, you'll never go back to the bucket-and-brush technique again!

If you do regular work in your shop, you might consider renting a solvent tank from a company like Safety-Kleen®. They will deliver a wash tank mounted on top of a 15- to 30-gallon drum of solvent to your shop/home. Then at prearranged intervals, they will clean the washer and install a new drum of solvent. If this fits into your budget, it solves both problems of parts cleaning and solvent disposal. Solvent tanks of this design are also available for outright purchase from several companies. The costs range from about $200 to $600.

Chemical Carburetor Cleaners

Solvent is an excellent choice for removing dirt, grease and oil. However, gum, varnish, and fuel residues that build up on carburetor surfaces and in internal passages are quite resistant to solvent. It's here that you need to bring in the "big gun": liquid carburetor cleaner. Composed of nearly everything that modern science can dream up (limited by the EPA, of course), *this stuff really works!*

Carburetor cleaner is available at most auto part stores. It can be purchased in various sizes, starting in one-gallon cans. McKay, Mechanic's Brand, Thyme, and several other manufacturers offer three-to-five gallon sizes (supplied in five-to-seven gallon cans having several inches of airspace above the solvent) that are often the minimum practical size for automotive carburetor rebuilding. Most carburetor cleaners are supplied with a dip basket that is essential for proper use. If the basket is not included with the cleaner, be sure to purchase one separately.

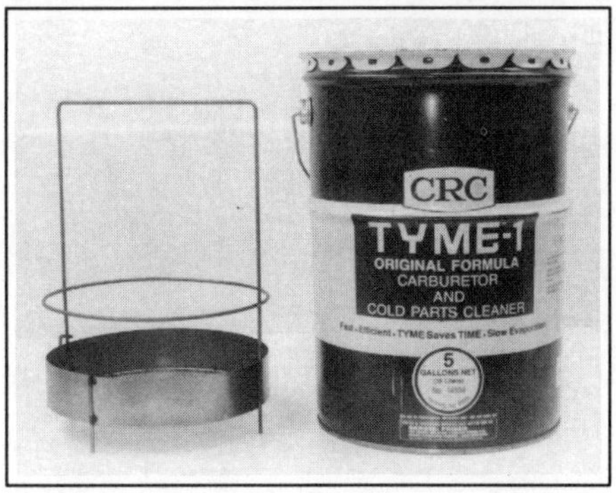

All the best intentions at the solvent tank won't budge tough carburetor deposits. It's here that you need to bring in the "big gun": liquid carburetor cleaner. Most carburetor cleaners are supplied with a dip basket that is essential for proper use. If the basket is not included with the cleaner, be sure to purchase one separately.

How long do parts need to be immersed in carburetor cleaner? That depends on the strength of the solution and the amount of accumulated varnish on the parts. A relatively unsoiled carburetor in fresh cleaner should be thoroughly clean in under twenty minutes. Very dirty carburetors, immersed in a weaker solution, may require several hours of immersion or even an over night soaking.

Remember, carburetor cleaner is not just a slightly stronger solvent. Some carburetor parts can be permanently damaged if placed in carburetor cleaner. Never immerse rubber O-rings, diaphragms, and rubber check valves, as they will quickly transform into a swollen, grotesque shape. Most plastic and nylon parts will withstand brief periods of immersion, but avoid dipping electrical parts, such as heated choke covers or solenoids. Some newer Holley carburetors use a green or black Teflon coating on the throttle shafts. Soak these parts just long enough to remove surface varnish; extended immersion can remove the Teflon coating.

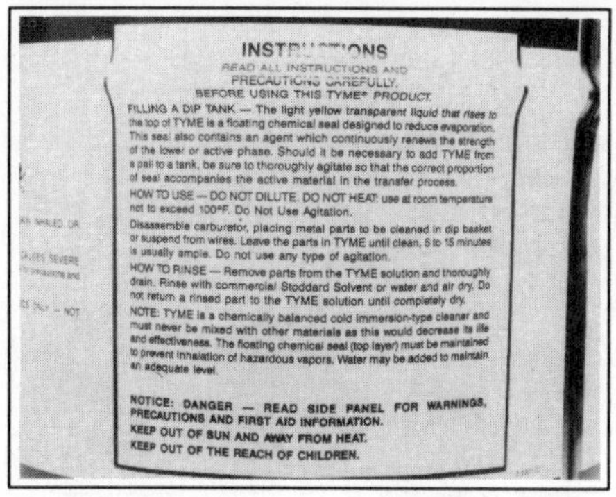

Carburetor cleaner is not cheap, but if you follow the instructions on the can and in this section you will be able to clean many carburetors and hundreds of other small parts before the cleaner becomes exhausted. And don't forget to read the precautions and warnings. Carburetor cleaner is easily absorbed through the skin; always wear gloves and goggles when you clean parts.

Extending Carburetor Cleaner Life

Carburetor cleaner is not cheap, but if you follow the instructions on the can and keep the following tips in mind, you will be able to clean many carburetors and hundreds of other small parts before the cleaner becomes exhausted. The most effective life extender is the simplest by far: keep the can tightly covered when not in use. Most brands of carburetor cleaner are supplied with an inch or two of chemical sealer that floats on top of the cleaner solution. This "cap" prevents air from reaching the cleaner and minimizes evaporation and deterioration. Over time this cap will diminish, but adding small amounts of water (usually water, but read the instructions on the can) will replenish this important air barrier.

Another way to extend carburetor cleaner life is to prevent contaminating the cleaner with dirt and sludge that can easily be removed with regular solvent. Never allow other cleaning solvents or gasoline to mix with your carburetor cleaner, as this can substantially reduce its strength.

If you plan to rebuild several carburetors, you may want to purchase an air-agitated immersion cleaner; the equivalent of a solvent tank for carburetor cleaner. They look similar to a standard 5- to 7-gallon pail, but have an air mechanism built into the lid. A piston-driven system raises and lowers (or rotates, or both) a parts basket through the cleaning

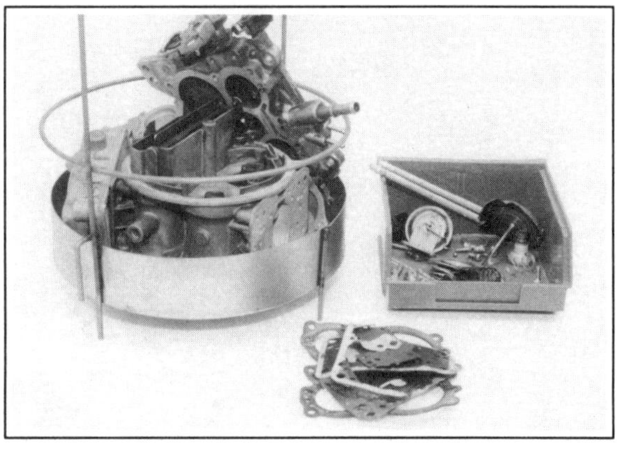

Carefully sort parts before immersion in carburetor cleaner. Never immerse rubber O-rings, diaphragms, or rubber check valves, as they will quickly transform into a swollen, grotesque shape. Most plastic and nylon parts will withstand brief periods of immersion, but avoid dipping electrical parts. Soak green or black Teflon-coated throttle shafts just long enough to remove surface varnish.

Gasoline contains a myriad of chemicals—and even small amounts of water—that can react with the metals used in Holley carburetors. The dichromate plating on this carburetor had worn off when it was left sitting on a shelf with fuel in the float bowls. The resulting white, powdery corrosion has clogged virtually every orifice in the metering blocks.

solvent at an adjustable speed. This substantially speeds cleaning and allows you to effectively use weak carburetor cleaner. Air-agitated cleaners can also be used with other chemical cleaners, degreasers and even a hot-water or soap bath (discussed in Chapter 7). Most units will attach to standard 10- or 12-inch diameter pails, permitting their use on several buckets in succession. They are available from auto parts and tool supply stores for between $100 and $200.

Finally, here's a carburetor cleaner storage tip that can save you a real headache. The chemicals in carburetor cleaners are dense and only slowly warm or cool to match room temperature. This can cause condensation to build up on the bottom of the can, particularly when it is allowed to stand on a concrete floor. *The problem is that condensation leads to rust!* Needless to say, finding five gallons of carburetor cleaner all over your workshop floor is not the best way to start off a Saturday morning. Store carburetor cleaner cans (and other solvents) on wooden blocks or on a sturdy shelf, away from moisture and direct sunlight.

Dichromate Refinishing

Gasoline contains a myriad of chemicals—and even small amounts of water—that can react with the metals used in Holley carburetors. It is for this reason that almost all cast metal components that come in contact with fuel, especially the float bowls and metering blocks, are treated with a dichromate "chemical film" plating that gives Holley carburetors their unique yellowish-gold appearance. This very thin coating effectively protects the interior passages and other surfaces from corrosion. Holley recommends against using any carburetor that no longer has a visible dichromate coating on the interior surfaces.

The protective dichromate film can be removed from carburetor surfaces by any type of abrasive cleaning, such as bead blasting, or even scrubbing with a wire brush. If the interior parts of your Holley look more silvery than yellowish, indicating that the base metal is exposed, you should consider having the components replated before reassembly, particularly if you live near the ocean (corrosion from salt in the air) or if the fuel you use contains high levels of moisture. If the dichromate plating is worn away from areas that do not come in regular contact with fuel (such as external surfaces), but the interior surfaces are intact, you do not need to replate the components. Instead, you may wish to apply a custom epoxy finish to the exterior of the carburetor. There are several fuel-resistant paints widely available in paint and auto part stores that can add an impressive custom look to your Holley. For more information on this procedure, refer to Chapter 8.

Every cast metal component in a Holley carburetor is treated with a dichromate "chemical film" plating that gives Holley carburetors their unique yellowish-gold appearance. This plating facility, located in Holley's Tennessee rebuilding plant, plates used carburetors scheduled for rebuilding and carburetors and components sent in by customers from all over the world.

If your carburetor is looking a bit worn because the dichromate plating is worn away from external surfaces, consider applying a fuel-resistant custom epoxy paint to the exterior. It can add an impressive custom look to your Holley. Further information on custom finishing can be found in *Chapter 8*.

If some or all of your carburetor components require redichromating, your choices are somewhat limited. A few years ago, it would have been possible to obtain redichromating services at many local plating shops, or even to do it yourself with a few cents worth of chemicals dissolved in water. But dichromate, like many other plating processes, involves the use of "heavy metals," like chrome, that are now considered hazardous waste by the EPA. Restrictions applied to the use of these materials throughout the U.S. have curtailed their widespread use, and finding local plating shops that will apply chemical films to small batches of carburetor components is quite difficult and/or expensive. Luckily, if you cannot find a local source, Holley will redichromate your carburetor for a minimal charge. For more information on redichromating, contact Holley Replacement Parts in Goodlettsville, Tennessee, 615-859-4924; see appendix.

Chemical Safety

A word of caution is in order. Carburetor cleaner, lacquer thinner, and other materials are effective, but powerful, chemicals. If you accidentally drop something in a bucket of carburetor cleaner, don't stick your arm in to fish it out and assume that it won't affect you. Many of these chemicals are toxic and can be absorbed through the skin. Keep them off your skin by using heavy rubber gloves (with "gripper" type fingers, since parts are very slippery when wet), a long-sleeved shirt, safety goggles, and extreme care. Avoid breathing any vapors. Follow all directions and precautions on the containers, and treat them like the dangerous, but effective, chemicals that they are.

ASSEMBLY LUBRICANTS

While it's true that some experts assemble Holley carburetors without lubricants of any kind, the appropriate lubricants used in the right places are similar to special tools: they make your job easier and the final product consistently higher in quality. But using incompatible lubricants on rubber O-rings can damage the rubber and lead to fuel leaks and other carburetor malfunctions, including the possibility of underhood fires! The nearby table indicates what lubricants to use and what lubricants to avoid when rebuilding Holley carburetors.

The fail-safe lubricant to use on Holley carburetors is petroleum jelly. It is available in nearly every market in America under the familiar Vaseline® brand name. This material will not damage O-rings or other rubber parts. In addition, it can be sparingly applied to bowl and metering block gaskets to facilitate their removal if you plan on disassembling the carburetor for subsequent tuning or modification. However, petroleum jelly is not the best choice for throttle- or choke-linkage lubrication. In these applications, a lithium-soap grease works much better, but lithium grease is incompatible with many rubber parts, so don't use it during assembly or anywhere inside a Holley carburetor. For this reason, petroleum jelly (or other products that primarily consist of petroleum jelly, like ChapStick®, etc.) is the only acceptable assembly lubricant for fasteners on Holley carburetors.

Another widely-known lubricant is silicone spray. Not to be confused with silicone-rubber adhesive, silicone lubricant is often a water-thin liquid that is compatible with O-rings, but unfortunately, the other chemicals used to dispense silicone from aerosol cans are sometimes not compatible with O-ring rubber. *Furthermore, very small amounts of silicone lubricant can cause engine oil to foam, potentially leading to bearing failure, particularly in racing applications.* Because of these concerns, Holley recommends that silicone never be used as an assembly lubricant on any engine components.

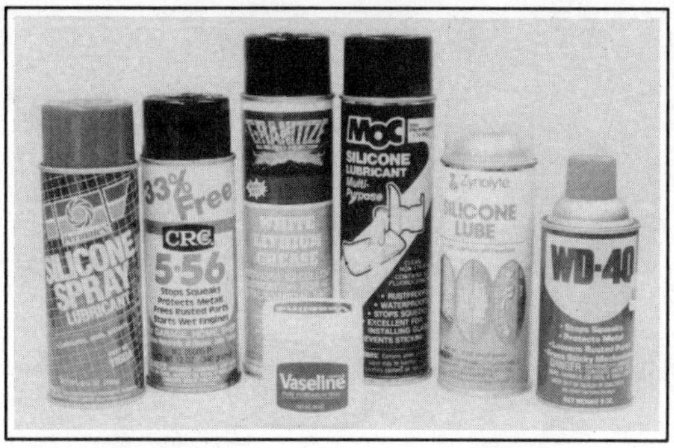

Some rebuilders assemble Holley carburetors without lubricants of any kind. However, I've found that the appropriate lubricants used in the right places are similar to special tools: they make your job easier and the final product consistently higher in quality. Review the application chart below before applying lubricants to carburetor components; the fail-safe lubricant to use on Holley carburetors is petroleum jelly.

WD-40® is a penetrating lubricant sold in virtually every hardware, auto parts, and office supply store. It is basically a thin oil and, as such, is not compatible with O-rings or other rubber parts. Don't use WD-40 anywhere inside a Holley carburetor, however, it can be sprayed on the exterior as a rust and corrosion preventative for long-term storage. WD-40 is too thin to be an effective lubricant for throttle or choke linkages. Again, use lithium grease on linkage systems.

Finally, a word about silicone rubber. Sold in tubes as a gel, silicone rubber is a remarkable, high-temperature adhesive with thousands of uses, but unfortunately, it is completely incompatible with gasoline. In the presence of fuel, it turns into a slimy mess that easily clogs carburetor passages. Don't use silicone rubber adhesive on any fuel tank, fuel line, mounting gasket, or carburetor component. If you do, I guarantee you'll only do it once. The trouble it will cause you and the mess you'll have to clean up will teach you a lesson you'll never forget!

Material	Trade Names	Application Notes
Petroleum Jelly	Vaseline Petroleum Jelly Petrolatum	Best lube to use on O-rings, allows them to slide with less chance of surface damage; OK to use on bowl and metering block gaskets; best thread lubricant. Pure petroleum jelly is a fail-safe assembly lube for Holley carburetors. Acceptable, but not the best choice for external linkage.
Rust penetrant/thin lubricants	WD-40 2-in-1 Oil CRC 5-56	Not compatible with O-rings or other rubber parts; don't use on bowl or metering block gaskets or as an assembly lube on fasteners, may seep out and come in contact with rubber parts; OK to spray on exterior of carburetor to prevent tarnishing during long periods of storage; Acceptable, but not the best choice for external linkage.
Lithium-Soap Grease Spray or Paste	Granitize White Lithium Grease	Not compatible with O-rings or other rubber parts; don't use on bowl or metering block gaskets; OK to lubricate external linkage.
Silicone Spray	Permatex Zynolyte	Don't use on O-rings or any internal parts during assembly; OK to spray on exterior of carburetor; not a good choice for linkage lubrication. Can cause engine oil to foam.
Silicone Rubber	Siliastic Silicone Gasket Bathtub Caulk	Silicone rubber is an adhesive gel not a lubricant. It is incompatible with gasoline. Don't use silicone rubber on any carburetor or fuel line components.

HOLLEY Rebuilding & Modifying

3—Component Identification

Many of the upcoming chapters refer to individual components that make up a Holley carburetor. In these chapters I've made the assumption that you have at least a familiarity with most major Holley parts, that you know most of their common names, and that you understand their basic function. If you are new to Holley carburetors, or if you would just like a refresher course, this chapter will take you on a quick tour of the Holley carburetor, its various models and characteristics, and its unique part numbering system.

HOLLEY MODULAR CARBURETORS

Reader note: The four-digit model numbers commonly used to identify various Holley carburetors should not be confused with the "part numbers" that Holley stamps on carburetor main bodies (Holley calls these stamped part numbers "list" numbers). The only way to determine the model number of a particular Holley carburetor is by evaluating its physical characteristics (e.g., by comparing it with photographs of known models). Additional information on Holley model/part/list numbers is presented later in this chapter.

The first modular-style Holley carburetor, introduced in 1957 on Ford's high-performance 312cid V8 engine, was a model 4150. The carburetor used similar metering blocks and removeable jets in both the primary and secondary sides. This basic design was later supplied with a non-jetable secondary metering plate and designated as a 4160, otherwise it is identical to the original model 4150. Both 4150 and 4160 carburetors are supplied with either mechanical or vacuum-actuated secondaries, depending on the specific part (list) number and the intended application.

A close cousin of the 4150/4160, the models 3150/3160 were introduced in the heady performance days of the 1960s, and they look, at first, identical to the 4150/4160. Just flip these carburetors over, however, and the difference is quite visible; they're three barrels! The models 3150/3160 use a large "flap" in place of two secondary throttle bores, but in almost every other respect, they're a standard, modular 4150/4160. The high-flowing three barrels were only produced for a short period of time, and they now are quite rare.

In 1971, Holley introduced two new modular carburetors, the models 4165 and 4175. These carburetors both had a spread-bore design, using small primaries and very large secondary bores. The 4165 has mechanically-actuated secondaries and, like the 4150, removeable jets in the secondary side. The 4175, on the

This book covers all Holley modular carburetors; like the 4150 pictured on the left. Other Holley carburetors not based on this modular concept, like the 4010 and the 1900 on the right, utilize many dissimilar components and, because of these variations, are not included in this book.

other hand, uses vacuum-actuated secondaries and, like the 4160, has a non-jetable secondary metering plate.

In the late 1960's, to meet the needs of factory-sponsored Nascar racing, Holley introduced a carburetor that looks very different from it's previous modular models. The "Dominator" 4500 series, however, is very much the same modular design, using 4150 metering blocks and float bowls. The major technical difference in the 4500, aside from its large main body and airflow capacity, is its unique mechanical secondary-actuation linkage located within the main body, between the throttle bores.

The last variations in Holley's modular carburetor series are the 2300 and the 2305. The 2300 is a high-performance two-barrel, and can, for the purpose of rebuilding and modifying, be considered the front half of a 4150. The 2305 is a unique, staged two-barrel designed for 4- and 6-cylinder engines. The 2305 opens one barrel first then, as the "rear-mounted" throttle linkage continues to the full-open position, progressively opens the secondary barrel. Both of these models use the same, familiar, modular float bowls and primary jets.

All of the carburetors mentioned so far have a modular design, operate very similarly, and allow the interchange of many individual components (with varying degrees of functional success). These basic models, and the hundreds of individual carburetors with unique part numbers (Holley calls them list numbers) based on these models, are covered in this book. The table on the left lists most Holley carburetor models, indicates the unique features of each, and designates those models covered in the rebuilding and tuning sections of this book.

Carburetor Type	Model Number	Identifying Features	Covered In This Book
Modular	4150	Original Holley modular design. Uses metering blocks and replaceable jets on both primary and secondary sides. Available with both vacuum and mechanical secondary actuation.	YES
	4160	Variant of original modular design. Uses thin metering plates that incorporate non-replaceable jets on secondary side. Available with both vacuum and mechanical secondary actuation.	YES
	3150/3160	Called a three barrel, it looks very similar to a 4150/4160. Uses large "flap" in secondaries instead of dual throttle bores. Now very rare, especially the 3150.	YES
	4165	Spread-bore design, with small primaries and very large secondary throttle bores. Has mechanical-secondary actuation, dual accelerator pumps. Uses dual metering blocks and replaceable jets in both the primaries and secondaries, similar to the 4150.	YES
	4175	Spread-bore design, with small primaries and very large secondary throttle bores. Has vacuum-secondary actuation, and incorporates non-replaceable jets on secondary side, similar to the 4160.	YES
	4180/4190	Similar to the 4160. Designed for Ford OEM applications. Have primary jets turned inward. Have unique bowl vent lines and idle-mixture screws located under lead plugs in the throttle plate rather than in the primary metering block.	YES
	4500	Float bowls and metering blocks similar to the 4150. Uses large main body with secondary-actuation linkage located underneath the main body, between the throttle bores.	YES
	2300/2830/2305	2300 and the 2830 are non-staged two-barrel carburetors; they look and function identically to the primary side of a 4150. 2305 is a unique, staged two-barrel available in street and competition versions; has throttle linkage on "rear" of carburetor.	YES
Non-Modular	1900/1901/6100	A variety of single-barrel carburetors, often used on six-cylinder engines. Many were manufactured with the Autolite tradename imprint. The model 1901 was a non-staged two-barrel carburetor designed for off-road VW applications.	NO
	2010/2011	New carburetor design for performance applications. Two-barrel versions of the 4010/4011, described below.	NO
	2100	Non-staged two-barrel carburetors designed for VW street applications.	NO
	2200	Non-staged two-barrel carburetors manufactured for Chrysler and Chevrolet OEM use.	NO
	4010/4011	New carburetor design for performance applications. Looks similar to Autolite four-barrel. Available with both vacuum and mechanical secondary actuation. 4011 is a spread-bore design, with small primaries and large secondary throttle bores.	NO
	4360/2360	Replacement for Rochester Quadra-Jet. Float bowl located in main body.	NO
	5200	A staged two-barrel. Called the Holley/Weber, it is made by Holley under license from Weber Carburetor. Commonly used on Pintos and Vegas.	NO
	6500/6700	Similar to above but with various electronic components and other minor design changes.	NO

The 3-barrel models 3150/3160 use a large "flap" in place of two secondary throttle bores, but in almost every other respect, they're a standard, modular 4150/4160 Holley. This particularly rare 1050cfm version, found at an auto swap meet in the Los Angeles area, uses discharge tubes instead of booster venturis in the secondary bore.

HOLLEY COMPONENT PARTS AND THEIR FUNCTION

A typical Holley four-barrel carburetor consists of over 200 individual parts. Although these pieces can make an intimidating pile on your workbench, there are only about 40 principal components that you need to be on a first-name basis with to proficiently work on your Holley. These main components are illustrated in the nomenclature drawing on the next page. Of the remaining component "pile," some are included in sub-assemblies that are not usually disassembled, such as

HOLLEY CARBURETOR NOMENCLATURE

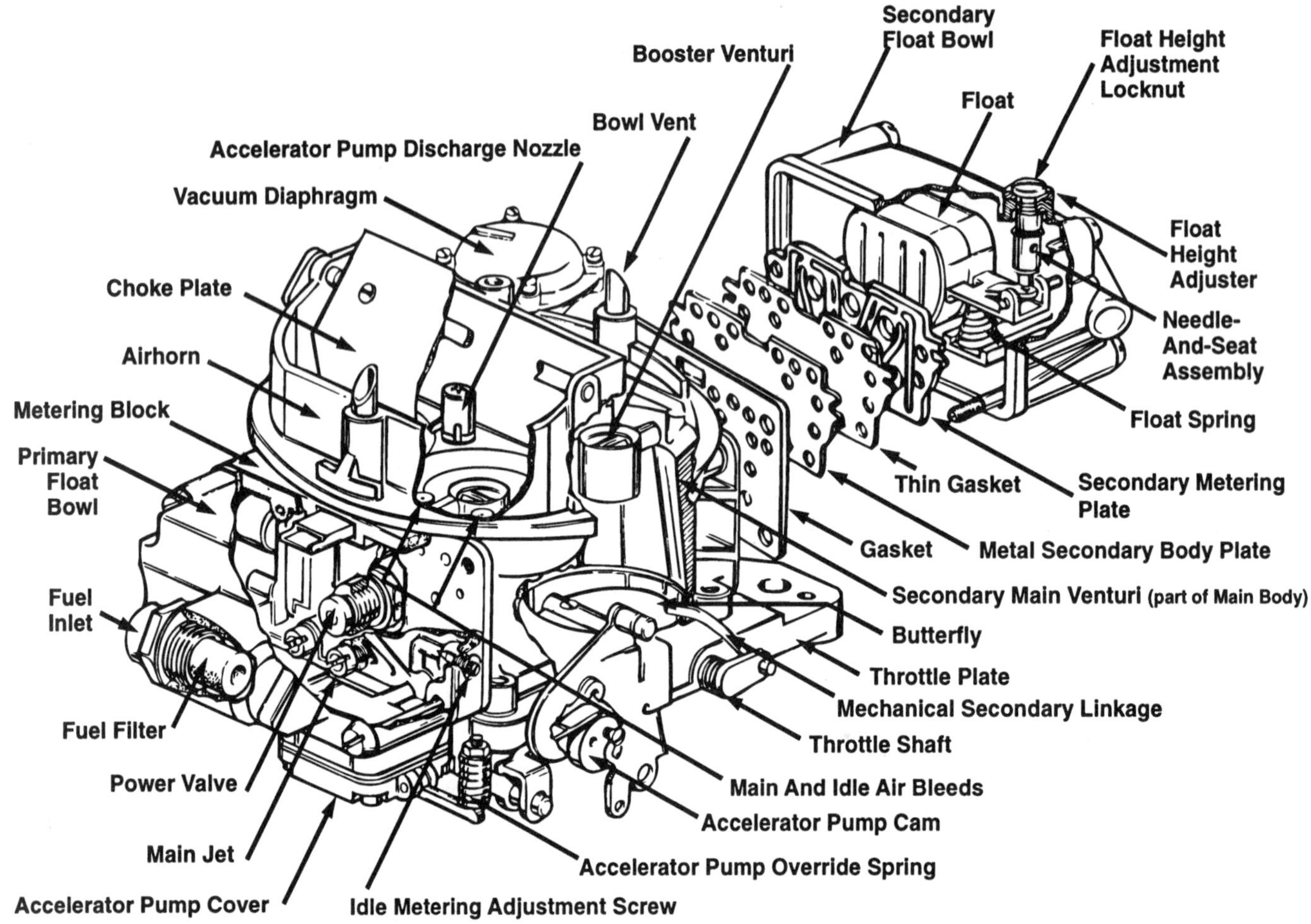

the throttle-plate and butterfly assembly, the choke plate and shaft, and the metering block interior components. The remaining parts are simply screws, clips, gaskets, O-rings, and miscellaneous metal stampings and brackets. So, once you're acquainted with the following important components and systems, you'll be well on your way to feeling like your Holley is just an old, familiar friend.

Accelerator Pump—A bellows-like plunger that pumps fuel from float bowl to a *discharge nozzle* located near the *venturi*. The pump is actuated by an *accelerator pump cam* mounted on the *throttle* shaft. When the throttle is opened, the pump discharges fuel through internal passages into the airstream to "cover up" the inescapable time lag in the activation of the *main metering* system. Without this temporary injection of fuel, the engine would suffer from lean misfire and backfiring. Holley uses two sizes of accelerator pumps: The standard 30cc pump, and a thick 50cc Reo pump for performance applications. For more information, refer to the sidebar *Accelerator Pump Basics* on page 92.

Accelerator Pump Cam—A plastic cam, located on the *throttle shaft*, that activates the *accelerator pump* through the pump-operating lever. Pump cams have two (usually) mounting holes and matching screw positions on the *throttle* linkage arm. These two positions, labeled "1" and "2," vary the pump discharge rate and volume. Holley also produces several different pump cams (each colored differently) that can be installed to tune pump discharge volume and timing to suit virtually any engine application.

Accelerator Pump Check Valve—A one-way valve located in the bottom of the *float bowl*—in the *accelerator pump* cavity—that allows fuel to fill the accelerator pump bellows but prevents reverse flow. Holley uses several different check-valve designs: A steel ball retained by a small metal "strap," a flexible red-plastic "umbrella," and a hard white-plastic "cap." Each of these designs allows fuel to enter the pump cavity when the *throttle* is closing and prevents reverse flow when the throttle is opened.

Accelerator Pump Discharge Check Valve—Most Holley carburetors use either a pump-discharge needle valve (located under the *discharge nozzle* in the *main body*) or a ball check (in the primary *metering block*) that prevents fuel from dribbling from the *discharge nozzle* after the pump shot is completed. Do not confuse this check valve with the *accelerator pump check valve* that allows the pump bellows to function.

Accelerator Pump Discharge Nozzle—A multiple-outlet nozzle mounted near the *venturi* that squirts fuel, pumped by the *accelerator pump,* directly into the airstream. The size of the orifices in the discharge nozzle determines <u>how fast</u> the fuel is injected. Holley manufactures four discharge nozzle designs—anti-pullover, tube, center-squirter, and standard straight type—and a wide selection of orifice diameters to adjust the pump discharge rate to a variety of applications.

Accelerator Pump Override Spring—An adjustable spring assembly located on the *accelerator pump* linkage arm that permits temporary compression in the pump linkage when the *throttle* is opened quickly. Because liquid fuel contained in the pump bellows (diaphragm) is not compressible, the linkage arm would bend without a method to absorb this load. Linkage damage is prevented when the override spring compresses, and while under compression, it continues to force the pump bellows to discharge fuel until the preset fuel volume (determined by the shape of the *accelerator pump cam*) is pumped from the *discharge nozzle*. The override spring assembly contains an adjustment that, when properly set, will eliminate both "free-play" and binding.

Accelerator Pump Transfer Tube—A short connecting tube used on some Holleys between the primary *metering block* and the *main body*. It is used to prevent internal leakage of pressurized *accelerator-pump* discharge fuel. The transfer tube fits into counterbores located in the main body and metering block and seals with rubber O-rings. Defective O-rings will allow fuel to directly enter the intake manifold through the *power-valve* vacuum passage, generating an over-rich condition each time the *throttle* is opened.

Air Bleeds (Idle & Main)—Part of the *idle and main metering systems*. The air that enters the idle and main metering systems through the air bleeds "pre-mixes" with the fuel and helps emulsification and atomization. The sizes of the air bleeds affect the overall air/fuel ratio and the start-up point for each system.

Airhorn—The upper extension of the *main body* over the primary *venturis* that supports the choke plate and shaft assembly. When the choke plate is closed, it forms a snug fit to the airhorn that increases signal strength (vacuum) at the *booster venturis* and *idle circuits*. This dramatically increases fuel flow and enriches the mixture during cold startup and running.

Booster Venturi (Primary & Secondary)—A small venturi positioned within the *main venturi*. The booster venturi forms a restriction within the main venturi and boosts the pressure drop in the *main metering system*. The increased signal strength improves fuel flow sensitivity and accuracy, resulting in superior throttle response, atomization, and fuel economy.

Bowl Vent—A passageway that allows atmospheric pressure to enter the airspace above the fuel in the *float bowl*. These pressure-equalization passages prevent pressure or vacuum from building up in the float bowls that would adversely affect the air/fuel ratio.

Butterfly (Butterflies)—Thin, round metal (sometimes steel, but also aluminum or brass) plates that are attached to the *throttle shafts* that control the flow of air and fuel from the carburetor to the engine. The butterflies are machined to form a nearly airtight seal to the bores in the *throttle plate*. The butterflies are secured to the throttle shafts with screws that have been crimped to prevent loosening. As an additional safety margin to help prevent engine overspeeding, the screw holes in the butterflies are always offset in such a way that engine vacuum tends to hold the *throttles* closed.

Choke Plate—See *Airhorn*.

Dominator Carburetor—The name given by Holley to a unique series of carburetors principally designed for racing applications. Dominators have large *main bodies*, *venturis*, and *butterflies*, and are rated at 1050 and 1150cfm. These high flow volumes are essential to producing maximum horsepower on large racing engines. Holley recently introduced a 750cfm Street Dominator that uses large *booster venturis* to maximize signal strength on street engines.

Float—A lever arm with a buoyant element that responds to fuel level in the *float bowl* and activates the *needle-and-seat* assembly to control the flow of inlet fuel. Some Holley floats are made from a buoyant phenolic material, however, Holley also uses floats made from either sealed brass or plastic "cans."

Float Bowl—A reservoir for fuel that supplies the *metering systems*. A float bowl contains the *float*, *needle-and-seat* assembly, and a linkage mechanism that maintains a constant fuel level, essential to delivering a predictable and stable air/fuel ratio. Holley has manufactured several float bowls for specific applications, however, they are all based on two design concepts: <u>dual-feed center-pivot</u> bowls are often used on large carburetors in drag-racing applications; <u>single-feed side-pivot</u> bowls, that use a fuel transfer tube to supply the secondary bowl, are often used on street carburetors.

Float Height—The extent to which the *float* rises in the *float bowl* before it forces the needle into the seat and prevents further fuel flow into the bowl. This adjustment controls the level of the fuel in the float bowl and, to some extent, the air/fuel ratio generated by the carburetor. Some Holley models use an externally-adjustable float (as shown in the Nomenclature drawing) while other models require bending the float arm to vary fuel height. Holley has specified a float height for every carburetor they have produced. These specifications can be found in rebuilding kits and in Holley's *Illustrated Parts & Specs Manuals*.

Float Spring—A "bumper" spring that is sometimes used to counterbalance part of the downward force of gravity applied to the *float*. The reduction of downward force helps dampen float movement and maintain a uniform fuel level in the bowls, especially when driving on rough roads.

Idle Mixture Adjustment Screws—Two small slotted screws, located one per side on the *metering block* or sometimes in the *throttle plate*. Many Holley carburetors use idle mixture screws only on the primaries, however, some carburetors have adjustable "4-corner" idle systems. Most Holley idle-mixture adjustments control the flow of fuel to the idle and off-idle discharge ports. A few "reverse-idle" carburetors use idle-mixture screws to control the flow of air mixed with idle fuel. Carburetors with a reverse-idle system are often labeled with stickers just above the mixture screws.

Main Body—This is the "heart" of every Holley. It consists of the *main venturis*, *booster venturis*, *air bleeds*, *airhorn*, and many interior passages. The main body determines, for the most part, the airflow capacity and overall functional characteristics of the carburetor. The main body is not available from Holley as a replacement part.

Main Jets (Replaceable)—Replaceable (screw-in) restrictions in the *metering block* that determine main-system air/fuel ratio while the *power valve* remains closed. When manifold vacuum drops and the power valve opens, additional fuel is metered into the main circuit through the *power-valve channel restrictions*. The main jets can be thought of as the principal restrictions that determine air/fuel ratio during part-*throttle* cruise. Also see *Secondary Metering Plate*.

Mechanical Secondary Linkage—A direct mechanical linkage between the *primary* and *secondary throttle shafts* that begins to open the secondaries when the *primary butterflies* open a preset amount, regardless of the airflow through the carburetor. The primary throttle shaft position at which the secondaries begin to open and how fast (the rate) they reach a full-open position are determined by the design of this linkage.

Metering Block—A casting that contains the *main jets*, *power valve*, and the idle and main-circuit channels and restrictions; it is positioned between the *float bowl*(s) and the *main body*. Most metering blocks contain two independent metering systems: *idle and main*. Idle circuit adjustment is usually provided by means of two adjustable screws located on each end of the metering block (note: some Holley carburetors for OEM applications use *idle-mixture screws* located in the *throttle plate* instead of the metering block). Removeable main jets, changeable power valves, and *power-valve channel restrictions* determine main-system air/fuel ratios. Some Dominator and other special-application carburetors use metering blocks that contain a third, intermediate metering system that provides separate transition fuel flow to maintain optimum air/fuel ratios. Holley models 4150, 4165, and the 4500 use metering blocks on the primary and secondary sides, while other Holley 4-barrel models use *secondary metering plates* with non-removeable jets.

Metering Systems (Idle & Main)—The combination of metered restrictions (*main jets, power valve, main & idle air bleeds,* etc.) and fuel and air passages that deliver the optimum amount of fuel into the airstream at various throttle positions. The primary idle system provides fuel at low engine speeds or when the throttle plates are nearly closed. Idle air/fuel ratio is variable with two adjustment screws commonly located in the primary *metering block.* The primary main metering system delivers fuel at higher engine speeds or when the throttle is opened more than about 20%. The secondary idle (if so equipped) and main metering systems function similar to those in the primary.

Needle-And-Seat—A valve assembly, activated by the *float,* to turn on and off the flow of fuel into the *float bowl.* When the fuel level rises in bowl, the float pivots upward and forces the needle—often tipped with Viton rubber—into the seat and closes off the flow of inlet fuel. When the float drops in response to a lower fuel level in the bowl, the needle moves away from the seat and fuel flow resumes. Needle-and-seats for performance applications have larger seat diameters to increase fuel flow rates. Some needle-and-seat assemblies use internal float springs to help dampen float movement and maintain a uniform fuel level in the bowls. Holley also manufactures needle-and-seat assemblies that do not use Viton-tipped needles for use with exotic fuels such as alcohol, benzine, or acetone.

Power Valve—A vacuum-actuated valve in the *metering block* that allows fuel to pass through the *power-valve channel restrictions* into the *main circuit* and enrichen the mixture. When the *throttle* is opened and manifold vacuum drops, the mixture established by the *main jets* is too lean to supply the power required. In response to low manifold vacuum, the power valve "opens" and permits additional fuel to enter the main circuit through two channel-restriction orifices. This additional fuel flow richens the mixture and allows the engine to produce maximum power. *Power-valve channel restrictions* can be thought of as the principal orifices that determine air/fuel ratio during heavy acceleration. Holley produces a variety of power valves that "open" at various vacuum levels. For further information on power valves see Chapter 11 and the sidebar *Power Valve Basics* on page 52.

Power-Valve Channel Restrictions (PVCRs)—Two drilled restrictions in the *metering block* (not shown in the Nomenclature drawing) that, when the *power valve* opens, allow additional fuel to enter the *main circuit* and enrichen the mixture. Some performance experts recommend tuning the main jets to provide good fuel economy and adjusting the PVCR diameters to obtain an optimum air/fuel ratio at wide-open throttle. This method (instead of just "tuning" the main jets) can optimize both fuel economy and power. For more information see Chapter 11, *Performance And Economy Tuning.*

Secondary Body Plate—See *Secondary Metering Plate.*

Secondary Metering Plate—Thin cast plate with air and fuel *metering* passages and restrictions, used only on the secondary side of some Holley 4-barrels and on the outboard pair of carburetors on three-two-barrel induction systems. The secondary metering plate performs many of the same functions as a *metering block.* The metering plate, however, does not have removeable *main jets.* If air/fuel ratio changes are required, the entire metering plate must be replaced (Holley manufactures several plates with various metering orifice sizes for this purpose). Metering plates are used either with or without a secondary body plate (see Nomenclature drawing), depending on the design of the main body. Body plates—and the thin gasket between the body plate and the metering plate—are only necessary when the secondary side of the main body has a cavity cast into it, similar in appearance to the primary side *power valve* cavity; if the secondary metering face is flat, no body plate is required.

Spread-Bore Carburetor—A carburetor that uses small primary and very large secondary butterflies. Based on the original Rochester Quadra-Jet concept, Holley 4165 and 4175 are spread-bore carburetors.

Square-Bore Carburetor—A carburetor that uses similar-sized primary and secondary butterflies. All 4150/4160/4180/4190 and 4500 Holley four-barrels are square-bore carburetors.

Throttle—A term that refers, variously, to the *throttle shaft/butterflies* in the carburetor, the "gas pedal," or even the linkage connecting the gas pedal to the carburetor. It usually refers to a device or devices that control the flow of fuel to the engine. So that, "opening the throttle" would mean "performing a procedure that in some way would result in the rotating of the throttle shafts and allowing the engine to draw in additional fuel."

Throttle Plate—A casting that contains the *throttle shafts, butterflies,* various vacuum passages, and attaches the *main body* to the intake manifold on all modular Holley carburetors except the 4500 (which has the throttle-plate components built into the main body). The throttle plate is an aluminum casting that is easily bent or cracked if the incorrect base gasket (too compressible) is used or if the attaching bolts are overtightened.

Throttle Shafts (Primary & Secondary)—A steel rod that rides in the *throttle plate* that is slotted or machined to accept the *butterflies* and has various linkage attachments on one or both ends. A four-barrel carburetor has individual *primary* and *secondary throttle shafts.* The throttle shafts may be coated with Teflon to reduce wear and help prevent sticking.

Vacuum Secondary Diaphragm—A vacuum actuated "motor" that opens the secondary *throttle plates* when flow through the primary *venturis* reaches a preset level. A small hole just below the minimum diameter of one primary venturi is the source of the vacuum that is used to actuate the diaphragm. This vacuum signal, proportional to the airflow in the primary venturi, is "conditioned" by several counterbalancing bleeds, restrictions, and springs to open the secondary throttles at the precise point in flow and at a precise rate to optimize performance.

Venturi (Main, Primary & Secondary)—A tapering tube cast into the *main body* above each *butterfly.* As air flows through the venturi, it creates a pressure drop near the *booster venturi* (greatest drop is just below narrowest portion of venturi). This pressure drop draws fuel from the *main metering system,* through the booster venturi, and into the airstream. As engine speed increases, airflow through the venturi increases causing an even greater pressure drop. The increased pressure drop draws additional fuel into the air-

You cannot locate the *model number* on any Holley carburetor. The *model number* can only be determined by evaluating the physical characteristics of the carburetor, i.e., having two, three, or four barrels; a metering block or metering plate on the secondaries; if it is a spread- or square-bore design; or in some cases if it is an emissions or racing carburetor. However, the *list (part) number* can always be found on Holley carburetors. It is located on the airhorn or, in the case of the Dominator and a few racing carburetors, on the top of the main body.

If you are interested in obtaining any of the Holley performance parts mentioned throughout this book—like chrome float bowls, main jets, power valves, etc.—first call Holley for their latest performance catalog (see appendix for phone number and address ordering information).

stream. This relationship between pressure drop and airflow maintains a uniform air/fuel ratio throughout a wide range of airflow levels and is the basic principle behind carburetor function.

UNDERSTANDING HOLLEY PART NUMBERS

Much has been written about Holley part numbers, and a good deal of it does little to help the average enthusiast understand the variety in the numbering system that Holley uses to identify their carburetors and components. Fortunately, all of this complexity can be reduced to a relatively simple concept consisting of two groups of numbers: *model numbers* and *list numbers* (*list numbers* are just another name for *part numbers*).

All carburetors have both a model number, identifying its general characteristics, and a list number (part number) by which you can order a replacement carburetor and component parts from Holley. As I mentioned earlier in this chapter, you cannot find the model number stamped anywhere on the carburetor. You can only identify the model number by evaluating the physical characteristics of the carburetor and comparing it with photographs in this and other books, including Holley's documentation. The model number identifies: if the carburetor is a two-, three-, or four-barrel; if it has a metering block or metering plate on the secondaries; if it is a spread- or square-bore design; or in some cases if it is an emissions or racing carburetor. The model number generally *does not* indicate: if the carburetor has mechanical or vacuum secondaries; its CFM rating; whether it has an automatic or manual choke; or if it will work well on any particular engine. In other words, the model number only identifies the general configuration; it's the list number that identifies a specific carburetor within any model group.

The list number is, for all practical purposes, Holley's part number for the carburetor. A list number is stamped on every Holley carburetor. As it applies to the modular carburetors covered in this book, list numbers can be found on the airhorn or, in the case of the Dominator and a few racing carburetors, on the top of the main body. When you want to order a rebuilding kit or secondary jet plate or any other replacement part, it's the list number that you'll need. While it's true that some list numbers have been superseded with new *sales numbers*, the new number should only be a concern for the individual behind the counter at the parts store. All that's required of you is to be sure you know the original list number stamped on your carburetor (and, of course, try to find a counter man that's at least partially awake).

Note: There are many idiosyncrasies in Holley's part numbering system. Those aware of these details can, in some cases, determine if a carburetor was designed for performance or street applications, identify unusual carburetors, etc. Because of the complexity involved and space required to cover Holley part numbers in detail, it must be considered beyond the scope of this book. However, for those interested in learning a little bit more about Holley numbering, refer to the sidebar *Dating Holley Carburetors*, on page 39.

PERFORMANCE PART NUMBERS

If you are interested in obtaining any of the Holley performance parts mentioned throughout this book—like chrome float bowls, main jets, power valves, etc.—first call Holley for their latest performance catalog (see appendix for phone number and address ordering information). This catalog is aimed at the enthusiast and is easy to understand and use.

Part numbers in the performance catalog consist of a prefix, that indicates the category of the part, and a suffix, that identifies a specific part within the category. For example, all main jets for modular Holley carburetors begin with the prefix **122-**. This number indicates that the part is a main jet, while the suffix indicates the specific jet and, in fact, shows the relative flow rate. A number 70 jet would be identified with the part number **122-70**. Other performance parts have their own identifying prefix: **0-** for performance carburetors, **12-** for fuel pumps, **34-** for performance accessories, **121-** for discharge nozzles, **134-** for secondary metering plates, etc.

As simple as this prefix-suffix part numbering system is to understand, all you *really* need to know is the part number itself or, if you want to purchase a replacement or performance carburetor, the carburetor list or part number. With the part number in one hand and the telephone number to Holley's parts division (or your local speed or carburetor shop) in the other, you can obtain any currently available carburetor, replacement part, or any performance part for the entire line of Holley carburetors. What could be simpler?

4—Before You Begin

The information in this chapter will guide you through the essential first step of the carburetor rebuilding process: *Isolating whether an engine problem is actually due to a carburetor malfunction.* By following the techniques presented here you should be able to locate almost any engine malady quickly with minimum expense. And if the trouble is eventually traced to your Holley, you will know what component parts deserve special attention during the rebuilding process.

HOW TO USE THIS CHAPTER

This chapter is composed of four parts: 1) an introduction to troubleshooting techniques that follows; 2) a troubleshooting chart on pages 24 and 25; 3) specific diagnostic procedures that cover pages 26 to 31; and 4) information on selecting carburetor kits to start your rebuild on the right foot.

The troubleshooting chart on pages 24 and 25 is really your starting point to pinpointing engine problems. The first Problem column lists five basic engine troubles. Select the problem that most closely matches your situation and refer to the next Possible Causes column. Then for each possible cause, refer to the Corrections and Section columns for the appropriate corrective measures and a reference to the detail diagnostics procedures that follow the chart. As you work your way down the rows, you first will be testing for non-carburetion problems. When you reach the final possible cause, you will have isolated the malfunction to your Holley. You can then refer to the carburetion diagnostics section on page 29 for advice on what carburetor components deserve special attention during your rebuild. Finally, proceed to the last two pages in this chapter where you will find helpful information on selecting carburetor kits.

The detailed diagnostic procedures that follow the chart will help you perform each of the indicated troubleshooting tests. Although this section contains a great deal of practical testing advice, because of the limited space in this book, there is only room to describe the most common procedures. You may find it helpful to obtain additional information about your specific vehicle. Service manuals that cover a range of vehicles over a span of several years of production are available from Motors and Mitchell. Haynes, Chilton, and Clymer manuals are more focused and cover only one make and model and range over about two years of production. The best and most expensive repair manuals are available from the vehicle manufacturer. They are comprehensive, accurate, and offer detailed troubleshooting information. They are often available at new car dealers.

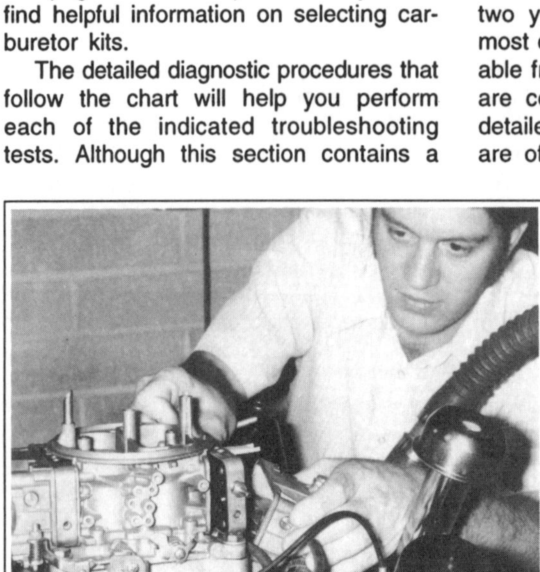

I've lost count of the number of times I've seen both amateur and professional mechanics routinely single out the carburetor as the culprit. While it's true that carburetor malfunctions can be the source of an engine problem, the trouble often can be traced to other engine systems.

You may find it helpful to obtain additional information about your specific vehicle. Service manuals that cover a range of vehicles over a span of several years of production are available from Motors, Mitchell, Haynes, Chilton, and Clymer. The most detailed and most expensive repair manuals are available from the vehicle manufacturer. They are comprehensive, accurate, and offer detailed troubleshooting information.

Timing lights are a fundamental piece of test equipment. This MSD light produces a brilliant beam, has an easy-to-use inductive hookup, looks terrific, and is rock-solid reliable.

TROUBLESHOOTING TECHNIQUES

The procedure for correcting engine malfunctions is a three-step process. The first is to *identify and pinpoint the cause* or causes of the problem. The second is to *fix the problem* by adjusting, repairing, or replacing the malfunctioning component(s). The third step is to make sure the repair has actually fixed the original problem–and not created new problems–by *performing a road test.*

You should be able to locate almost any defective component that can affect engine performance by using the troubleshooting chart on the next two pages and referring to the detailed diagnostic procedures that follow it. Unfortunately, not all defective parts are defective all the time. These intermittent problems are the most difficult malfunctions to fix. In these situations, it pays to look for the less than obvious, such as loose or damaged electrical connections. Search in places where the insulation of a wire may have worn through, causing intermittent shorts. Inspect moving components, such as heat riser valves and choke mechanisms, for sticking, since these are often the source of sporadic problems.

Some intermittent problems are even more difficult to locate because they provide no visible clues. For example, a coil that occasionally gets hot enough to malfunction, but then cools down and begins working properly before you can perform diagnostic tests, can be very difficult to pinpoint. A good strategy to use in these situations is to substitute a part that is known to be good for the suspect component. And I can't emphasize this enough: *Make sure you know that the substitute part is in perfect working order!* More than once I've come across new parts that were just as "dead" as the 50,000-mile old part they replaced.

The final step of any quality repair requires confirming that the problem was fixed and that no other problems were created. The only way to accomplish this is to run the vehicle through a complete road test. A road test should always include the following two major modes: *1) Engine start-up and cold operation.* The engine should crank over strongly and start quickly. It should not stall when cold. *2) Acceleration, cruise performance, and idle quality after warm up.* The vehicle should accelerate smoothly without hesitation or surging under light, medium, and heavy throttle. The engine should run at low and high speeds under light throttle without missing or surging.

TROUBLESHOOTING TOOLS

In order to perform the diagnostic procedures and tests discussed in this chapter, you will need some basic test equipment. Here are some of the most commonly needed items:

Hand Tools–In addition to the tools discussed in Chapter 2, the following tools are also helpful for engine diagnostic work: 1/2-inch drive ratchet and socket set, a distributor wrench, slotted and Phillips screw starters, ignition wrenches, an Allen wrench set, sparkplug socket(s), sparkplug wire pliers, a battery terminal puller, and a battery terminal cleaning brush.

Timing Lights–The timing light is a fundamental piece of test equipment. Inexpensive timing lights require a direct connection at the distributor or sparkplug, while more costly models have an inductive pick-up that clamps around the plug wire. However, all timing lights will accurately indicate ignition timing, providing they have no internal malfunctions.

Compression Gauge–The compression-test gauge is an indispensable engine diagnostic tool. The best gauges (usually not much more expensive than cheap models) are equipped with a flexible hose and adapters, allowing compression tests on almost any engine.

Vacuum/Pressure Gauge–Vacuum readings reveal information about the valvetrain, the intake manifold, carburetor, cylinder condition, and the exhaust system. (Note: *Avoid highly damped gauges.* Rapid needle movement can help diagnose engine problems, such as valvetrain malfunctions).

Electrical Test Equipment–An extensive array of automotive electrical test equipment is currently available, ranging from simple volt meters to sophisticated digital oscilloscopes. If you can afford only two basic test devices, they should be a volt-ohm meter, or VOM, and a simple 12-volt test light. General purpose VOM's are available at electronics supply stores such as Radio Shack. Test lights are very inexpensive and can even be home made by soldering wire leads to a 12-volt light bulb.

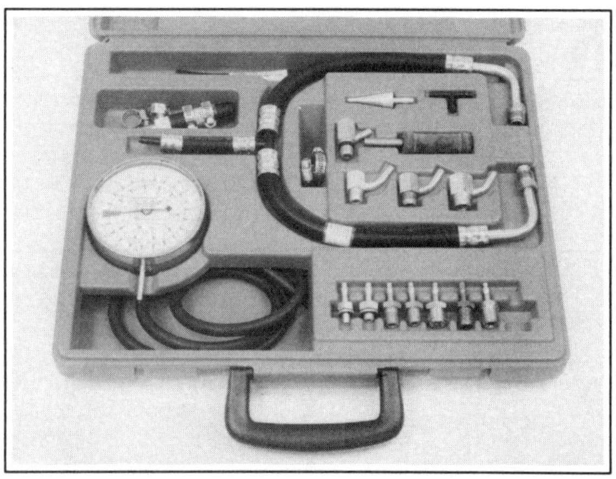

The vacuum gauge is so inexpensive and valuable as a diagnostic tool that I find it incomprehensible that so many tune-up technicians either don't own one or rarely use it. This Snap-on vacuum gauge kit includes a wide variety of fittings for general-purpose and specific-vehicle testing.

On-Car Troubleshooting Chart

Problem	Possible Causes	Corrections And Comments	See Section
Engine Will Not Crank Or Cranks Slowly	Defective Or Discharged Battery	Charge battery, load test, replace if necessary.	Starting System
	Corroded, Loose, Or Damaged Battery Cables, Missing Or Damaged Engine Ground Straps	Clean or replace corroded or damaged battery and starter cables. Check engine ground.	Starting System
	Damaged Or Burned Out Fusible Link	Check fusible link with ohmmeter or test light, repair or replace as necessary.	Starting System
	Improperly Adjusted Shift Linkage Or Defective Neutral Safety Switch	Check ignition switch and starter circuit for continuity; check neutral safety switch and shift linkage for proper operation.	Starting System
	Defective Starter Motor Or Starter Solenoid	Check starter for excessive current draw or intermittent operation; check solenoid for open or burned contacts.	Starting System
	Starter Drive Gear Jammed In Flywheel	Remove starter, inspect flywheel for damaged or missing teeth, inspect starter drive gear for damage.	Starting System
	Engine Overheated Or Internally Damaged	Check for internal engine damage by rotating engine manually (make sure starter isn't engaged--see above).	Engine Mechanical
	Timing Too Far Advanced Or Distributor Mechanical Advance Malfunction	Check distributor for proper mechanical advance operation; check and adjust engine timing.	Ignition System
	Improper Grade Of Engine Oil	Check for proper engine oil viscosity, particularly in cold climates.	Engine Mechanical
	Coolant Hydrolocking	Remove all sparkplugs; check for presence of coolant in cylinders (leaking head gaskets, cracked heads or block, etc.).	Engine Mechanical
	Carburetor Malfunction; Fuel Hydrolocking	Possible in extreme cases of carburetor flooding caused by defective needle-and-seats or ruptured power valves, etc. Remove all spark plugs; check for presence of fuel in cylinders .	Carburetion
Engine Cranks Over Normally But Will Not Start Or Hard Starting	Primary Ignition System Malfunction	Check points and condenser (reluctor and pick-up coil); check coil primary wires and ballast resistor wiring for integrity; check distributor shaft for excessive play; check distributor shaft for rotation when engine is cranking; check ignition module (electronic ignition); check for proper cranking dwell (points and condenser ignition).	Ignition System
	Secondary Ignition System Malfunction	Check sparkplugs for excessive gap, wear, or fouling; check coil output; check coil wire and sparkplug wires; check distributor cap and rotor for carbon tracking, cracks, or moisture; check for proper firing order; check for correct initial timing.	Ignition System
	Insufficient Fuel Delivery To Carburetor	Check for fuel in gas tank; check for proper fuel volume and pressure at carburetor inlet; check fuel pump for correct operation; check fuel lines and hoses for cracks, kinks, or other damage; check all fuel filters in system for restrictions .	Fuel System
	Contaminated Or Wrong Fuel In System	Check for contamination in fuel; stale, or wrong type of fuel.	Fuel System
	Intake System Vacuum Leak	Check all vacuum hoses for damage or incorrect connections; Check for damaged power brake booster or leaking vacuum reservoir cannisters; check for cracked intake manifold or leaking intake gaskets; check for damaged heat riser passages or faulty EGR valve .	Intake/Vacuum System
	Engine Mechanical Problem	Perform compression and/or leak down test; check for broken cam or timing chain; check for damaged pistons or rings.	Engine Mechanical
	Carburetor Malfunction	Check carburetor choke for proper operation and for sticking closed; check accelerator pump operation; check float level and for presence of fuel in bowls.	Carburetion

Problem	Possible Causes	Corrections And Comments	See Section
Idles Rough Or Stalls	Faulty Distributor, Incorrect Timing Or Secondary Wiring	Check distributor points for correct adjustment or burning; check distributor cam for excessive lobe wear; check for loose distributor shaft or breaker plate; check for correct sparkplug wire routing.	Ignition System
	Burned Or Faulty Sparkplug Wires	Test all sparkplug and coil wires with an ohmmeter or scope, replace entire set if any test bad.	Ignition System
	Intake System Or EGR Malfunction	Check for intake manifold vacuum leaks; inspect EGR system for proper operation or stuck EGR valve.	Intake/Vacuum System
	Valvetrain Problem	Check for worn cam lobes, bent pushrods, broken rockerarms, or incorrect valve lash adjustment.	Engine Mechanical
	Carburetor Malfunction	Check and adjust idle-speed setting, idle-mixture setting, choke adjustment and function, fast-idle setting, idle solenoid or dashpot function; check float level setting; check idle circuit and transfer circuit; check for damaged or incorrect carburetor gaskets; check for excessively worn primary or secondary throttle shafts; check for loose carburetor-to-manifold fasteners or a damaged carburetor mounting gasket.	Carburetion
Missing Or Poor Performance On Light Throttle Acceleration Or Cruising	Faulty Vacuum Or Mechanical Advance	Check for malfunctioning vacuum advance; check for frozen or sticking mechanical advance; check for loose breaker plate or distributor shaft.	Ignition System
	Incorrect Timing	Check and adjust static timing; check for abnormal timing fluctuation at low and high engine speed.	Ignition System
	Broken Pick-Up Coil Wire Or Points Wire	Check distributor pick-up coil wires for damage (electronic ignition); check points and condenser wires for damage and tight connections.	Ignition System
	Low Secondary Voltage Or Faulty Sparkplugs	Check coil output; check for defective ballast resistor or wire; check spark plugs for broken insulators, loose or damaged plug wire boots, or fouling.	Ignition System
	Insufficient Intake Manifold Vacuum	Check for intake manifold vacuum leaks; missing, damaged, or incorrect vacuum hose routing; stuck or malfunctioning EGR valve.	Intake/Vacuum System
	Carburetor Malfunction	Check carburetor float level; check power valve for damage and proper number; check for loose or damaged booster venturi; check main jets for obstructions and correct size; check for damaged or incorrect carburetor gaskets; check accelerator pump system for proper operation; check secondary throttle return linkage and springs for proper operation .	Carburetion
Poor Performance Or Missing Under Heavy Acceleration Or High RPM's	Ignition Timing Incorrect	Check and adjust timing, make sure mechanical and vacuum advance mechanisms are operating properly.	Ignition System
	Secondary Ignition System Malfunction	Check for sparkplug and coil wires "breaking down" under load; test for adequate coil output voltage; check sparkplug condition.	Ignition System
	Inadequate Fuel Supply	Check fuel lines and hoses for damage or kinking; check all fuel filters for proper flow; check fuel pump for pressure and volume supply.	Fuel System
	Dirty Or Restrictive Air Filter	Inspect air filter and replace if necessary; be sure air filter is of adequate size for engine size and rpm range.	Intake/Vacuum System
	Restrictive, Plugged, Or Damaged Exhaust System	Inspect heat-riser valve for free movement; check exhaust system for plugged catalytic convertor(s), damaged muffler(s) or other restrictions.	Exhaust System
	Low Compression In One Or More Cylinders	Perform compression and/or leakdown tests.	Engine Mechanical
	Worn Cam Lobes Or Incorrect Cam Timing	Inspect cam and valvetrain for worn or broken components; check timing chain and gears for excessive wear or jumped timing.	Engine Mechanical
	Carburetor Malfunction	Check for full throttle opening and secondary operation; check choke operation for complete opening; check for correct float level; check power valve system for proper operation; check main jets for obstructions or incorrect sizes.	Carburetion

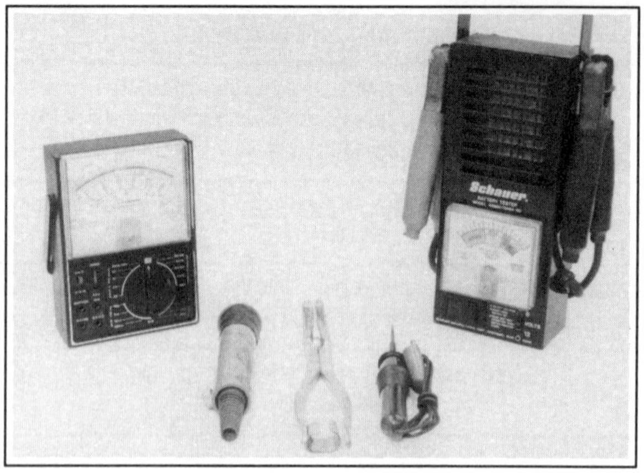

A test light and a VOM (volt-ohm meter) are necessary for testing the starter circuit. Battery terminal service tools are also helpful. If the battery cables and the starter pass diagnostic tests, the battery should be checked with a high-current load tester, like the unit on the right.

THE STARTING SYSTEM

Battery and Cables—Most engine cranking difficulties can be traced to the battery cables or the battery. Remove, inspect, clean, or replace both battery cables. Battery terminals are easily damaged, so use only minimal force—or use a puller—to remove stuck battery cable ends. And always remove the ground cable first and reconnect it last. The engine and chassis ground straps are potential problems that are often overlooked. Carefully check for damaged, loose, or missing straps and repair or replace as necessary. If servicing the cables does not correct the problem, the battery should be removed, *fully charged*, load tested and replaced if necessary.

Starter Switch and Wiring—If, after checking the battery and cables, the starter fails to spin and no audible "clunk" is heard, the ignition switch and its related components should be examined. A wiring diagram, a test light, and a VOM are necessary for testing the starter circuit. Two items to check carefully (if your vehicle has them) are the neutral safety switch and the main fusible link. You can find the fusible link location in your service manual. It should be checked for continuity with a test light or ohmmeter. The neutral safety switch prevents starter operation except in park or neutral. The starter will not operate if the neutral safety switch is defective or the shift linkage is out of adjustment.

Starting-system diagnostics should include a check of the neutral safety switch and the main fusible link with a test light or VOM. The starter will not operate if either of these items is defective.

The Starter Solenoid and Relay—If, after the above tests, there is still no "clunk" from the starter, the solenoid and/or starter relay may be defective. Refer to a shop manual for the procedures for locating and checking the starter solenoid and relay.

Starter—The starter should be removed and tested if all of the above tests fail to crank the engine. While the starter is removed, examine the flywheel and starter for damaged or missing teeth. Worn starter teeth occasionally jam, locking-up the engine.

Engine—Before reinstalling the starter remove all of the sparkplugs and check for the presence of coolant or large amounts of fuel in any of the cylinders by turning the engine over with a breaker bar and socket on the front dampener bolt. Liquid in a cylinder can "hydrolock" the engine and cause severe damage if enough torque is supplied by the starter.

Ignition—Ignition system malfunctions can sometimes cause slow engine cranking. You should suspect the ignition system if the engine kicks back or cranks slowly, but turns over normally when the coil wire is removed. Make sure initial timing is set properly and that the mechanical advance mechanism is not stuck. Crossed or mis-routed sparkplug wires will cause engine kickback, as can a cracked or carbon-tracked distributor cap.

THE IGNITION SYSTEM

Distributor—Most primary ignition problems are related to the distributor. Points, if used, should be checked for wear, alignment, pitting, burning, and correct adjustment at least every 5,000 miles and replaced, along with the condenser, at least every 10,000 miles. Check the distributor cam for worn lobes and/or excessive roughness; then **sparingly** lube with a light grease. The points or reluctor mounting plate is an area of trouble that is often overlooked. If this plate is loose or worn, it will be impossible to keep the points (reluctor) in proper adjustment. Worn distributor shaft bushings will also produce this effect.

Cap and Rotor—Remove the distributor cap and carefully inspect it and the rotor for the following: cracks, condensation, carbon tracks, and burned contacts. Also closely inspect the center carbon contact for wear or burning. When in doubt replace the cap and rotor.

Electronic Ignition—There is an air gap measurement between the reluctor and pick-up that should only require setting if these components have been serviced or replaced. *A special non-magnetic (brass) feeler gauge must be used to perform this operation.* The magnetic pick-up has one or more small, fragile, wires connected to it. These wires can break (internally) causing erratic ignition operation or even a no-start condition. Electronic control module problems are often intermittent and, since they

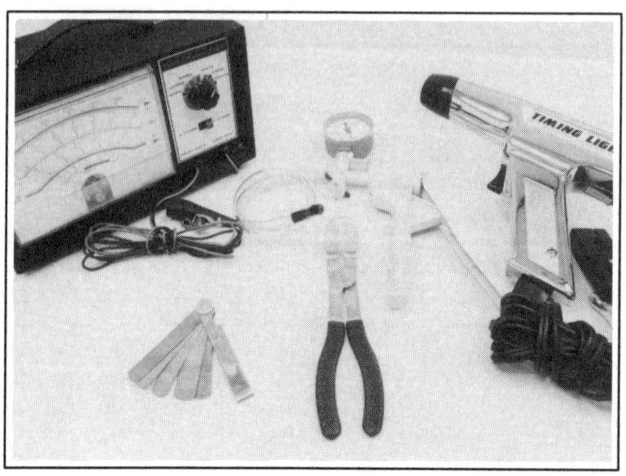

A dwell meter, a feeler gauge set, a sparkplug wire puller, a timing light, and a vacuum tester such as the *Mity-vac* are useful for ignition troubleshooting. A special non-magnetic—brass—feeler gauge must be used to set the air-gap between an electronic reluctor and pickup.

are sealed, they cannot be repaired. The easiest way to check a module is to substitute one that is known to be good in place of the suspect unit. Unfortunately, since most parts stores don't allow the return of electrical parts, this can be expensive unless you can borrow a "test" module.

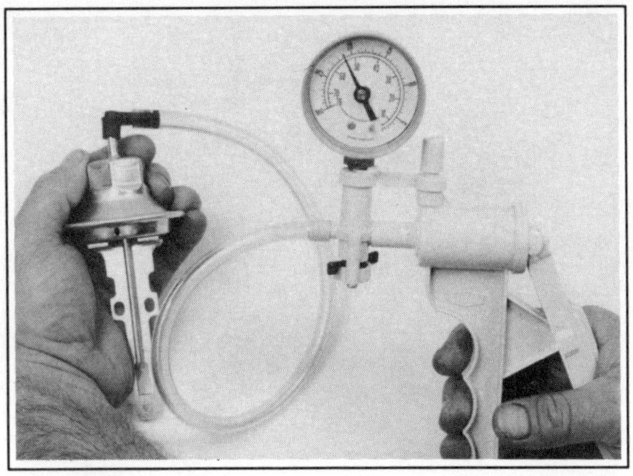

A defective vacuum advance can dramatically reduce fuel economy or cause surging. Test the vacuum advance with a vacuum tester, like the *Mity-vac*, available at many auto parts stores. Use a timing light to verify that the timing advances as vacuum is increased.

Mechanical and Vacuum Advances—Frozen, worn, or binding advance mechanisms can cause one or more of the following problems: hard starting and engine kick-back, lack of power, poor fuel economy, engine stalling, and surging during light throttle operation. The mechanical advance can be checked with a timing light by first disconnecting and plugging the vacuum advance hose. Check to see that the timing mark advances as engine speed increases. To test the vacuum advance use a vacuum tester, such as the *Mity-vac*, and verify that the timing advances as vacuum is increased.

Coil Resistor—During start-up, the ballast resistor is by-passed to allow a full 12 volts to the coil. A defective resistor or wiring is indicated if the engine starts but stalls immediately upon releasing the ignition switch.

Coil—Coils used in negative ground electrical systems are *always wired with the positive terminal connected to the ignition wire.* Always check the coil to make sure the primary leads are not reversed. Modern coils are very reliable and seldom fail. Output voltage can be checked by holding the distributor end of the coil wire (with a suitably insulated tool) about 1/2 inch from a grounded metal object and cranking the engine. An intense blue-orange spark should jump from the wire.

Plug And Coil Wires—Examine high-voltage plug and coil wires for missing or torn boots, cuts, abrasions, or burned spots. These conditions will cause arcing and engine misfire. *Arcing plug wires sometimes can be located by running the engine in the dark and looking for any visible sparking.* Resistor-type plug and coil wires often visually appear to be in perfect condition but are internally defective (burned). Test them with an ohmmeter. Wire resistance should measure approximately 1000 to 10,000 ohms per foot. Higher values than these indicate worn out wires. If one or two wires test bad, replace all of them.

Sparkplugs—Remove and inspect sparkplugs for correct heat range, proper gap, worn electrodes, excessive contamination, and cracked insulators. Engines that are modified and driven hard usually need one or two heat ranges colder plugs than stock. Engines that are worn (burn oil) may need "hotter" than stock plugs to prevent fouling. *Note: the correct heat range plugs will still foul if the carburetor mixture is too rich.*

THE INTAKE/VACUUM SYSTEM

Warning! Vacuum leaks can be hazardous to your engine's health. Rough engine operation or misfire that disappears as engine speed increases points to a vacuum leak. Unfortunately, if a vacuum leak is severe, it can lean the air/fuel mixture enough to cause severe engine damage. A malfunctioning EGR (exhaust gas recirculation) system can also act in a similar fashion.

Vacuum Hoses—Pinpointing a vacuum leak starts with a careful visual inspection of all vacuum hoses and hard lines. Check for cracking, splitting, swelling, and collapsed hoses. Examine steel lines for cracks or damage caused from abrasion. Refer to your service manual to verify vacuum line routing. Check each attachment point for snugness.

Vacuum Components—A defective vacuum-operated component is usually not revealed by a visual inspection. Pinch closed the vacuum line that leads to each component with needlenose *Vice-grips®*. If the idle noticeably improves, the tested component is defective. Test and repair items with the largest hoses (i.e. brake boosters, PCV valves) first and work down to the smaller sizes. A vacuum tester like the *Mity-vac* is useful for checking diaphragm-operated devices like the EGR valve. If the EGR valve is stuck open or opens too easily, it will cause a rough idle or poor low-speed performance.

Intake Manifold—*The intake manifold and gaskets should be tested only after all vacuum lines and components have been checked.* Leaking intake and carburetor mounting gaskets can sometimes be located using a mechanic's stethoscope or by squirting light oil around the suspected leak. Also check the carburetor mounting gasket. Although rare, the manifold may be cracked, causing a vacuum leak. Examine all the intake runners for visible cracks, paying particular attention to areas near the heat cross-over and EGR passages.

Air Filter—Don't overlook the air filter and element when checking the intake

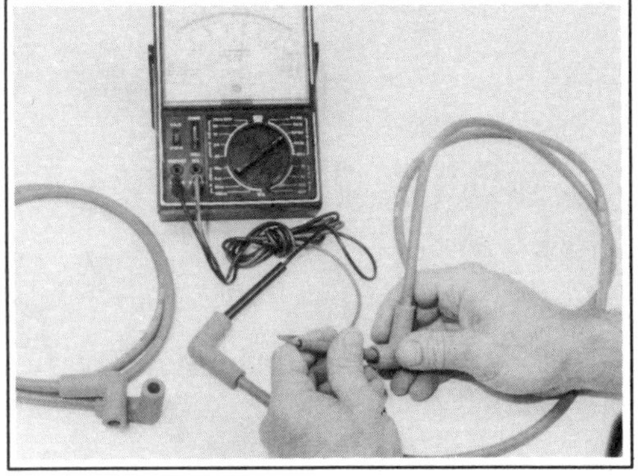

Resistor-type ignition wires often appear to be in perfect condition but can be defective (burned) on the inside. They can easily be tested with an ohmmeter. Wire resistance should measure approximately 1000 to 10,000 ohms per foot. Higher values than these indicate damaged conductors. If one or two wires test bad, replace them all.

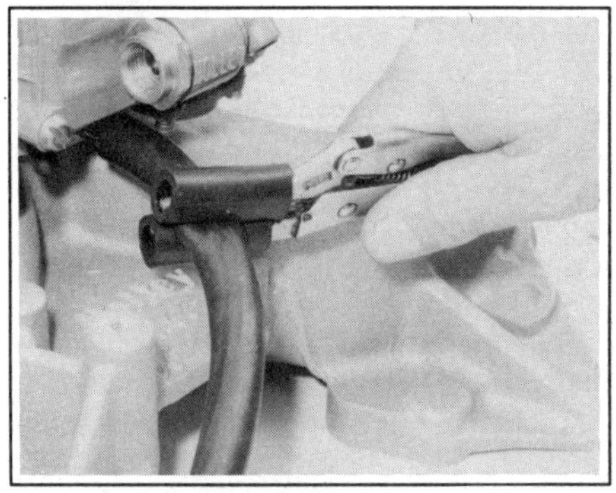

Defective vacuum-operated components are usually not found with a visual inspection. Try pinching the vacuum line closed with needlenose *Vice-grips®* that have pieces of rubber hose on the jaws to protect the vacuum lines. If the idle noticeably improves, the component under test is defective.

Don't overlook the air filter and element when checking the intake system. Installing a tall, high-flow element, like this K&N cotton-wire filter, will not only ensure optimum airflow capacity, but it will last many years, since it can be cleaned and reused.

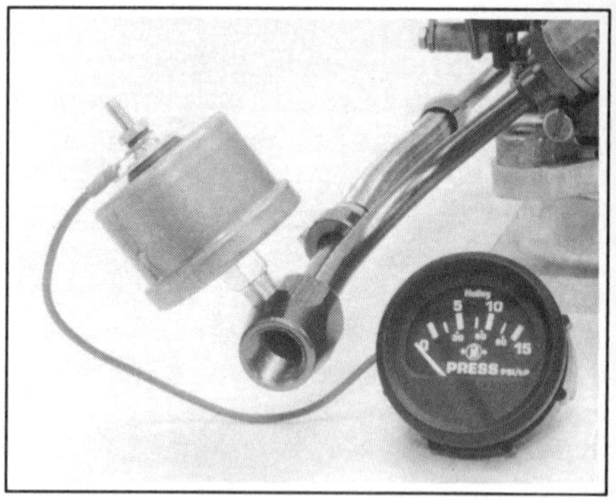

Monitoring fuel pressure can be very helpful in troubleshooting, however, routing a pressure takeoff for a mechanical pressure gauge into the driver's compartment is extremely dangerous. Holley offers this quality fuel pressure gauge that requires only an electrical connection from the pressure sensor to the dash.

system. A dirty, clogged filter will usually not affect idle or low-speed operation, but at higher speeds the engine will run rich and lose power. This over-rich condition can foul sparkplugs and accelerate piston ring wear.

THE FUEL SUPPLY SYSTEM

Contamination—Obtain a fuel sample and inspect it for contaminants, freshness, and purity (no water present). Gasoline begins to lose its volatility after a few months and will not provide the octane necessary for proper combustion. Many mysterious engine problems have been caused by nothing more than contaminated, old, or improper fuel.

Fuel Delivery Pressure—An adequate supply of fuel to the carburetor depends on two things: pressure and volume. Troubleshooting the fuel supply system requires both pressure and volume tests. *Caution: Exercise extreme care when performing the following tests. Avoid spilling gasoline on hot engine parts or near the ignition wires. Always have a fully charged fire extinguisher on hand when testing the fuel system. (Note: Clean or replace all fuel filters in the system including the filter located in the carburetor inlet before performing these tests.)* A pressure test is done by disconnecting the carburetor inlet line and installing a fuel pressure gauge on this line. Normal pressure for street vehicles should be between 4.5- and 7.0-psi. If the pressure is not correct and you have a pressure regulator installed, connect the gauge directly to the pump outlet. Adequate pressure at this point indicates a faulty or improperly adjusted pressure regulator. If the pressure is still too low, attach a hose from the pump inlet to a container of clean fuel. If the pressure remains low, the fuel pump is defective. Normal pressure after this test indicates a restriction or air leakage in the fuel-tank line. The fuel pick-up should be removed from the tank (when possible) and inspected for cracks or a plugged filter. Check all rubber hose connections to metal fuel lines. And don't forget to check the gas cap; never use a non-vented cap on a system designed for a vented one.

Fuel Delivery Volume—*A test for adequate fuel volume to the carburetor must be performed after the pressure test is completed.* Many times the fuel pressure is sufficient, but the fuel flow (volume) is inadequate. Fuel delivery volume is often rated in gallons-per-hour (gph). Gph can be calculated easily by timing how long it takes the fuel pump to deliver one pint of fuel. Disconnect the fuel inlet line from the carburetor and run a hose from it to a container marked at one pint. Divide 450 by the number of seconds it takes to fill the container to the pint level. This is the gph output of your system. The minimum gph depends on several factors: engine size, vehicle weight, maximum rpm, the type of carburetor(s), and engine modifications. If, as the container was filling, the fuel was foaming and pump volume was low, there may be air entering the suction side of the pump. If the fuel did not foam, there may be a restriction in the fuel-tank line or pickup or the pump itself may be defective.

Vapor Lock—Vapor lock is the interruption of fuel delivery to the carburetor caused by vaporization of fuel in the fuel lines or fuel pump from extreme heat. Minor cases of vapor lock can often be cured by routing fuel lines away from exhaust manifolds and other hot engine parts. Replacing a defective fuel pump can sometimes solve the problem. Installing an electric fuel pump in place of the mechanical unit is often the only way to cure chronic vapor lock on heavy RV or towing vehicles.

THE EXHAUST SYSTEM

Restrictions—Exhaust system restrictions, unless extreme, will cause problems only at speeds above idle. Engine overheating, lack of power, and a lowered rpm range are all symptoms. A visual inspection of the entire exhaust system is the first step. Check the heat-riser valve for free movement. If it is operated by vacuum, make sure the diaphragm and rubber lines are not damaged. Check all exhaust tubing, mufflers, and catalytic convertors for obvious damage.

Vacuum Test—Mufflers, catalytic convertors, and even exhaust tubing (double wall type), can appear fine and yet have internal restrictions. A vacuum test can reveal hidden problems. Connect an undamped vacuum gauge to a direct engine-vacuum source. A normal steady reading at idle will be about 16- to 22-in(Hg). Even a restricted exhaust system should give this reading at idle. Raise the engine speed to about 2500rpm and

Vapor lock is caused by gasoline becoming so hot in the fuel lines that it boils (vaporizes), preventing the fuel pump and carburetor from functioning. Installing a high-volume electric fuel pump, like this Holley unit, in place of the engine-mounted mechanical pump can often cure chronic vapor lock, especially on heavy RV or towing vehicles.

observe the gauge reading. While an unrestricted exhaust will show a slightly higher vacuum reading, an engine with a defective exhaust system will indicate a decreasing engine vacuum value, approaching zero. Since this low reading can also be caused by a malfunctioning EGR valve, test and/or replace the EGR valve and repeat the test. If the reading is still abnormal, each exhaust component must be removed and checked.

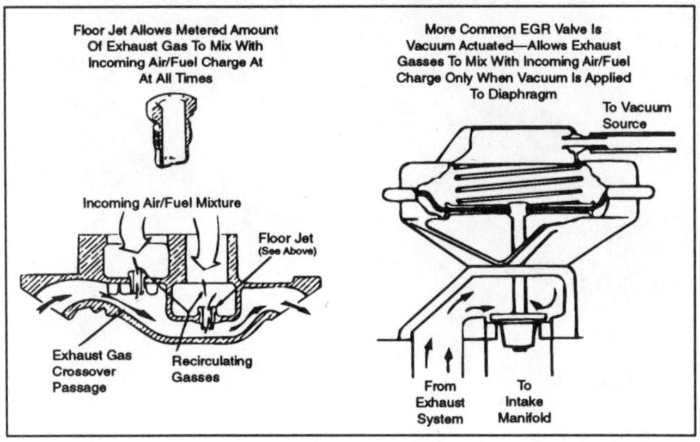

Exhaust gas recirculation (EGR) is either accomplished with a vacuum-actuated valve or a fixed orifice in the floor of the manifold. Either system directs exhaust gas into the induction system to reduce nitrogen-oxide emissions at moderate acceleration and at cruise when NO_x formation is the highest.

ENGINE MECHANICAL CONDITION

Power and economy both depend on the basic mechanical condition of the engine. No amount of carburetor, ignition, or exhaust work will make up for burned valves, leaking head gaskets or worn piston rings. The following tests and inspections will help you locate mechanical problems in the engine that may cause rough running, lack of power, or hard starting.

Vacuum Gauge Tests—Use a quality *undamped* vacuum gauge for these tests. Begin by connecting the gauge to a direct manifold vacuum source. The following are some typical gauge readings and what each indication means:

1) A steady reading between 16- and 22-in(Hg) at idle: This is normal. Radically cammed engines have lower, less steady readings.
2) Normal range at idle with sporadic drops below normal: This indicates a sticking valve.
3) Normal range at idle with needle vibration of about 2-in(Hg): This indicates an ignition problem. Check plug gap, dwell, coil, distributor cap, and plug wires.
4) A steady reading slightly higher than normal: This can be caused by a dirty air filter or overly advanced ignition timing.
5) A steady reading 3- to 12-in(Hg) lower than normal. This indicates one or more of the following conditions: Intake or carburetor vacuum leak, late ignition or cam timing, worn piston rings.
6) Gauge needle drifts slowly over a range of 4- to 5-in(Hg) at idle: This indicates an idle mixture that is too rich or lean.
7) Gauge needle fluctuates rapidly between 10- and 21-in(Hg) at idle: This occurs when one or more valve springs are weak or broken.
8) Gauge reads normal at idle but drops slowly as engine speed is increased to 2500: This indicates a restricted exhaust system.
9) Gauge reads below normal and fluctuates rapidly over a range of about 3-in(Hg) at idle, then the needle becomes steady as engine speed is increased: Worn intake guides usually cause this reading.

A compression test indicates how well the piston rings, valves, and head gaskets are sealing. Cylinder-to-cylinder balance is the most important information. The lowest cylinder reading should be at least 80% of the highest reading.

Compression Test—A compression test indicates how well the piston rings, valves, and head gaskets are sealing. The cylinder-to-cylinder balance is the most important information the compression test gives. The lowest cylinder reading should be at least 80% of the highest reading. Before checking engine compression, make sure the valves are properly adjusted (mechanical camshafts) and all the sparkplugs are removed. The engine should be at normal operating temperature. Disable the coil and make sure the throttle and choke plates are fully opened. The compression reading for a mechanically sound engine can vary from as low as 110psi to over 180psi, depending on the mechanical compression ratio and camshaft profile; it should, however, be consistent between all of the cylinders.

Uniformly worn piston rings can give consistent, but low, readings. To check for ring wear in a cylinder, remove the gauge and squirt some motor oil through the sparkplug hole. Crank over the engine to distribute the oil and recheck the compression. *The new reading should* ***not*** *increase more than 5- or 10-psi.* A larger increase indicates worn piston rings.

TROUBLESHOOTING CARBURETION

Before examining the carburetor, the other systems and components discussed in this chapter should be checked, repaired, replaced, and eliminated as sources of poor engine performance.

Carburetor Flooding—Flooding is one of the most common carburetor problems. It is caused by an excessively high fuel level in the float bowl(s). Fuel dripping from the booster venturis, running out the bowl-vent tubes, or excessively wet butterflies indicate flooding. The float level should be the first item checked when flooding occurs (see the sidebar *Fuel Bowl Basics* on page 55 for adjustment procedures). The fuel pressure should

Flooding is one of the most common carburetor problems. It can be caused by high fuel level in the bowls, often the result of contamination in the needles-and-seats, defective floats, or excessive fuel pressure. Checking the float level is the first thing to do when flooding occurs. Externally-adjustable float bowls are easily checked by removing the sight plug. After float adjustment, make sure the fuel level remains stable by observing the bowl level for a minute or two before replacing the plug.

The most common automatic choke failure is due to a defective external pull-off. If the diaphragm does not open the choke plate when the engine starts, an excessively rich mixture will be delivered to the engine. The pull-off is easily tested with a hand pump, like the *Mity-vac.* Apply vacuum for several seconds to check for slow leaks.

also be checked. Flooding is likely to occur if pressure exceeds 7.5psi at the carburetor.

If adjusting the float level(s) fails to eliminate flooding and the fuel pressure is not too high, the following items should be examined during carburetor disassembly: 1) needle-and-seat assemblies for dirt, wear, or damage to sealing gaskets or O-rings, 2) float weight, obvious physical damage, or leaks (see the sidebar *How To Weigh Floats* on page 84), 3) the float arm and pivot for correct alignment and free movement.

Hard Starting/Poor Cold Performance—The automatic choke and its related components are often the cause of cold starting and performance problems. Automatic chokes most often malfunction when the choke pull-off mechanism is not adjusted correctly or is inoperative. The choke cover may also be incorrectly adjusted. *The choke plate should be fully closed when the engine is cold and not running and fully opened after it is warmed up.* Replace the choke-cover/heating-element if the choke plate cannot be adjusted for full opening and closing.

Check the following items carefully when overhauling a carburetor that has poor cold running performance: 1) too-lean main jets, 2) the accelerator pump circuit, including the check valve and squirter nozzle size, 3) restrictions in the choke assembly vacuum-feed circuit, 4) low float level, 5) bent or binding choke-plate shaft, 6) restricted idle and main air bleeds, and 7) burned out heating element in the choke cover.

Hard Hot Starting—Difficult starting of a hot engine is often caused by fuel "percolation" or boiling in the float bowls. Make sure the heat-riser valve isn't stuck closed and that there is a restrictor in the crossover passage, if required. An insulator gasket under the carburetor will often reduce percolation problems.

Other items to check: 1) choke adjustment and operation, 2) needle-and-seat assemblies for leaks, 3) float height and operation, 4) external or internal float bowl leaks (bowl draining after shut off), sometimes due to a defective power valve, and 5) vapor lock from an overheated fuel pump or lines.

Rough Idling and Stalling—Check and adjust idle speed and idle mixture as described in Chapter 10. Make sure there are no vacuum leaks and the EGR valve is not stuck open. Check the carburetor mounting gasket for damage and make sure the mounting bolts are secure. Check and adjust the float level and make sure the carburetor is not flooding. If the carburetor is equipped with a hot-idle compensator valve, make sure it is operating correctly. It should be closed when the engine is cold and open above approximately 140 degrees(F). Also look for a dirty air filter element, clogged air bleeds or bowl vent tubes.

During disassembly pay close attention to: 1) idle-feed circuit for debris or incorrect size, 2) ruptured power valve, 3) flatness of main body mating surfaces, 4) incorrect or damaged metering block gaskets, 5) dirty idle-well passages, 6) loose or badly worn throttle shafts and butterflies, 7) cracked or warped throttle plate, and 8) damaged or incorrect carburetor mounting gasket.

Surges at Light Throttle—An EGR valve that is stuck or opens too soon can often cause surging during light throttle operation. Also check the PCV valve and crankcase breather for clogging. Poor fuel delivery caused by a weak fuel pump or restrictions in the fuel tank pickup or venting system can also cause engine surging. Double check for vacuum leaks. And don't overlook the air filter element; change it if there is any doubt.

Check the following carefully during rebuilding: 1) clogged or improper size main jets, 2) float level, 3) power valve for damage and correct number, 4) restrictions in the main and idle-air bleeds or bowl vents, 5) contaminated fuel, and 6) flatness of main-body mating surfaces.

Hesitates On Acceleration—A defective accelerator pump system is most often the cause of stumbling or hesitation on initial acceleration. With the engine shut off, remove the air cleaner and visually check the accelerator pump squirter. A quick movement of the throttle should produce a steady stream of fuel from both nozzles aimed toward the center of the venturi. The flow of fuel should begin as soon as the throttle is moved. There should be no free play between the pump cam and lever when the throttle is closed. At wide-open throttle there should be at least .015 to .020-inch additional movement possible. The engine may still hesitate if the amount or rate of fuel discharge is incorrect. See Chapter 11 for

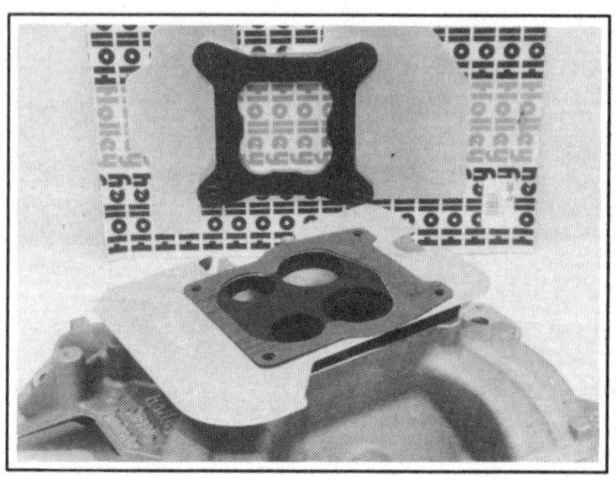

If the engine is hard to start when hot, the problem may be due to carburetor overheating. Installing an insulated base gasket, like this one from Holley, can often reduce hard starting and improve performance.

The main-body mating surfaces can become so warped that they will not properly seal, even with new gaskets. This can cause internal leaks that produce all sorts of performance problems, like rough idling, surging, poor fuel economy, and hesitation. If you measure warpage of more than .008-inch, you should consider filing or machining the surfaces flat.

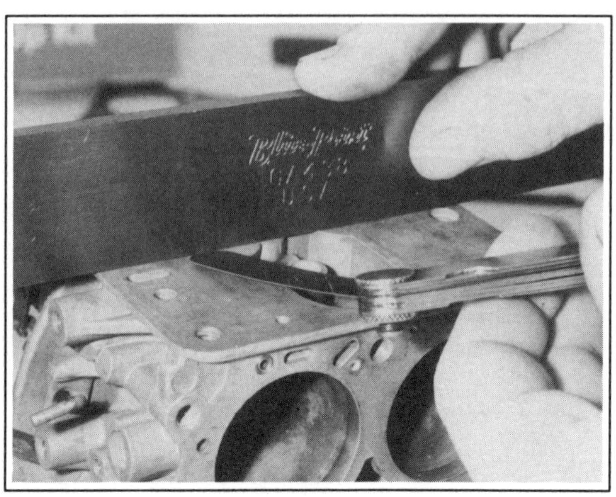

tips on fine tuning the accelerator pump.

Off-idle hesitation problems can also have several other causes. A few of the most common are:

1) Low float level, since the main circuit will have a delayed start up.
2) An off-idle flat-spot can be caused when the idle-stop adjustment is advanced too far, uncovering too much of the idle-transfer slots. See the sidebar *Idle Problems With Racing Cams* on page 119 for more on this problem and its remedy.
3) If the throttle blades are completely closed (can be necessary to obtain correct idle speed when there is a vacuum leak) an off-idle stumble is also likely. If everything else is normal, unscrewing the idle-speed adjustment should kill the engine.
4) Power valve malfunctions. Check for a valve that is stuck closed, the wrong number, or ruptured.
5) Vacuum secondaries that open too soon. A similar condition occurs when too large of a carburetor is used. See Chapter 11 for tips on tuning vacuum secondaries.

Carefully inspect the following areas during carburetor disassembly: 1) the entire accelerator pump circuit, including the discharge check valve, pump check valve, discharge nozzle, transfer tube and O-rings, metering block gaskets, pump diaphragm, and all accelerator pump passages, 2) the power valve and all passages, 3) the vacuum secondary assembly for missing opening restrictor, check ball, or incorrect spring, 4) throttle blades too far open or closed at idle, 5) float level, and 6) main-body mating surfaces.

No Power Under Heavy Acceleration—Incomplete throttle opening caused by improperly adjusted linkage is often the reason for lack of full power. With the engine shut off, have a helper operate the accelerator and check for full throttle opening and closing. Also check

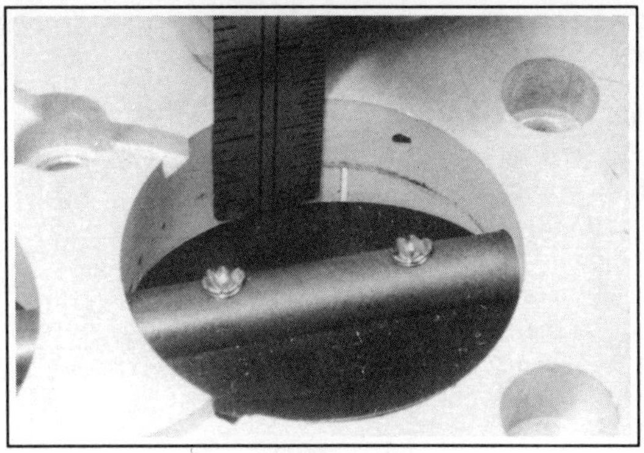

An off-idle flat spot can be caused when the idle-stop adjustment is advanced too far, uncovering too much of the idle-transfer slots. Similarly, if the throttle blades are completely closed (can be necessary to obtain correct idle speed when there is a vacuum leak) an off-idle stumble is also likely. If everything else is normal, unscrewing the idle-speed adjustment should kill the engine.

the following: 1) clogged main jets, 2) plugged power valve channel restrictions or a power valve that is not opening, 3) low float level or a float that is not dropping far enough, 4) fuel filters and the fuel delivery system, and 5) make sure the air filter is clean and of sufficient size.

Vacuum secondaries that aren't fully opening will also decrease maximum power. Incomplete secondary opening can be caused by: 1) ruptured secondary diaphragm, 2) clogged diaphragm opening restrictor or vacuum feed, 3) secondary spring too stiff, 4) stuck or binding secondary throttle shaft, and 5) carburetor too large for application.

Poor Fuel Economy—There are many factors that affect fuel economy besides the carburetor. Before checking the carburetor make sure the engine is "tuned-up"; that includes fresh plugs, points adjusted, timing set, etc. Also check the following chassis components that may affect fuel economy: 1) tires for proper inflation, 2) front-end alignment, 3) brakes for dragging, and 4) bad wheel or axle bearings. Some performance modifications will also reduce fuel economy: 1) long-duration and/or high-overlap camshafts, 2) numerically high rear-axle ratios, 3) high-stall torque convertors, 4) open-plenum type intake manifolds, and 5) blocked heat-riser passages. The most important factor influencing fuel economy is, of course, driving technique. Smooth acceleration and braking, reduced cruising speeds, and steady throttle use will help reduce fuel consumption.

Optimum fuel economy is achieved when the fuel mixture is as lean as possible and still provides crisp throttle response without surging or flat spots. Anything that overly richens the fuel mixture will decrease fuel economy. Improper choke operation is frequently the cause of an over-rich fuel mixture. Verify that the choke is fully open as soon as the engine is warmed up and that it is properly adjusted.

A float setting that is too high, leaking needle-and-seat assemblies, or too much fuel pressure will also affect fuel economy. Remove the float bowl sight plug and adjust the float level to just below the sight plug. This setting should remain stable after several minutes of idling. When disassembling the carburetor, check for debris in the float bowls or fuel-saturated floats.

The accelerator-pump discharge check valve can decrease mileage if it is missing or not functioning. Check for this condition by removing the air cleaner and running the engine at a high enough speed for the booster venturis to start flowing fuel. Observe the accelerator pump squirter. If fuel dribbles from the nozzle, check for a missing or damaged check valve. *No fuel should discharge from the nozzle at steady throttle.*

The power valve circuit plays an important role in fuel economy. A power valve that is stuck open, leaking, or the wrong number will increase fuel consumption. Don't expect good fuel economy if the power valve has been removed and blocked.

A defective accelerator pump system is most often the cause of stumbling or hesitation on initial acceleration. A quick movement of the throttle should produce a steady stream of fuel from both nozzles aimed toward the center of the venturi.

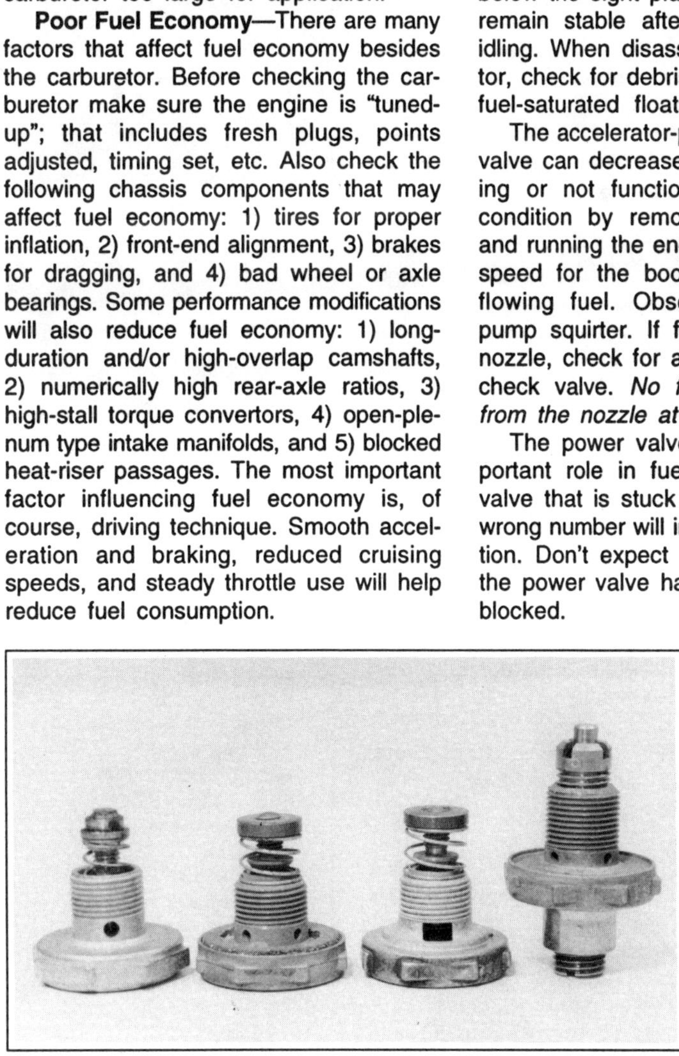

The power valve circuit plays an important role in fuel economy. A power valve that is stuck open, leaking, or the wrong number will increase fuel consumption.

There are many brands of carburetor rebuilding kits on the market. Some are excellent; some are mediocre. Since the quality of the components in a carburetor kit will affect the final results of your rebuild as much or more than any other single factor, selecting a carburetor kit is an important step. The below-average parts found in some overhaul kits can offset the careful preparation, assembly, and fine adjustments that you perform during the rebuild. The most consistent way to obtain quality is to purchase a genuine Holley *Renew* or *Trick Kit.* These rebuilding packages contain only Holley-approved components that are the same or equal to original parts. Some carburetion specialists, like The Carburetor Shop, Braswell Carburetion, Mike Jones Carburetion, etc., also offer top quality overhaul components (see appendix).

When it is not possible to purchase a Holley kit or its equal, make sure you carefully examine the contents of the kit before you make your purchase. Most overhaul kits are packaged with cellophane so the contents may be inspected without actually breaking the seal. Don't rely on past experience with a particular brand as a guarantee of quality components inside. Surprisingly, kit manufacturers often buy from other vendors and re-box items when they cannot supply parts out of their own inventory. I have found completely different parts of varying quality supplied in two identical kits (from a well-known manufacturer) that were purchased several months apart. Always look before you buy!

But what should you look for as you peer through the cellophane? The most important items are: 1) the composition of the metering block and bowl gaskets, 2) the construction of needle-and-seat assemblies, and 3) the construction of power valves.

GASKET MATERIAL

Float-bowl and metering-block gaskets and their sealing surfaces are especially important on Holley modular carburetors. Since the float bowls are located on the side of the main body, effective gasket seal prevents both external leaks and fuel/air leakage between internal passages. Holley originally used cork composition gaskets on these surfaces. These gaskets have several drawbacks. They tended to shrink, crack, and split, especially on carburetors that are stored for long periods of time. They also easily lost their resiliency and require retorquing the float bowl screws to prevent leaks. If that's not enough, cork may also swell and can restrict passages in the metering block. Holley stopped using cork material several years ago, and all cur-

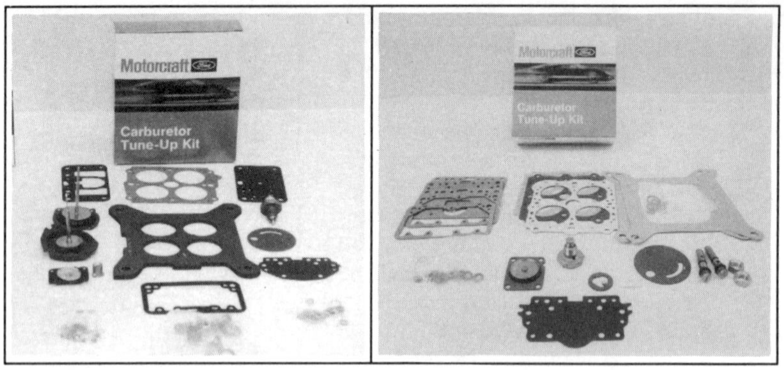

Kit manufacturers often buy from other vendors and re-box items. Here are two Motorcraft® carburetor kits, both for Holley four-barrels. The one on the left has Holley-type paper gaskets, needle-and-seats, and power valve; the one on the right uses inferior cork gaskets, reverse style needle-and-seats, and an unmarked power valve. The moral is: Always look before you buy!

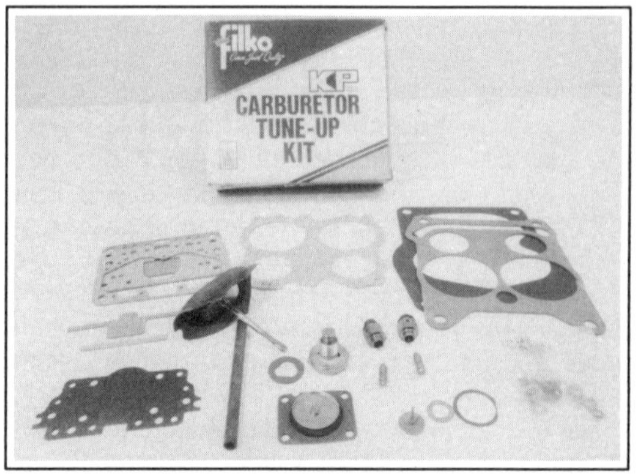

There is no logical way to predict what you'll find in a carburetor rebuilding kit. This package by Filko® includes a new vacuum secondary diaphragm (great!), but it also had cork gaskets and an unmarked limited-flow power valve (not so great).

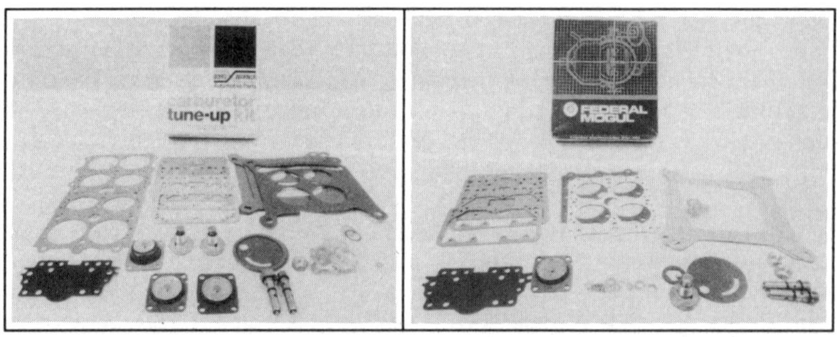

Here are some more surprises! A close examination of these kits by Federal Mogual® and Borg Warner® reveals identical types of gaskets and components; they even included the same instruction sheets. They were, in fact, re-boxed by the same (unnamed) vendor.

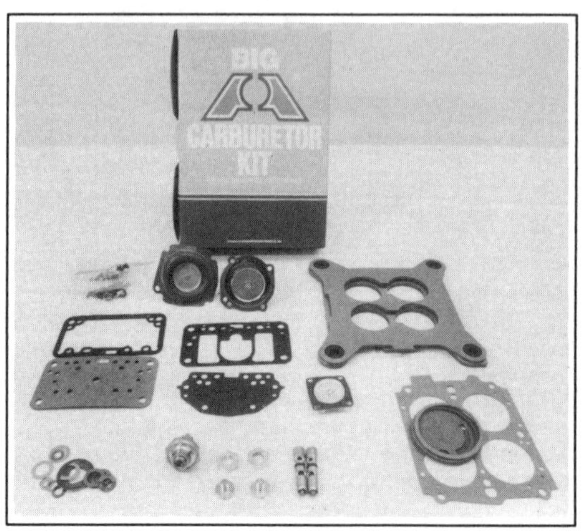

The components of this kit from Big A® were found to be well above average. Paper-type gaskets, factory-style needle-and-seats, and vacuum-secondary and choke pull-off diaphragms were provided. This kit also included replacement needle-and-seat screws and hex nuts, and a very nice insulating base gasket.

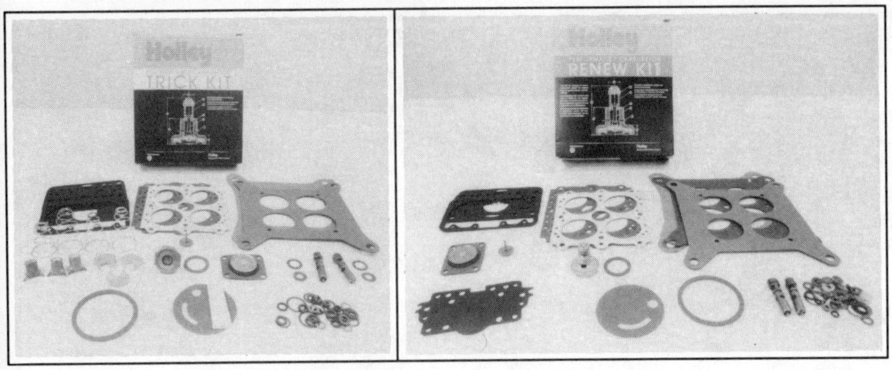

The purchase of genuine Holley *Renew* or *Trick Kit* rebuilding kits assures you of receiving top quality, replacement gaskets and components. The *Trick Kit* (left) offers a variety of additional components for fine tuning carburetor performance.

Cork composition gaskets (left) tend to shrink, crack, and split, especially on carburetors that are stored for long periods of time. Holley stopped using cork material several years ago, and all current float-bowl and metering-block gaskets are either a brown or black treated paper. Unfortunately, many manufacturers still use cork gaskets in their rebuilding kits.

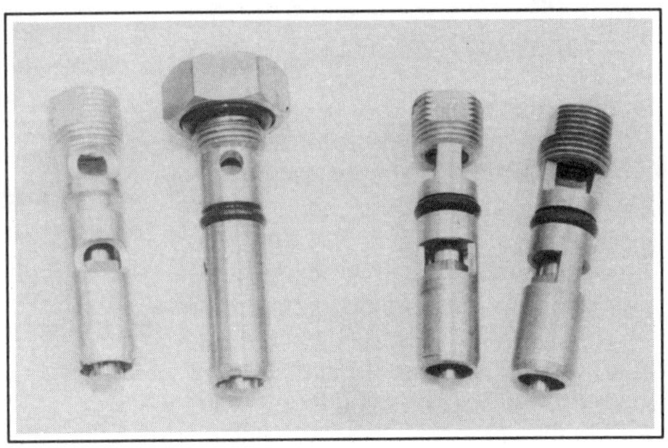

The needle-and-seat assemblies found in most kits will work fine in low-performance, street-type applications (two left assemblies). However, Holley needles-and-seats have large, square "window" openings (two right assemblies) and incorporate other subtle improvements that improve fuel flow, essential for racing or high-performance use.

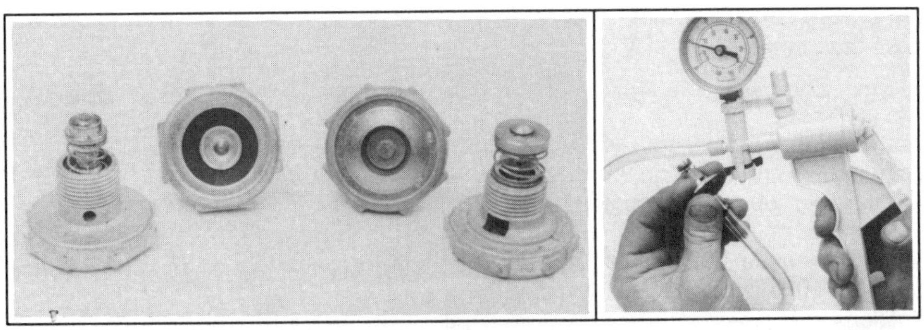

Holley power valves are individually calibrated and numbered (two right valves). Many carburetor kits contain generic, unnumbered power valves (two left valves) that may not function the same as the original Holley part. Unmarked power valves can be checked with a hand vacuum tester, such as the *Mity-vac*, to determine their opening point. By using a "suction-cup" adaptor and observing at what point the valve just begins to open, you can determine the precise opening point within plus or minus 1.5-in(Hg).

rent float-bowl and metering-block gaskets are either a brown or black treated paper that is much more stable. Unfortunately, many manufacturers still use cork gaskets in their rebuilding kits. I advise against using cork gaskets, however, if it is unavoidable, make sure to retorque the float bowl screws after a few hours of operation and again in a week or so.

NEEDLE-AND-SEAT

The needle-and-seat assemblies found in most kits will work fine in normal (low-performance, street-type) use. However, be aware that there are two different styles of needles-and-seats and each require a different adjusting procedure (see **Step 14** in the reassembly chapter; page 85). Some non-Holley needles-and-seats are quite restrictive and may not provide adequate fuel flow for racing or high-performance applications. One identifying feature of these "low-performance" parts is small fuel discharge holes instead of the large square "window" openings found in genuine Holley assemblies. The seat diameter and the machined taper approaching the seat may also not meet Holley's standard. If you are working on a performance carburetor, you shouldn't settle for less than genuine Holley parts or their equivalent from a company specializing in high-performance Holley accessories.

POWER VALVES

Holley power valves are individually calibrated and numbered. The number indicates the opening point (in inches of vacuum) that the valve will begin to richen the fuel mixture in the main metering system. This opening "timing" will affect fuel economy and throttle response. Many carburetor kits contain generic, unnumbered power valves that may or may not function the same as the original Holley part. Fortunately, all power valves can be easily checked with a vacuum tester, such as the *Mity-vac*, to verify the opening point. By using a "suction-cup" adaptor and observing at what point the valve just begins to open, you can determine the opening point to within plus or minus 1.5-in(Hg). You should purchase the correct Holley power valve if the opening point of the valve in the kit is not within the measured limits of the recommended value listed in the chart on pages 120 to 122.

The best words of advice I can give you are: Whenever possible, obtain a genuine Holley rebuild kit. This is especially important for racing or heavy-duty application. Your rebuilt carburetor will only be as good as the component parts that you install.

HOLLEY Rebuilding & Modifying

5—Carburetor Removal

This chapter provides detailed procedures that will guide you through the step-by-step removal of virtually any Holley modular carburetor. In order to properly apply the information in this chapter, you should have a basic familiarity with hand tools (refer to the sidebar, *Removing Stubborn and Hidden Fasteners* on page 37 for more information). In addition, you should be able to identify the individual component parts of your Holley, as discussed in Chapter 3.

Testing Before You Begin

Many engine problems are incorrectly diagnosed as carburetor malfunctions. This often results in carburetor rebuilds that waste time, money, and do not fix the original problem. If the malfunction you are repairing is not due to an obvious carburetion problem (like flooding or other fuel leaks), I urge you to perform the on-car diagnostic tests illustrated in Chapter 4 before you begin removal. These tests will pinpoint the source of the trouble, and if the carburetor is at fault, you will have a good idea of the specific components that should receive extra attention when you disassemble your Holley. If it is not possible to do these tests, subsequent chapters will help you locate and correct many carburetor maladies both before and after you rebuild your Holley.

Work-A-Long Sheet

Some of the steps in this and subsequent chapters instruct you to note information on your Work-A-Long sheet. Make a copy of the blank Work-A-Long sheet provided on pages 123 and 124 now. As you proceed through removal, disassembly, cleaning, and reassembly, your completed Work-A-Long sheet will become a valuable record of what problems you encountered, the solutions you found, and the individual components you installed in your Holley. This information can make tuning easier, and it can even add value to your carburetor if you sell it.

Step-By-Step Procedures

Not all of the step-by-step procedures in this chapter and subsequent chapters apply to every job. For example, if your vehicle does not have an EGR plate under the carburetor, steps related to this feature should be skipped. In many cases, instructions are provided in the step-by-step captions that direct you to "jump ahead to..." when upcoming steps do not apply to a specific carburetor model. However, all steps are clearly titled and in the proper sequence, so if you come upon a step that does not relate to your project, just skip ahead to the next step.

Several steps in the step-by-step chapters in this book are labeled with icons—symbols—that represent important information or special help, such as **IMPORTANT STEP**, **MASTER MECHANIC TIP**, or **SAFETY STEP**. Carefully read these steps unless you are specifically directed to skip over them. These steps contain important information that can often make the difference between a successful rebuild and disappointment.

Finding Additional Information

If you would like additional information on the following subjects before you begin removal, refer to the indicated chapters.

How This Book Is Organized .... Chapter 1

What Holley Models Are Covered In This Book .... Chapter 1

Shop/Workspace Considerations ...... Chapter 2

Required And Optional Tools .... Chapter 2

Identifying Holley Carburetors And Components .... Chapter 3

Understanding Holley Part Numbers .... Chapter 3

Diagnostics Before You Begin And How To Select A Carburetor Kit ...... Chapter 4

1 Remove Air Cleaner

1—Holley carburetors have been installed as standard equipment on Ford, Chevrolet, and Chrysler vehicles. They have been used on high-performance, big-block V8s producing over 500 horsepower, and they have been used on small-displacement engines designed for good gas mileage and minimum emissions. This wide variety of vehicles require an equally wide variety of carburetor removal and installation procedures, however, many of the steps are similar. Regardless of vehicle, the first step is almost always: Remove the air cleaner.

MASTER MECHANIC TIP PRO TIP

2 Container For Fasteners

2—Locate a suitable container, like this Holley parts tray, to hold all of the nuts, bolts, clips and other hardware that you remove. *Don't set them down in the engine compartment.* If you don't heed this advice, you'll likely waste your time crawling around on all fours looking for a clip, spring, or bolt that "just disappeared."

SAFETY STEP

3 Wear Eye Protection

3—Consider eye protection as just another essential tool while servicing your carburetor. Many mechanics vividly recall the first time a drop of gasoline splashed into their eye. Don't take any chances; goggles are cheap and are only a minor inconvenience. Keep a pair in your toolbox.

IMPORTANT STEP

4 Locate Vacuum Lines

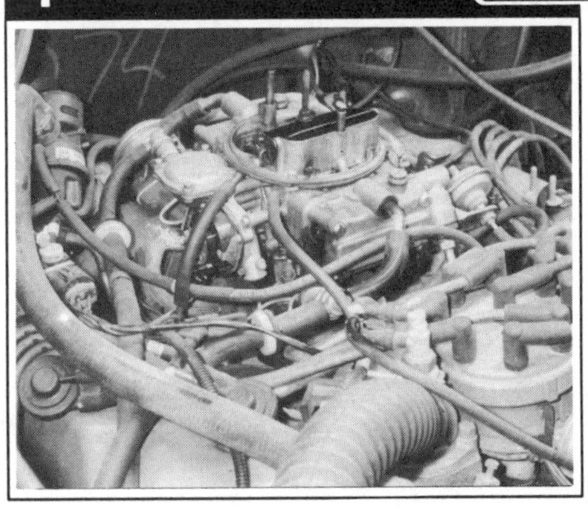

4—Emission requirements have increased the complexity of carburetor induction systems. While the latest computer-controlled vehicles may actually be more "complex" in their function, this complexity is virtually lost in software programming within the electronic controls. On a carbureted vehicle like this Ford truck, this complexity is very visible in the form of sensors, dashpots, switches, and hoses; lots of hoses! But don't pull them off yet. First, familiarize yourself with the overall layout.

NOTATION REQUIRED

5 Use Work-A-Long Sheet

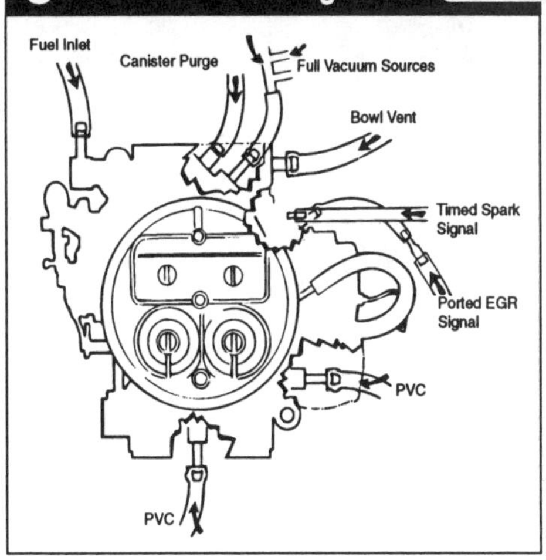

5—If you haven't already, tear out or make a copy of the Work-A-Long sheet on pages 123 and 124. The vacuum line identification drawing on the Work-A-Long sheet illustrates some of the hose connections used on Holley carburetors. Use these label names as is, rename them, or add your own to match the connections on your engine. As explained in the next step, transfer the names to masking tape, and mark all hoses and electrical connections leading to or from your Holley.

6 Label All Vacuum & Electrical Connections

6—Referring to your Work-A-Long sheet, label each vacuum line and electrical connection with a number or a functional name, e.g., "1," "2," or "VACUUM ADVANCE," "CHOKE," etc. You may also find it helpful to identify some linkage connections or make a few removal notes as you proceed. Simpler installations, like this Dodge Dart 340 (right), require fewer labels.

7 Remove Springs And Linkage

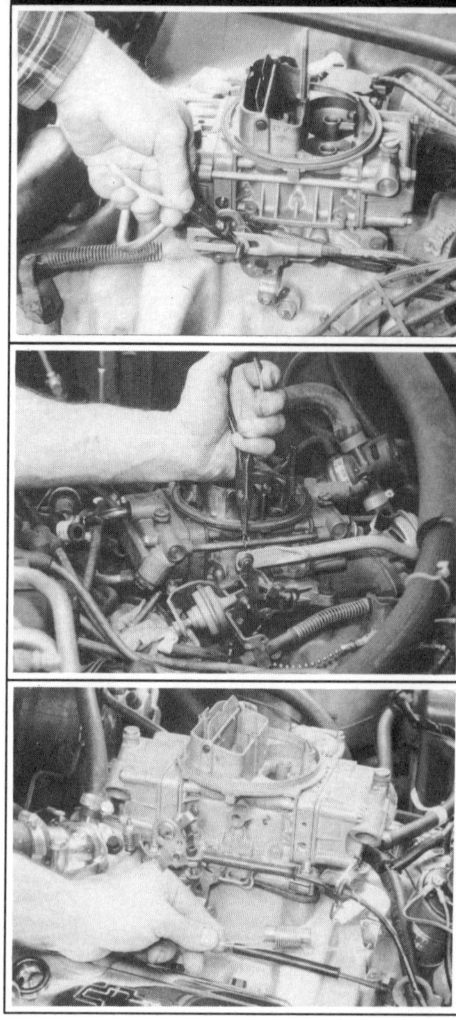

7—If the vehicle is equipped with an automatic transmission, the next step is usually to remove a horseshoe or "E" clip and, perhaps, a washer or two that retain the kickdown linkage (top and middle photos). Next unclip or unbolt the throttle-pedal linkage. Then remove all throttle-return springs connected to the throttle arm (lower photo). Finally, some carburetors may have electronic switches or sensors located near or on the throttle arm. Unplug or disconnect these devices (make sure to note these connections on your Work-A-Long sheet).

MASTER MECHANIC TIP PRO TIP

8 Remove Fuel Line

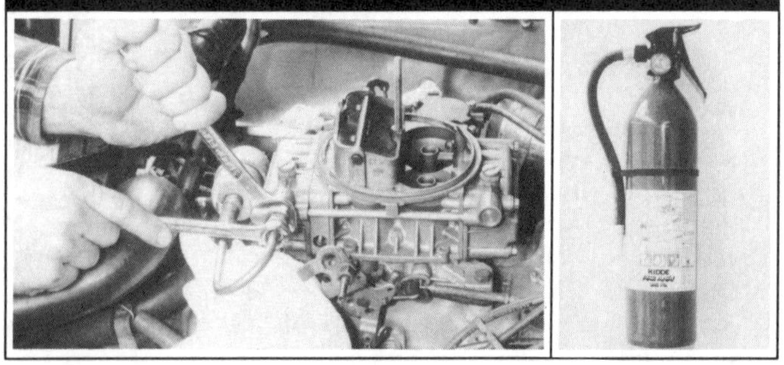

8—Place a rag under the fuel line fitting to catch dripping fuel. Most vehicles have a steel fuel inlet tube that is removed with a 1/2- or 5/8-inch tubing wrench and a 1-inch open-end wrench. The larger wrench prevents the carburetor fitting from rotating. If your vehicle has dual-fuel inlets, remove the fuel line to the secondary bowl also. Tip: Save yourself a step during carburetor disassembly by loosening the bowl inlet fitting(s) now, while the carburetor is still firmly held by the intake manifold.

SAFETY STEP

9 Cap Off Fuel Line

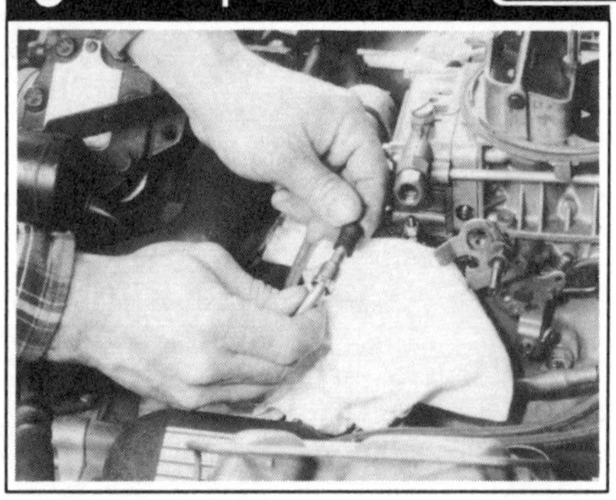

9—After you have unscrewed the fuel line, slide a rubber cap plug over the end to prevent fuel from running out. This not only makes the job safer, it's a great way to keep fuel off engine paint, etc. And make sure you have a fire extinguisher nearby. A puddle of fuel combined with an inadvertent spark (dropping a wrench on the battery terminals is a common source) can result in a nasty fire.

10 Unbolt Carburetor Base

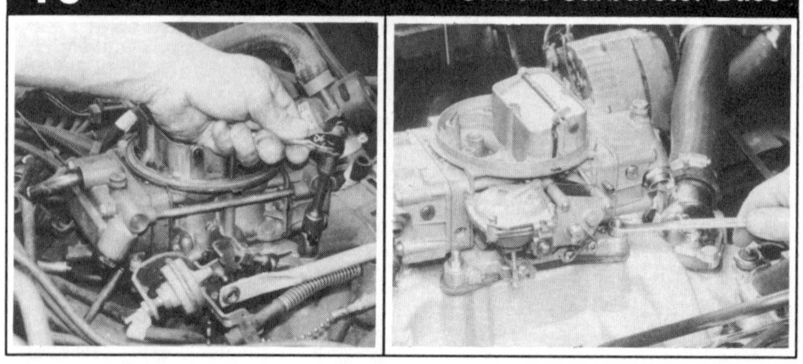

10—Remove all remaining vacuum lines and electrical connections (check your Work-A-Long sheet and note any you may have missed the first time). If your vehicle uses a water-heated choke, drain the radiator into a drain pan—to prevent water and antifreeze from spilling over the engine—before you remove the hoses. Finally, unbolt the carburetor from the intake manifold. Some Holleys can be unbolted with a 1/2-inch socket, others will require an open-end wrench (or some of the special tools illustrated in the sidebar on the next page) and some patience.

Removing Stubborn And Hidden Fasteners

Removing very tight or frozen fasteners requires special techniques to avoid damaging the bolts or component parts. Carburetors are particularly susceptible to stuck screws because of a process called electrolysis. This electrochemical phenomenon causes corrosion between dissimilar metals, like the aluminum alloys in a carburetor body and the steel in fasteners. Carelessly applying brute force usually results in stripped Phillips or straight-slot heads and broken fasteners.

Here are some tips that will help you successfully remove all but the most stubborn fasteners. The most important tip is: be patient! When you encounter a stuck screw or bolt, STOP! Don't apply excessive force! Use the following tools and chemicals to put the odds on your side.

Chemical Wrenches—Applying a liberal amount of a chemical penetrant, such as WD-40®, will often make the difference between success and failure. Allow the penetrant to soak in for at least five to ten minutes. When possible, strike the center of the bolt or stud with a light hammer and drift several times. This tends to break down the corrosion between the threads and helps the penetrant to work its way in. But don't deform or damage the head of the fastener.

Special Tools—Use only a high quality tool that is not worn and fits the fastener tightly. *This is essential! Removing stuck fasteners with cheap tools is almost always doomed to failure.* If possible, use a six-point socket or wrench, since twelve-point sockets don't allow as much torque before "rounding off" the fastener. Never use an open-end wrench on a frozen bolt or nut, unless a socket, box-end, or line wrench simply cannot be used. Long-stem, six-point, box-end wrenches are usually effective for breaking loose stuck bolts when a socket can't be used. Top quality line wrenches are a must for working on stubborn flared fittings.

Techniques and Tricks—Make sure that the tool is fully engaged on the fastener before you apply force. Always use deep sockets on studs with nuts to ensure full engagement on the hex surfaces of the nut. If a fastener starts to unscrew but then locks up, screw it in slightly and apply penetrant to the threads. Continue screwing the fastener in and out while adding more penetrant until it eventually can be removed.

Fasteners that are in hard-to-reach areas often require special tools or unique combinations of wrenches, sockets, and other tools to obtain access and full fastener engagement. Some "trick tools" for tight spots include the following: 1) flex-drive sockets, 2) swivel-head wrenches, 3) offset wrenches 4) various length extensions and wobble-head extensions, 5) crows-foot wrenches, 6) offset Phillips and straight-slot screwdrivers, and 7) flex-head and bent-handle ratchets (see photos).

A little patience and the right tools will make removing stubborn fasteners much easier. Leave the hammers and chisels, hacksaws, and cutting torches to the less patient "mechanic."

A liberal application of penetrating lubricant should be your first step when you encounter a stubborn fastener. Allow the penetrant to soak-in for five or ten minutes. Lightly striking the fastener several times with a hammer and drift can help breakdown corrosion and allow the lubricant to enter the threads.

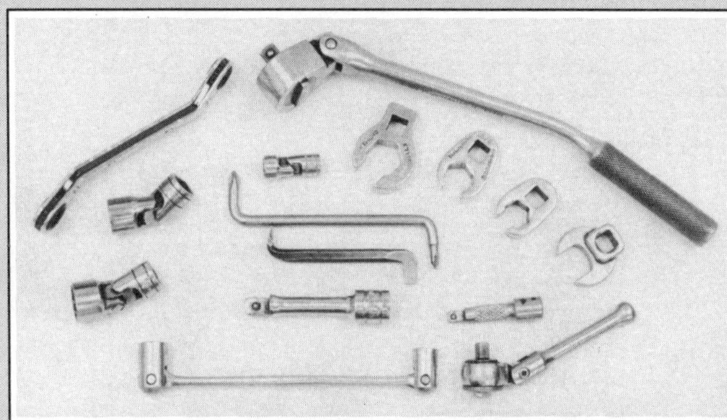

Here are a few of the "special" tools that professional mechanics use to allow full torque to be applied to fasteners in difficult locations. Many of these tools can only be purchased from professional tool trucks, like Snap-on and Mac.

Using unique combinations of tools will allow you to reach virtually any fastener. Here a crows-foot line wrench is attached to a wobble-head extension and bent-handle ratchet. Never use open-end wrenches on stubborn fasteners.

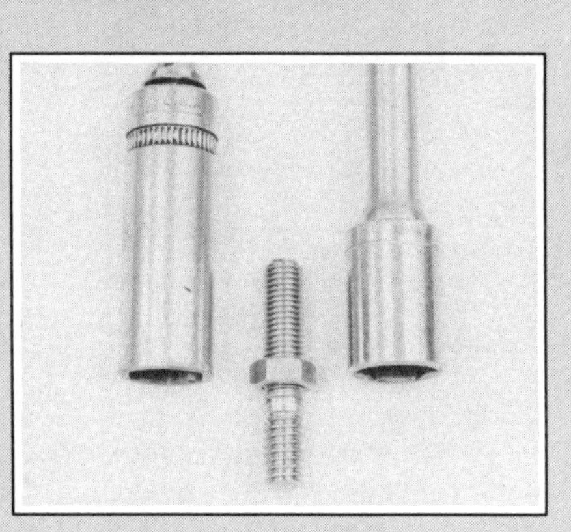

Studs with stubborn nuts can really cause problems. Make sure to use a six-point socket that allows full engagement on the nut, such as the deep socket on the left.

11 Remove Carburetor

11—Now attempt to lift the carburetor free. Sometimes the carburetor can stick to the manifold gasket and may have to be pried loose. But before you start prying, double check to make sure all the carburetor attaching bolts have been removed. Once free, lift the carburetor slowly to make sure it's still not connected to a wire or hose.

NOTATION REQUIRED

12 Fuel/Vacuum Leaks

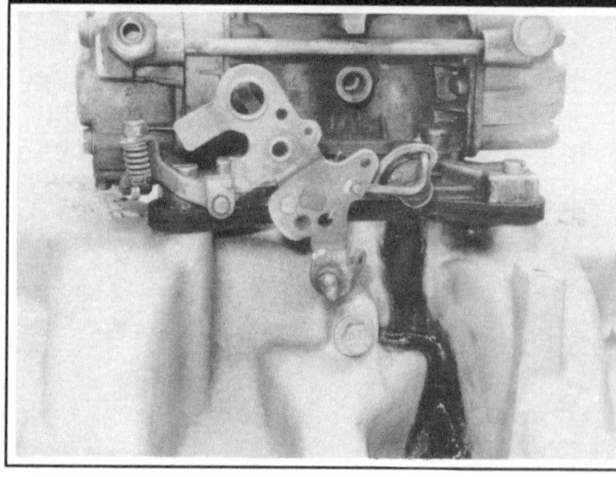

12—Take a close look at the carburetor base gasket. Does it look like it might have caused a vacuum leak? If so, and you didn't check for vacuum leaks earlier (as explained in Chapter 4), you may have found the source of the malfunction. Also look closely at the intake manifold and note any signs of fuel leakage on your Work-A-Long sheet; this information can help you locate and confirm problems as you proceed with the rebuild.

13 Inspect EGR System

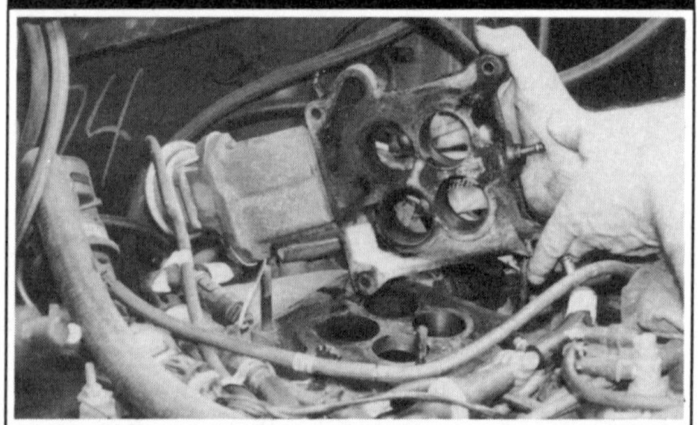

13—Many vehicles use an Exhaust-Gas-Recirculation (EGR) system that feeds a controlled amount of exhaust gas into the induction system to help the engine meet emissions requirements. When these systems fail, they can cause the engine to run extremely rough. Some engines, like this Ford truck, have an EGR plate under the carburetor; inspect for burned out gaskets, etc. Other engines use a metered exhaust "jet" in the base of the manifold plenum. Refer to Chapter 4 for EGR testing and troubleshooting information.

IMPORTANT STEP

14 Cover Intake Manifold

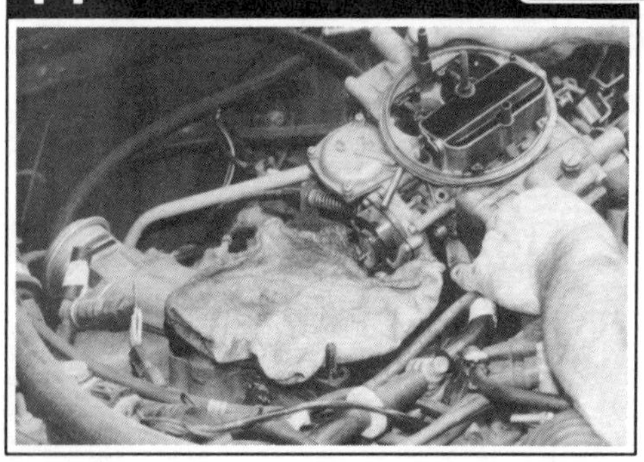

14—Cover the intake manifold with a clean rag. If a socket or bolt finds its way into the induction system, you unwittingly may be beginning a complete engine overhaul, not a simple carburetor rebuild!

NOTATION REQUIRED

15 Locate The List Number

15—Locate the carburetor list number and note it on your Work-A-Long sheet. The list number is almost always located on the airhorn, except on 4500's and some race-only carburetors, where it is stamped on the main body. Now, look up the list number in the *Numerical Parts Listing* on pages 120 to 122, and record the configuration specifications. When disassembly is started in the next chapter, you'll record additional facts, like the jet sizes and power-valves *actually found* in your carburetor.

16—If the list number on your carburetor is not included in the parts listing on pages 120 to 122, you'll have to refer to Holley's *Illustrated Parts And Specs Guide*, **part 36-51**, to obtain component part numbers and decipher the information. You can also contact Holley's technical support for stock specifications and recommended rebuilding kit numbers (see appendix). If you decide to obtain an *Illustrated Parts And Specs Guide*, you will find it to be loaded with information, but it can be a bit difficult to use. Here are a few tips. First, locate the list number of your carburetor; refer to information in Chapter 2 in this book (and How To Read This Catalog in the *Parts And Specs Guide*) for help. Next, find your list number in the Variable Specs And Parts listing beginning on page 1. The list number is the first number in each information "box" (the second number is the current part number for the complete carburetor). The information box contains a group of specifications labeled with "IV," "XII," "VII," etc. Refer to the Reference Notations page (usually on page iv) for a key to the meaning of these notations; for example, "I" indicates the dry-float setting; "IV" indicates the pump-cam position; "XIII" indicates pump-spring adjustment and pump travel, etc. Next, look for the T.V. number at the top of the information box. This number is the Typical View number; a reference to the exploded-parts view that can be found in the back of the book. Find the typical-view page for your Holley. You will see a list of part numbers for the various parts indicated on the drawing. To find the part number for parts marked with an "X," first note the index number (from the drawing), then match this number with the listing in the bottom of the information box; the part number will be indicated.

16 Unlisted List Number?

Dating Holley Carburetors

Various identification numbers are stamped on the front of the airhorn, or when no airhorn exists, as with Dominators and some racing carburetors, on top of the main body. The *list number*, required for ordering overhaul kits, accessories, or replacement parts, is easily identified; it is always preceded by the letters **LIST** or the single letter **L**. The list number is, in effect, the part number of the carburetor (for more information on list numbers, refer to Chapter 2 and 5).

Almost all Holley carburetors have a *date-code number* stamped near (usually under) the list number. Early models used three- and four-digit codes to indicate the date of manufacture, while carburetors built after 1972 exclusively used four-digit coding. Refer to the accompanying tables for the meaning of each date-code digit

Many Holleys that were built for OEMs (sold directly to auto manufacturers) carried an additional number: the *auto manufacturer's part number*. This number is usually found above the list number. The recent interest in the restoration of factory muscle cars has made OEM numbers quite important. A rare manufacturer's part number found on the airhorn can instantly transform a twenty dollar "junk" carburetor into a prized possession worth several hundred dollars.

Happy hunting!

Holley Four Digit Date Codes

First Three Digits—Day Of Year:

Day Code Non Leap Year	Day Code Leap Year	Month Of Production
001-031	001-031	JANUARY
032-059	032-060	FEBRUARY
060-090	061-091	MARCH
091-120	092-121	APRIL
121-151	122-152	MAY
152-181	153-182	JUNE
182-212	183-213	JULY
213-243	214-244	AUGUST
244-273	245-274	SEPTEMBER
274-304	275-305	OCTOBER
305-334	306-335	NOVEMBER
335-365	336-366	DECEMBER

Fourth Digit—Year:

1	1961/1971/1981/1991
2	1962/1972/1982/1992
3	1963/1973/1983/1993
4	1964/1974/1984/1994
5	1965/1975/1985/1995
6	1966/1976/1986/1996
7	1967/1977/1987/1997
8	1968/1978/1988/1998
9	1969/1979/1989/1999
0	1970/1980/1990/2000

Holley Three Digit Date Codes

First Digit—Year:

1	1961/71
2	1962/72
3	1963
4	1964
5	1965
6	1966
7	1967
8	1968
9	1969
0	1970

Second Digit—Month:

1	JANUARY
2	FEBRUARY
3	MARCH
4	APRIL
5	MAY
6	JUNE
7	JULY
8	AUGUST
9	SEPTEMBER
0	OCTOBER
A	NOVEMBER
B	DECEMBER

Third Digit—Week:

1	FIRST WEEK
2	SECOND WEEK
3	THIRD WEEK
4	FOURTH WEEK
5	FIFTH WEEK

Carburetor identification numbers are located on the front of the airhorn or on top of the main body. The first number is the *auto manufacturer's part number*. The second line is the Holley *list number* and the last line is the *date-code number*. This three digit date code indicates that the third week of July, 1969, was the date of manufacture.

This model 4180 carburetor was originally installed on a new Ford truck. The first number on the airhorn is the Ford part number. The four-digit date code indicates that this carburetor was manufactured on February 9, 1973 (1963 is too early; 1983 is too late).

HOLLEY Rebuilding & Modifying

6—Carburetor Disassembly

This chapter provides detailed workbench procedures that will guide you through the step-by-step disassembly of virtually any Holley modular carburetor. In order to properly apply the information in this chapter, you should be familiar with the tools needed to work on Holley carburetors; see Chapter 2. You should also be able to identify the individual component parts of your Holley; see Chapter 3.

Testing Before You Begin

Many engine problems are incorrectly diagnosed as carburetor malfunctions. This often results in carburetor rebuilds that waste time, money, and do not fix the original problem. If the malfunction is not due to an obvious carburetion problem (like flooding or other fuel leaks), I urge you to perform the on-car diagnostic tests illustrated in Chapter 4 before you begin disassembly. These tests will help pinpoint the source of the trouble, and if the carburetor is at fault, you will have a good idea of the specific components that should receive extra attention when you disassemble your Holley. If it is not possible to do these tests, this and subsequent chapters will help you locate and correct many carburetor maladies both before and after you reinstall your Holley.

The Step-By-Step Procedures In This Chapter

Not all of the step-by-step procedures in this chapter apply to all Holley models. For example, if you're rebuilding a 2-barrel, steps related to secondary disassembly are not applicable and should be skipped. In many cases, instructions are provided in the step-by-step captions that direct you to "jump ahead to..." when upcoming steps do not apply to a specific carburetor model. However, all steps are clearly titled and in the proper sequence, so if you come upon a step that does not relate to your Holley, review the subsequent steps until you come to the next relevant procedure.

Several steps in this chapter are labeled with icons—special symbols—that represent a unique Holley model, like 4500 or 4165. Only perform these steps if you are working on that model carburetor. On the other hand, some steps are labeled with icons that represent important information or special help, such as **IMPORTANT STEP**, **MASTER MECHANIC TIP**, or **SAFETY STEP**. Carefully read these steps unless you are specifically directed to skip over them. These steps contain important information that can often make the difference between a successful rebuild and disappointment.

Finding Additional Information

If you would like additional information on the following subjects before you begin disassembly, refer to the indicated chapters.

How This Book Is Organized .... Chapter 1

What Holley Models Are Covered In This Book .... Chapter 1

Shop/Workspace Considerations .. Chapter 2

Required And Optional Tools .... Chapter 2

Identifying Holley Carburetors And Components .... Chapter 3

Understanding Holley Part Numbers .... Chapter 3

Diagnostics Before You Begin ..... Chapter 4

Buying The Right Carburetor Kit .... Chapter 4

Typical Step-By-Step Procedures For Removing A Holley Carburetor From A Vehicle .... Chapter 5

1 Ready To Disassemble

1—The principal carburetor used in this step-by-step disassembly is a "Premium" 600cfm model 4160 with externally-adjustable float bowls (list 81850). But regardless of which modular Holley carburetor you're working on, the rebuilding procedures are covered in this—and in the remaining—step-by-step chapters.

2 Remove Surface Dirt

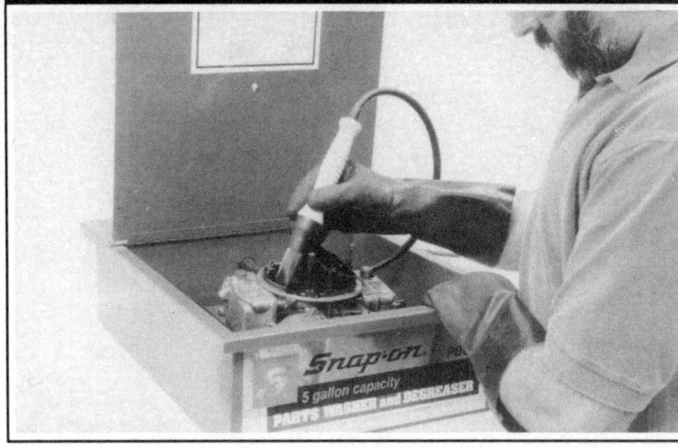

2—Before you begin disassembly, wash off excess surface dirt and grease with solvent and a stiff-bristled brush. A benchtop solvent tank, like this quality Snap-on unit with a built-in pump, is perfect for the job. In a pinch, you can use spray-on carburetor cleaner, engine degreaser, or brake cleaner. If necessary, rinse the carburetor off with water. *Don't forget to wear eye protection while you're working on your Holley. Springs, clips, and even fuel can fly into your eyes!*

NOTATION REQUIRED

3 Record List Number

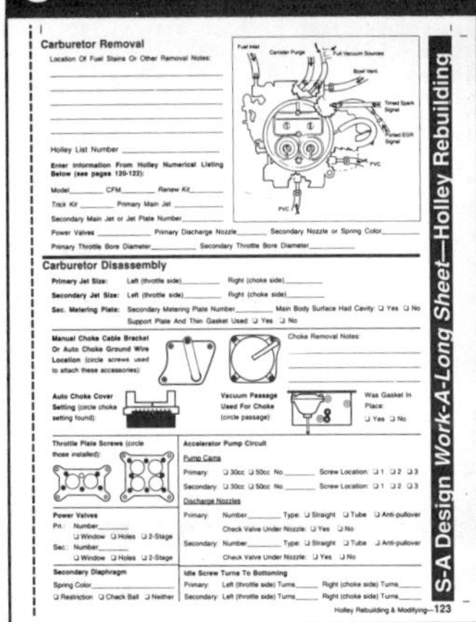

3—If you haven't already, remove (or make a copy of) the Work-A-Long sheet on pages 123 and 124. Locate the carburetor list number and note it on your sheet (refer to the final steps in removal on page 38 for help locating the list number). Locate the list number in the *Numerical Parts Listing* on pages 120 to 122 and record the configuration specifications. If the list number on your carburetor is not included in the *Numerical Parts Listing* in this book, refer to *Holley's Illustrated Parts & Specs* book (**part 36-51**) for the stock specifications and recommended rebuilding kit numbers (see page 39 for tips on using *Holley's Parts & Specs* book). Finally, there have been many variations in Holley carburetor design, and some disassembly steps may vary from those illustrated in this and subsequent chapters. Note any important differences on your Work-A-Long sheet; the information can really come in handy later on.

IMPORTANT STEP

4 Mark Metering Blocks/Float Bowls

4—Distinguishing the primary and secondary metering blocks and/or float bowls after disassembly can be confusing on some Holley double-pumpers, 4150s, or 4-barrels, especially those with four idle-adjustment screws. To be on the safe side, regardless of which Holley 4-barrel you're working on, scribe a small "s" on the secondary float bowl and metering block (if so equipped). If you don't want the marks to show, scribe them on the bottom.

5 Disassembly In A Drip Pan

5—If you want to keep your work surface clean and prevent small parts from rolling off into oblivion, try disassembling your Holley in a small drip pan. You may also find your rebuilding job easier when you use a carburetor stand, like this Peters model that holds the carburetor either by the throttle plate (rightside-up) or by the airhorn (upside-down).

6 Drain Fuel From Carburetor

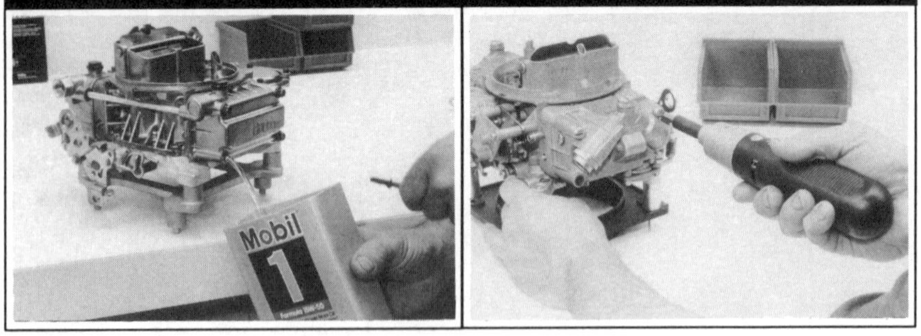

6—First, drain any remaining fuel from the carburetor by either removing a lower bowl screw from both the primary and secondary bowls or turning the carburetor upside-down over a bucket, allowing fuel to drain from the bowl vents. Then, remove the remaining primary bowl screws. Holley carburetors use either straight-slot or hex-head 12-24 bowl screws. Regardless of which style your Holley uses, the Black & Decker Screwdriver Plus (right) is a very handy tool for carburetor work; it's inexpensive, powerful, and comes with just the right bits.

MASTER MECHANIC TIP PRO TIP

7 Organize Parts In Bins

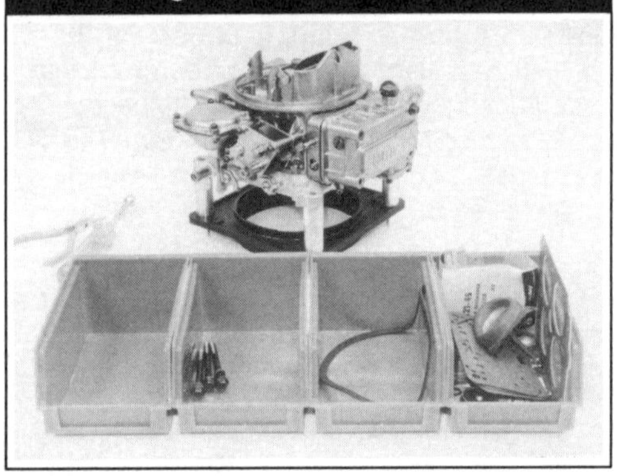

7—Placing carburetor parts in plastic storage bins will keep your workspace organized. I commonly use four bins: I place plastic parts, floats, electrical components, and very small parts that will not be immersed in carburetor cleaner into bin #1 (if you're not sure, put the part in this bin anyway, additional cleaning information is given in the next chapter); place all metal components in bin #2; place old gaskets, used rubber parts, etc. in bin #3 (old parts can be helpful during inspection and reassembly); and if you have opened your carburetor rebuilding kit, put the new parts and gaskets in bin #4.

IMPORTANT STEP !

8 Proper Way To Remove Float Bowl

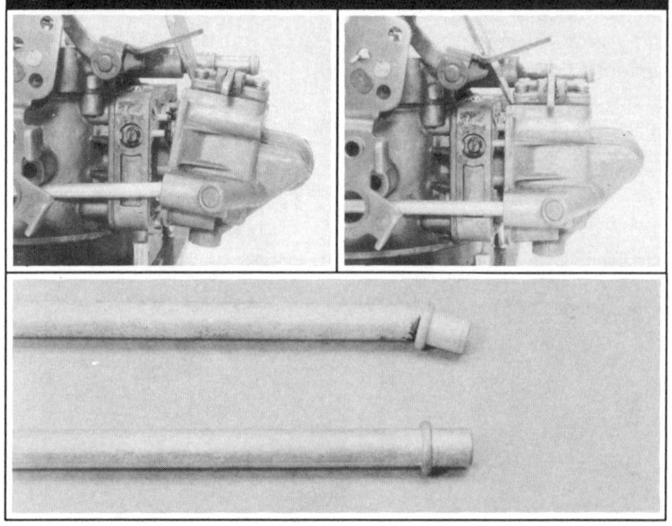

8—Loosen the primary bowl by prying the bowl, not the metering block, from the top and bottom. If the metering block comes off attached to the float bowl, refer to **Step 9.** Some gaskets can really stick to metal surfaces, so you may have to use considerable force. Don't insert the screwdriver between the bowl and the metering block, you'll damage sealing surfaces. IMPORTANT: *If your Holley has an aluminum fuel-transfer tube, don't pry the float bowl from just one side.* If you do, the bowl will tip, as shown on the left, bending the aluminum tube. Although it may take several attempts of careful prying from both the top and bottom of the bowl, the primary bowl should come off parallel to the metering block, as shown on the right. *When the bowl has been removed, pull the fuel-transfer tube out of the secondary float bowl and place it in bin #2.* Now, skip to **Step 10**, *Don't Scrape Gaskets.*

9 Separate Metering Block/Float Bowl

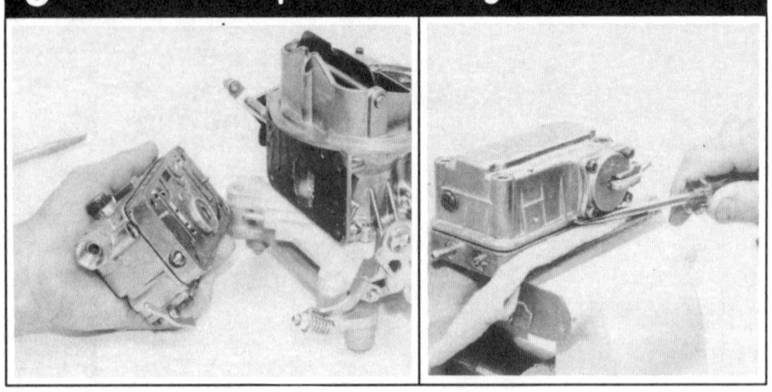

9—If, despite your best efforts, the metering block came off stuck to the float bowl, clamp the metering plate in a bench vise—use brass jaws or a rag to prevent damaging the exterior surfaces. Pry the bowl loose with a screwdriver as shown.

10 Don't Scrape Gaskets

10—The float bowl (and, later, the metering block and throttle plate) gaskets will usually tear during disassembly. Don't scrape off stuck-on gasket remnants; leave gasket removal for carburetor cleaner and/or the alternate techniques described in the next chapter. Scraping off gaskets with a putty knife or screwdriver can easily shave off cast-in sealing ridges or cut deep scratches in sealing surfaces.

11 Remove Fuel Vent Baffle

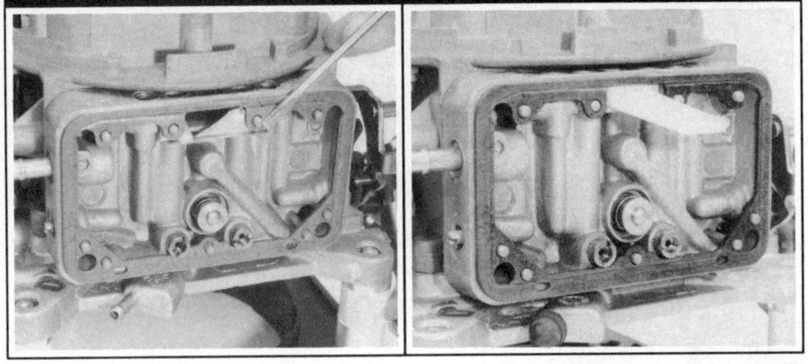

11—Many Holleys are equipped with a small metal or plastic fuel baffle held in place by the bowl gasket (left photo). Push the gasket aside and remove the baffle and place it in bin #2. If your Holley has an extended plastic bowl vent "whistle," (right photo) do not attempt to remove it. It is retained by a small rivet and is normally not removed during a rebuild.

IMPORTANT STEP **!**

12 Proper Way To Remove Metering Block

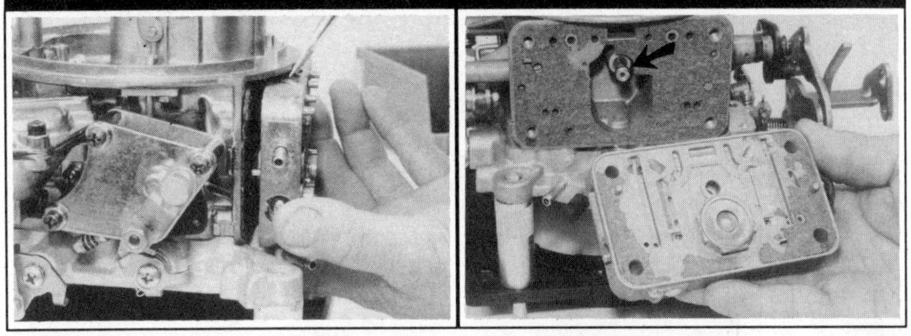

12—Using the same technique described for the primary bowl in **Step 8**; pry loose the primary metering block by using a screwdriver on both the <u>top and bottom</u> of the plate. Many Holleys are equipped with an accelerator pump fuel-transfer tube. This brass tube and O-ring assembly can be damaged if the metering plate is not pried off parallel with the main body. NOTE: If you insert a screwdriver into the idle- or main-well plug bores, make sure you don't inadvertently loosen the plugs; air leaks will adversely affect fuel metering.

When Not To Rebuild

By applying the rebuilding, refinishing, and component repair procedures outlined in this book, you can refurbish even the most "sick-looking" Holley carburetor. However, when a replacement for a severely damaged part is not available (like a main body) or a replacement carburetor is cheaper than the needed part and repair procedures, rebuilding doesn't make much sense. On the other hand, if you're restoring a rare carburetor, almost any cost or "heroic" repair procedure may be worth the investment.

The condition of the main body is often the most important rebuilding consideration, since replacements are not available from Holley. Main bodies with loose, broken, or missing booster venturis, cracks or breaks in the casting, severe corrosion, or fire damage can sometimes be repaired by a professional carburetor or machine shop. Unfortunately, the cost of these repair procedures may compare with, or exceed, the price of a new or rebuilt carburetor. If the main body can't be economically repaired—or replaced with a used part—you may be better off "junking" the carburetor and obtaining a replacement (in addition to new carburetors, Holley also sells an extensive line of quality, rebuilt carburetors, called *Holley II).*

Other than the main body, almost all component parts of current model carburetors are available from Holley. Float bowls, metering blocks, throttle plates, and hundreds of other small parts can be purchased through any Holley dealer or directly from *Holley Replacement Parts.* Most parts are inexpensive, however, some—like throttle-plate assemblies—are quite costly. If your carburetor needs any new component parts, carefully evaluate the costs before you proceed with your rebuilding plans. And don't forget that you can often find suitable used parts at a reasonable price from carburetor rebuilding shops or salvage yards.

Most Holley carburetors can be economically refurbished with only the parts from a quality rebuilding kit. But a careful inspection of all the component parts (see Chapter 7 for more help in parts inspection) just might save you from wasting time and money on a carburetor that really belongs in the junk pile.

A severely damaged main body—note the melted booster venturis—may make rebuilding or restoration impractical. Obtaining a new or replacement carburetor is often a better choice when the main body or the throttle plate has been cracked or badly damaged.

The replacement cost and/or rarity of some carburetors can determine how much time and money should be expended on repair. The cracked throttle plate on this 3-barrel (new plates are not available from Holley) was welded and remachined. Although this repair and other rebuilding costs exceeded $100, the owner considered this investment worthwhile.

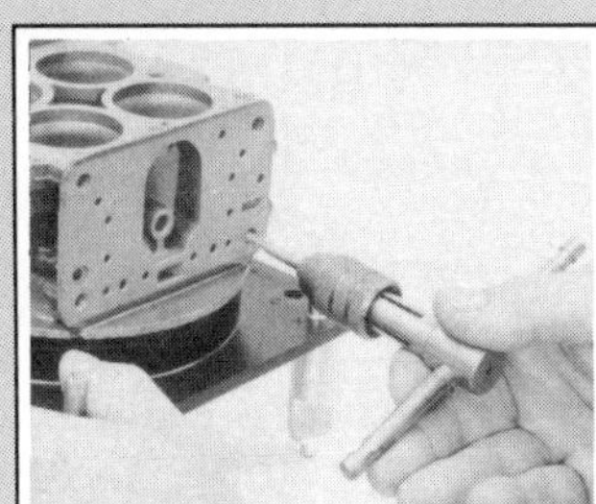

Not all damaged components need to be sent to a machine shop or replaced. Minor repairs, like restoring damaged threads or correcting warped surfaces, are inexpensive procedures that almost anyone can do.

NOTATION REQUIRED

13 Log Primary Main Jet Size

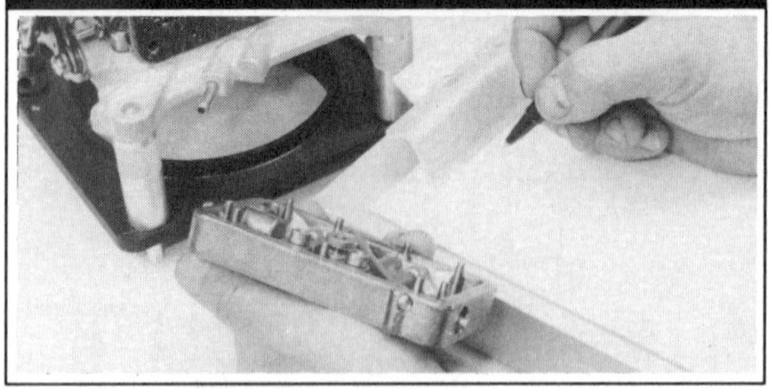

13—Rather than wait until later—when the primary and secondary metering blocks can be confused—note the size of the left (throttle-arm side) and right (choke-arm side) primary main jets on your Work-A-Long sheet now. If you're rebuilding a Holley 2-barrel, skip ahead to **Step 23**, *Begin Removing Choke*.

14 Remove Secondary Bowl

14—Now remove the secondary bowl screws and pry the bowl loose. If it doesn't easily come off, don't forget to turn the carburetor upside down and pry on the underside of the bowl also. If you haven't already done so, mark a small "s" on the secondary bowl (and metering block, if so equipped) to prevent any confusion during reassembly.

15 Secondary Metering Block

15—If you have a 4150, 4165, or 4500, your carburetor is equipped with a secondary metering block instead of a metering plate, and you should perform the procedure in this step, otherwise skip to **Step 16**, *Metering Plate Screws*. Pry off the metering block using the same technique you used on the primary side. Remember, never stick a screwdriver between the metering block and the main body; you'll damage sealing surfaces. Now, skip ahead to **Step 18**, *Stuck-On Secondary Gaskets*.

Two- and Four-Barrel Myths

The primary throttle blades of a spreadbore or small squarebore four-barrel are often smaller than those of many two-barrel carburetors. This is the key to improved fuel economy.

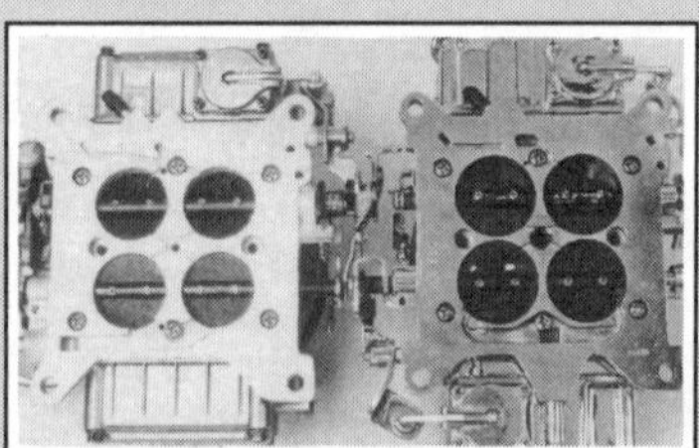

Large-diameter primary throttle bores improve high-speed power on racing engines, but are guaranteed to reduce fuel economy. This is because the larger bores require higher capacity main and intermediate circuits that are not able to meter part-throttle fuel flow as accurately.

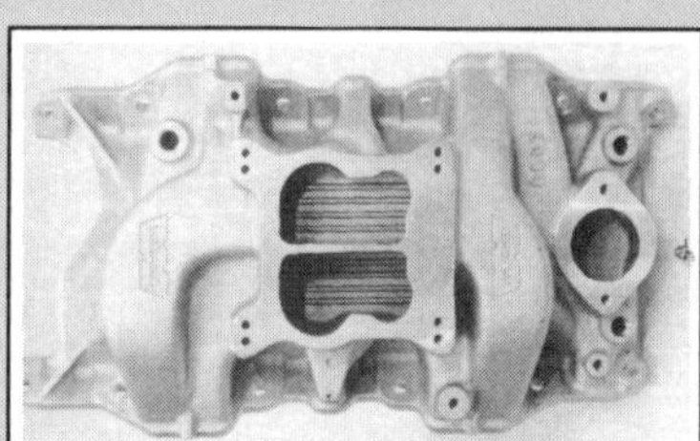

A dual-plane manifold, like this Weiand design, has minimum runner volume and will increase signal strength at the carburetor. This improves fuel-metering accuracy, an essential step for fuel economy.

Many enthusiasts may think it ridiculous to suggest that a four-barrel carburetor can provide the same or better fuel economy than a two barrel. Surprisingly, in most situations where a two-barrel is replaced with the right four-barrel carburetor, fuel economy actually does improve!

The reason a four-barrel can optimize fuel economy is really very simple: the primary throttle bores on a small squarebore or spreadbore carburetor with vacuum-secondary actuation are smaller than most two-barrels. The small primary bores often provide better throttle response, fuel metering accuracy, and improved fuel economy; and when the secondaries kick in, the horsepower potential leaves a two-barrel in the dust!

If you're interested in optimizing fuel economy with a Holley four-barrel carburetor, selecting the right model carburetor is absolutely essential. The most important things to look for are: 1) vacuum-secondary actuation—mechanical secondaries are great for racing but they will almost certainly decrease fuel mileage and throttle response, and 2) a spreadbore or a squarebore model with small primary throttle bore diameters—a spreadbore model 4175 Holley is an excellent choice since it combines the "best of both worlds" with small primaries for good economy and large vacuum-actuated secondary bores for optimum power.

Intake manifold selection is also an important factor when replacing a two-barrel or optimizing fuel economy with a four-barrel carburetor. *Do not install a four-barrel carburetor on a two-barrel manifold with an adaptor.* This will inevitably cause fuel-distribution irregularities and limit power and economy. Use a dual-plane manifold with small-volume intake runners. Avoid open-plenum, large-runner manifolds whenever fuel economy is a consideration. Contact Edelbrock, Weiand, Holley, or Offenhauser for advice on selecting a manifold that will optimize fuel efficiency in your specific application.

By carefully choosing a Holley four-barrel carburetor, installing the right manifold, and topping it all off with methodical carburetion and engine tuning, it is possible to develop an induction system that produces better fuel economy, throttle response, and top-end power than virtually any two-barrel installation.

16 Metering Plate Screws

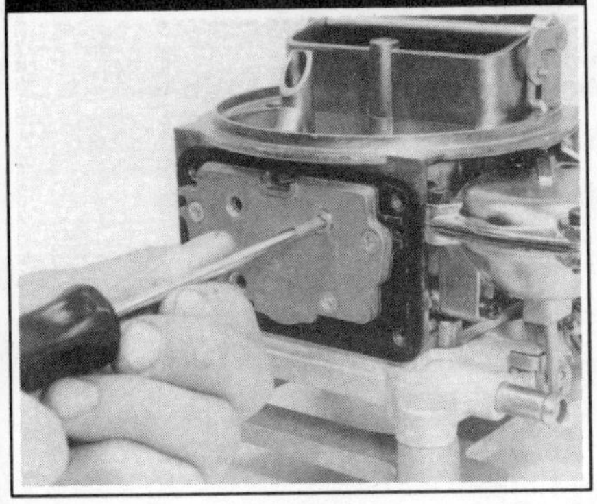

16—Many Holley carburetors are equipped with secondary metering *plates* instead of metering *blocks*. Metering plates are held in place with six clutch-head screws. Remove these screws with a .145-inch clutch-head-tip screwdriver (Snap-on part SDDC5).

17 Prying Secondary Metering Plate

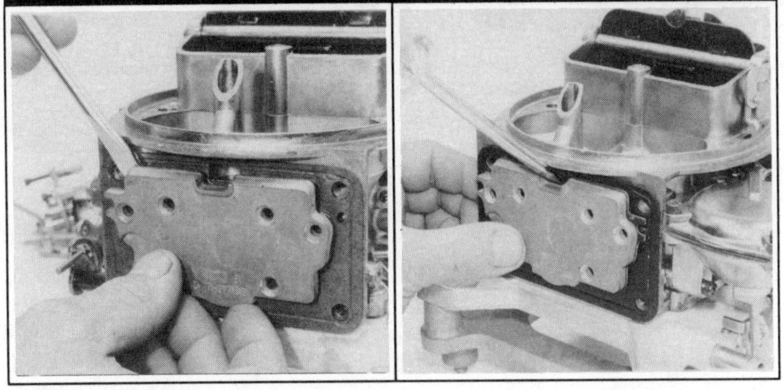

17—Metering plates can stick like glue to gaskets. Although you may be tempted, never insert a screwdriver between the metering plate and the gasket (or steel support plate) as shown on the left. The screwdriver can easily damage the sealing surfaces and cause internal fuel or air leaks. Carefully pry the metering plate on the *top surface* as shown on the right.

MASTER MECHANIC TIP PRO TIP

18 Stuck-On Secondary Gaskets

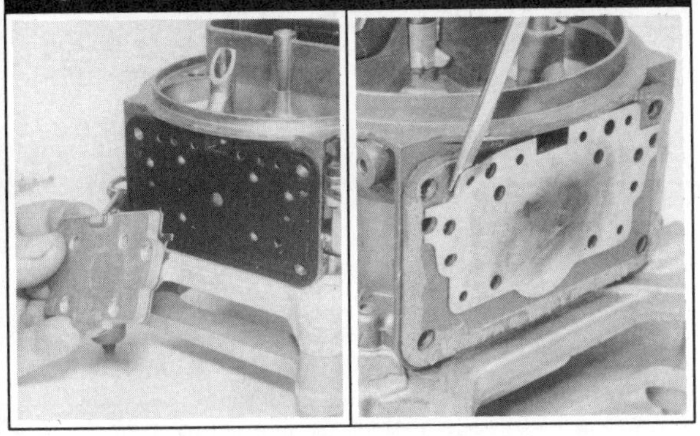

18—When the metering plate or metering block is removed, gasket material (or the steel support plate, if so equipped) may remain stuck to the main body or metering plate. *Don't scrape off stuck-on gaskets or attempt to pry off the metal support plate*; let carburetor cleaner or spray-on gasket remover loosen them later.

THREE-BARREL 3 bbl

19 3-Barrel Plate

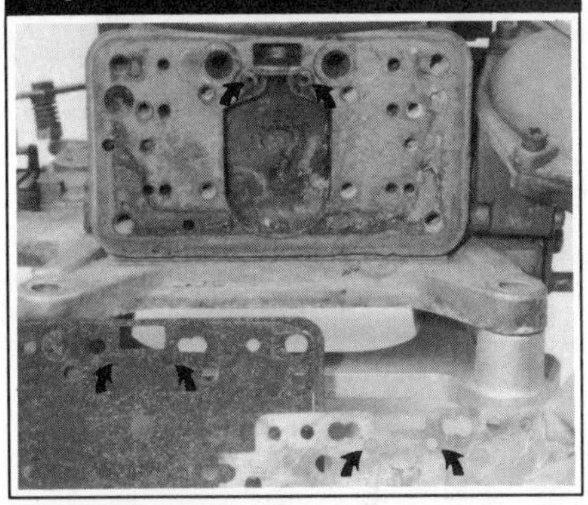

19—The three-barrel uses a large "flap" rather than two individual throttle bores. Holley engineers discovered that signal strength and fuel metering could be improved if the booster venturis were moved inward. This booster and air-bleed inset (arrows) requires a special support plate and gaskets. However, the secondary metering plate for 3-barrels are interchangeable with other 4160s. Not many parts are available for 3-barrels, but Holley still supplies rebuilding kits: Carburetor list R-3916 uses kit **3-405** while 3-barrel list R-4604 uses kit **3-475**. A new secondary diaphragm, **part 135-6**, will work, but the rubber diaphragm and stem must be modified (see *reassembly* **Step 30** on page 88).

NOTATION REQUIRED

20 Secondary Jet Sizes

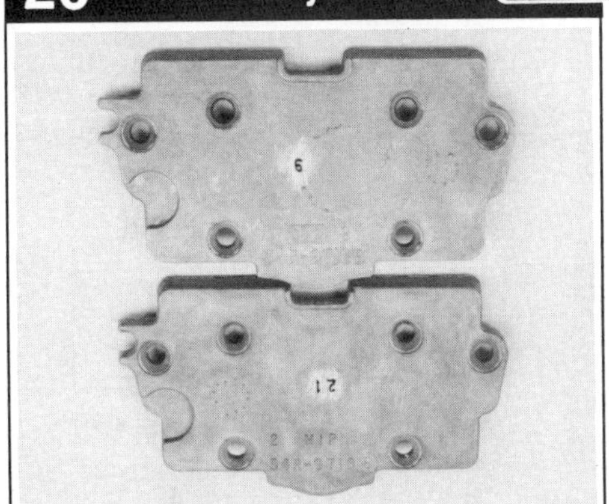

20—Note the size of the left (throttle-arm side) and right (choke-arm side) secondary main jets on your Work-A-Long sheet. If your carburetor is equipped with a metering plate, note the (upside down) stamped number in the center of the plate on your Work-A-Long sheet.

4500 DOMINATOR 4500

21 Remove Pump Arms

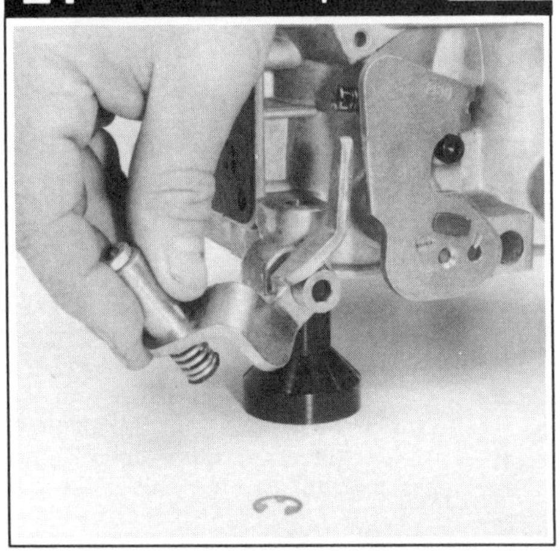

21—You can remove the accelerator pump arms from the main body of your Dominator (if you do, mark the underside of the secondary arm with a small "s"). Arm removal is not required for rebuilding, but they can become a nuisance if left in place. Since Dominators do not have vacuum secondaries or chokes (usually), skip ahead to **Step 40**, *Remove Accelerator Pump Cam*.

22 Remove Fuel Balance Tube

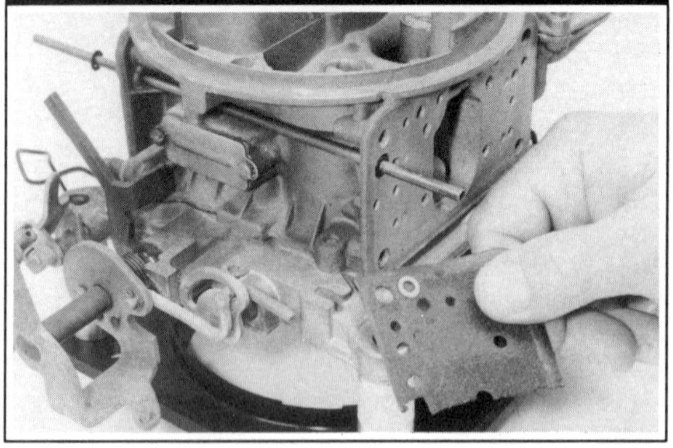

22—If your Holley uses a brass fuel-balance tube, perform this step, otherwise jump to the next step. This tube transfers excess fuel and vapors from the secondary bowl to the primary side to help maintain the correct air-fuel ratio on emission-calibrated carburetors. Each side of the tube is sealed to the main body with an O-ring and brass washer located under the metering block and plate gaskets. Slide the tube out of either end of the main body. Remove the O-rings (unless they're stuck on) and brass washers. If you can't find the washers, they may have stuck to the inside of the gaskets.

IMPORTANT STEP **!**

23 Begin Removing Choke

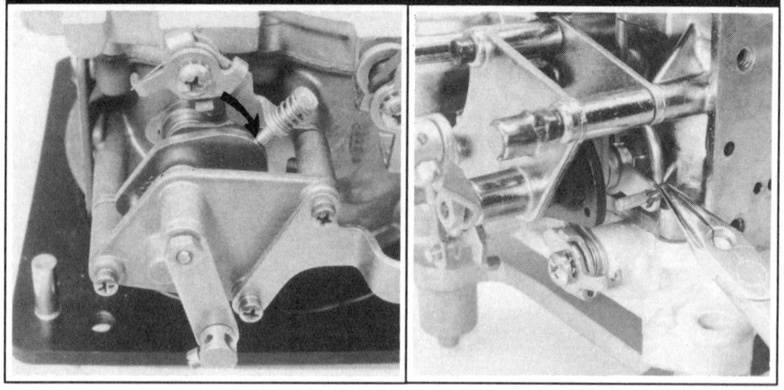

23—Look over the choke linkage carefully before you remove it in the following steps. Holley carburetors are equipped with many different choke mechanisms. Try to familiarize yourself with the overall design (like the contact between the fast-idle screw and cam—arrow); you may want to make some notes on your Work-A-Long sheet. If your carburetor is equipped with an automatic choke, skip ahead to **Step 26**, *Removing Automatic Choke* now. Otherwise, begin disassembly by pulling the lower choke-link "R" clip with needle-nose pliers or hemostats. Put the clip in bin #1.

Fuel Filter Basics

Efficient fuel filtering is critical for the proper functioning of any carburetor. A dirty filter can restrict fuel flow and cause hard starting, lack of power, or even prevent the engine from running. Just as bad, but much more difficult to diagnose, a defective fuel filter can allow microscopic particles to enter the carburetor and cause flooding, surging, lack of power, stalling and a variety of other subtle and major problems. Considering how important fuel filters are to proper operation, and, generally, how easy they are to change, it is amazing how often they are ignored until it is too late.

In a nutshell: Fuel filter(s) should be cleaned or replaced at every tune-up, *and they should always be replaced when the carburetor is rebuilt.*

Use only quality, brand-name fuel filters. The slight additional cost is always offset by superior filtering, better construction, and increased safety. Don't consider the paper, sintered brass, or a wire-screen type filter used in the bowl inlet of some Holley carburetors to be sufficient. *These elements are back-up filters only, and every carburetor installation—regardless of whether or not a filter is installed in the float bowl—should include a quality, external filter element.*

High-performance and heavy-duty vehicles require optimum fuel-filtering and increased flow capacity; that means use a large-diameter filter. Inadequate filter volume can cause high-speed starvation and, under extreme conditions, detonation and internal engine damage. Holley, Moroso, and several other specialty manufacturers offer high-volume filters that can satisfy the needs of even the "hottest" bigblock.

Always install filter(s) in the fuel line from the pump to the carburetor, rather than in the line from the fuel tank. Any restriction in the delivery line from the tank will substantially reduce the fuel-delivery capacity of the fuel pump. Don't locate filters near hot exhaust-system components; you'll avoid fires and help prevent vapor lock. And find a location that is easy to get at; you'll thank yourself when you need to change the element.

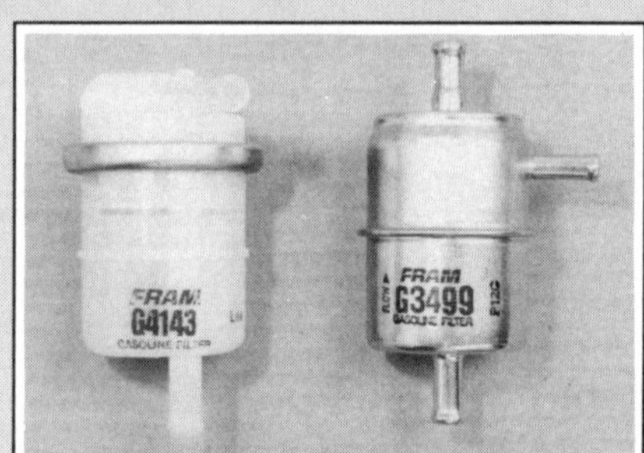

These in-line filters have large filter elements, are easy to install and change, and flow enough fuel for virtually any street machine. Steel cased elements are a better safety choice.

Filters with glass-type housings are very popular. Unfortunately, the small element size limits its filtering capacity, and the glass is easily broken. Avoid this filter for performance use.

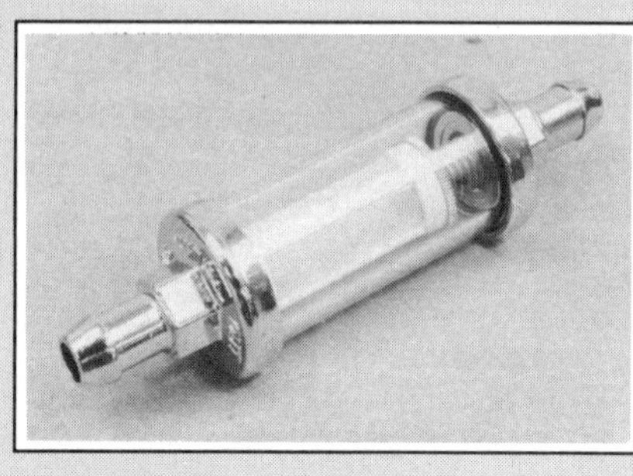

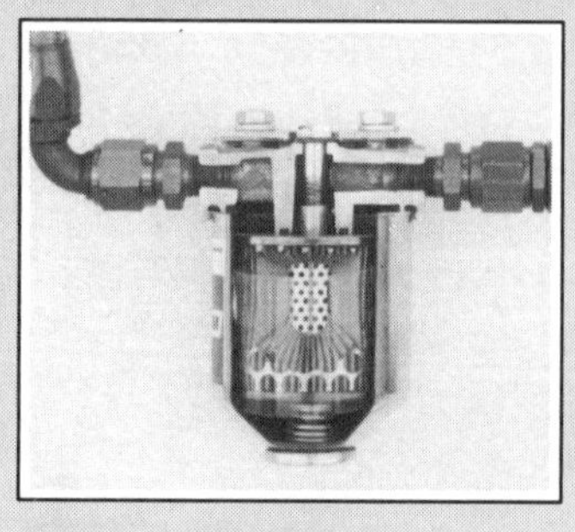

Race engines or highly modified street engines require high-capacity filters, like this Fram competition unit. Regardless of application, always install the filter on the pressure side of the pump.

24 Remove Choke Arm Assembly

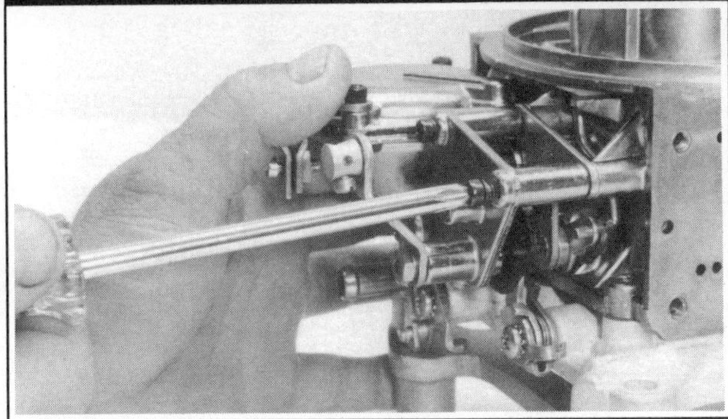

24—Remove the three screws that hold the choke-arm assembly. If you can, try to remove the backup plate and the arm assembly as one unit so you can study the orientation of the components. Note: Some Holleys have a manual choke cable bracket attached by screws on the choke assembly. If your carburetor is so equipped, note the location of the bracket on your Work-A-Long sheet.

25 Remove Backup Plate Assembly

25—If the choke backup plate separates from the arm assembly (it's difficult to hold them together), don't worry; reassembly tips are provided later on. The backup plate often consists of four or more parts that are held together on a stud with a hex-head nut. Unless the choke spring is broken or the stud is loose in the plate, there is no need to disassemble these components. Skip ahead to **Step 32**, *Secondary Diaphragm.*

26 Removing Automatic Choke

26—Holley has manufactured several types of integral automatic chokes. Some earlier models used water-heated choke covers, many used hot-air actuation, some used electrically-heated covers with hot-air assist, but many late-model carburetors (and choke conversion kits) use an electric choke mechanism—like the one shown here—that does not require an external hot-air connection. If your carburetor has a ground wire attached to the secondary diaphragm cover, remove it now.

NOTATION REQUIRED

27 Log Choke Cover Position

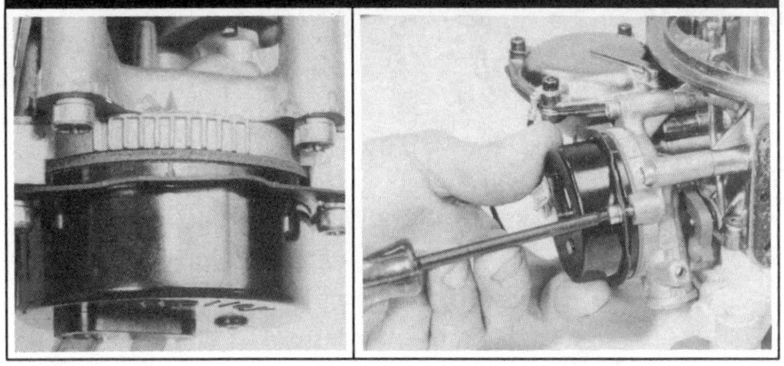

27—Note the position of the choke cover indicator mark on your Work-A-Long sheet. Returning the cover to this position during reassembly often provides a close initial adjustment. Now remove the three screws that hold the cover clamp in place. Some late-model cover screws require a Torx T-20 bit. Remove the cover and clamp; place the cover in bin #1 (no immersion) and the remaining parts in bin #2.

28 Remove Choke "R" Clip

28—Now pull the lower choke-link "R" clip with needle-nose pliers or hemostats. It is often partially hidden behind the plastic fast-idle arm and can be difficult to get at. Put the clip in bin #1.

29 Remove Choke Housing

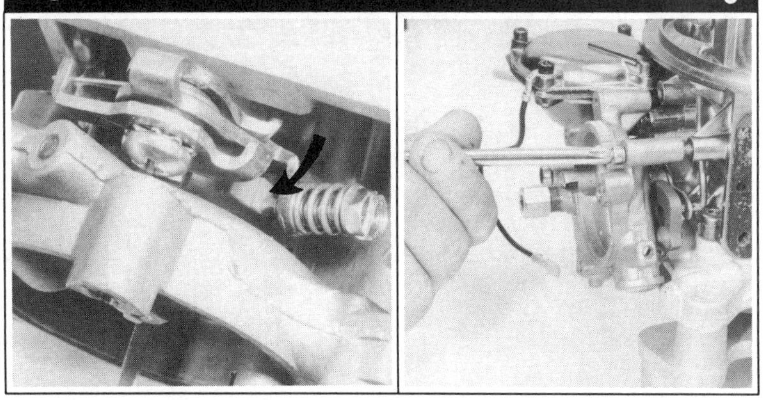

29—Turn the carburetor upside-down and examine the relationship between the fast-idle arm (attached to the primary throttle shaft) and the plastic fast-idle cam on the choke housing (arrow). If you haven't done so already, you may want to make some notes on your Work-A-Long sheet to help you during reassembly. Now turn the carburetor back over and remove the three screws that hold the choke housing to the main body. Finally, remove the housing by wiggling it off the choke link.

NOTATION REQUIRED

30 Is Gasket Present?

30—Many automatic choke assemblies use an external or internal vacuum source. If your choke uses internal vacuum (the vacuum passage in main body is open), check the housing or main body for the presence of a circular gasket that seals the vacuum passage to the choke housing. If the gasket is missing, the internal choke pulloff will malfunction (causing cold-engine stalling) and the choke cover heating element may burn out. Note its presence or absence and location on your Work-A-Long sheet.

31 Remove Choke Pulloff

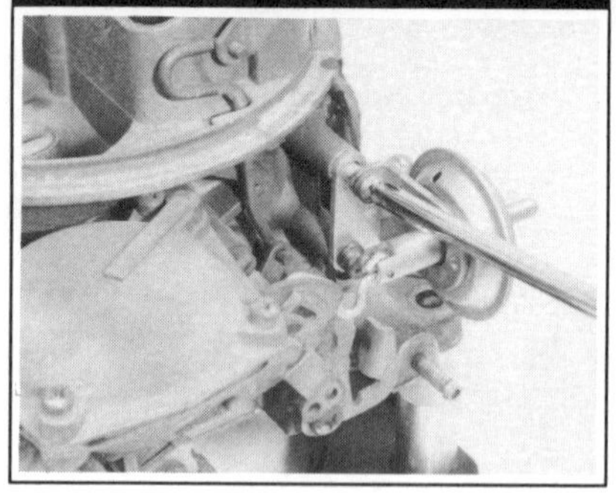

31—Some Holley carburetors use an external (divorced) choke mechanism located in a well in the intake manifold. On these carburetors, and even on some integral-choke carburetors, a small vacuum diaphragm may be mounted on the side of the carburetor that functions as a choke pulloff (more information on choke pulloff function and adjusting is provided in Chapters 9 and 10). If your Holley has this feature, remove the pulloff diaphragm and linkage now. You may wish to note the orientation of the linkage on your Work-A-Long sheet to help you during reassembly.

32 Secondary Diaphragm

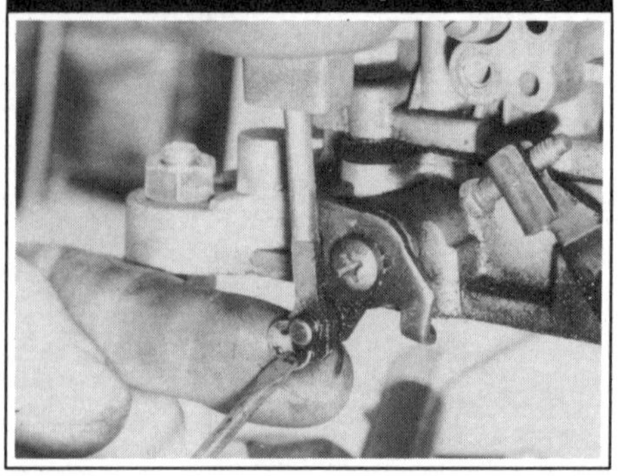

32—If you are working on a mechanical-secondary Holley, skip ahead to **Step 36**, *Removing The Throttle Plate*. Remove the "C" or "E" clip that holds the diaphragm rod to the lever assembly. A small screwdriver works nicely to remove the larger "E" clip, however, the smaller "C" clip is a bit more tricky...

33 Removing The Small "C" Clip

33—Since few of us own the special tool designed to remove small "C" clips, the next best method is to use two small screwdrivers. To minimize the chance of loosing the clip, orient the open-end up and place a rag under the carburetor. If all goes well, the clip will drop onto the rag. But if the clip flies off into the "wild-blue," don't panic; most good carburetor kits include a replacement.

NOTATION REQUIRED

34 Remove Diaphragm Housing

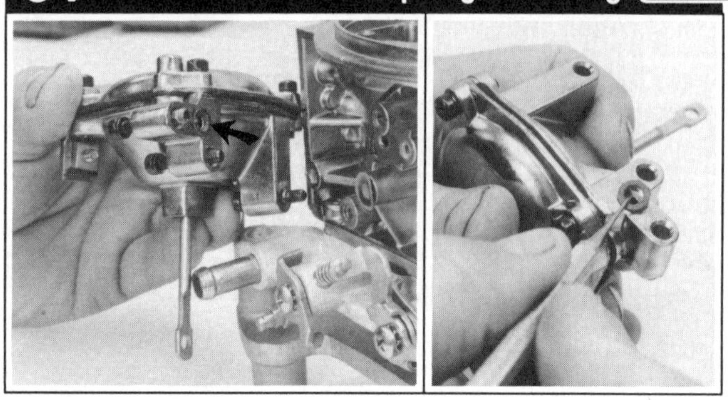

34—Remove the three screws that hold the vacuum-diaphragm housing to the main body. As you remove the housing, see if the circular gasket that seals the vacuum source to the housing is either stuck to the housing or the main body. If the gasket is missing, it may explain why the secondaries didn't open or opened too late. Note its presence or absence on your Work-A-Long sheet.

35 Throttle Plate Accessories

35—Some Holley carburetors, particularly 4180s and 4190s, have dashpots and electronic controls mounted on a bracket attached to the throttle plate or elsewhere. Remove these controls and set them aside in bin #1; they will not be immersed in carburetor cleaner.

NOTATION REQUIRED

36 Removing The Throttle Plate

36—Before you remove the throttle-plate attaching screws, note their position on your Work-A-Long sheet. Some Holleys use eight screws, many others use only six, and most 2-barrels use five screws.

PERFORMANCE STEP

37 Special Secondary Linkage

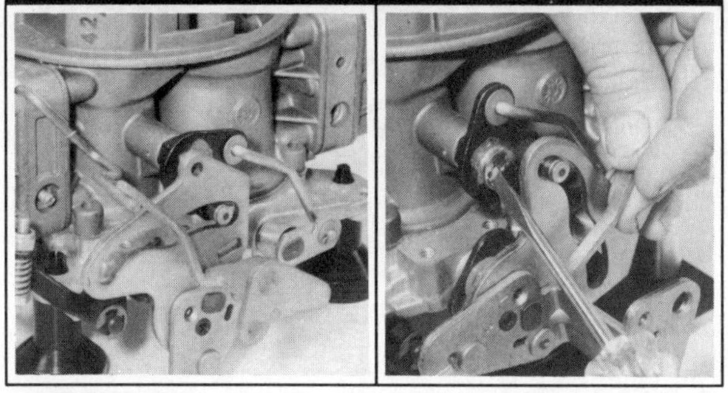

37—The 4165 and some other mechanical-secondary, spread-bore carburetors have a rather elaborate secondary linkage system. This linkage must be disconnected in order to remove the throttle plate. Instead of removing the secondary actuation link (arrow), just pull the "E" clip on the intermediate arm. Don't remove the arm yet.

38 Remove Throttle-Plate

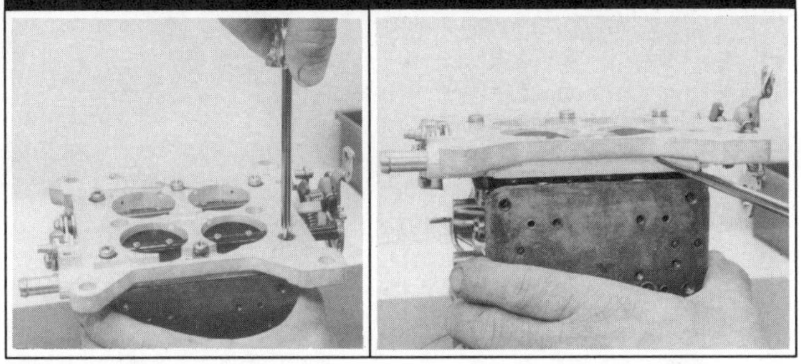

38—Now loosen and remove all the throttle-plate screws. These screws can difficult to remove if they have become corroded. Don't apply excessive force. For help, refer to the sidebar *Removing Stubborn And Hidden Fasteners* on page 37. Most throttle plates will lift off with only a little effort. But some are firmly stuck. Stubborn plates can be pried loose by positioning a screwdriver between the plate and the main body (make sure all screws have been removed before you begin prying).

PERFORMANCE STEP

39 Bellcrank Linkage

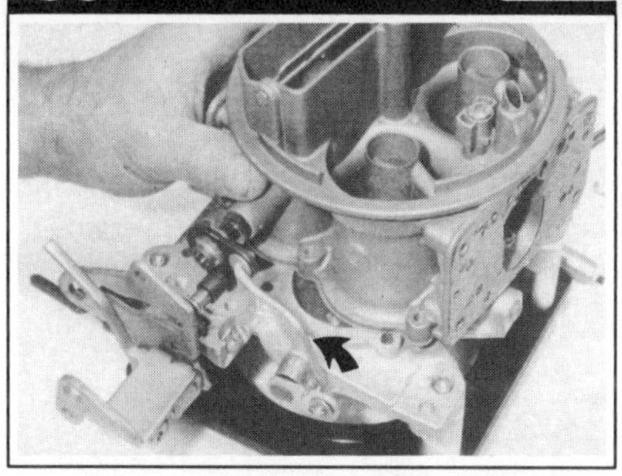

39—To remove the throttle plate on manual-secondary carburetors equipped with bellcrank linkage, just lift (or pry) the main body slightly and slide it sideways while removing the intermediate arm from its stud. Unless the throttle plate will be disassembled, there is no need to remove the secondary actuation link (arrow) from the throttle linkage.

NOTATION REQUIRED

40 Remove Accelerator Pump Cam

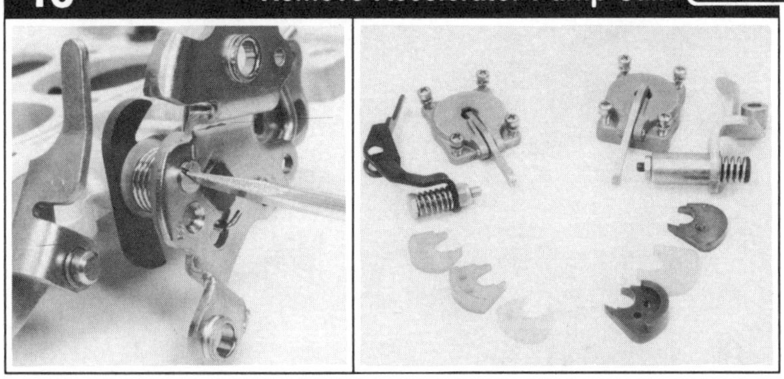

40—Now remove the primary—and secondary, for double pumpers—indexing screw(s) and cam(s) from the throttle shafts. Note the shape (30cc or 50cc type), color, ID number, and the indexing screw position on your Work-A-Long sheet. The photo shows the difference between the standard pump, lever, and cams that deliver 30cc, and the larger pump, actuating lever, and cams that deliver 50cc.

4180/4190 EMISSIONS CARBURETOR

41 Idle Screws In Throttle Plate

41—Some Holleys, particularly strict emissions models like the 4180, have the idle-mixture screws located in the throttle plate rather than the metering blocks. If your throttle plate has these casting extensions and tamper-resistant plugs pressed in them, see the sidebar *4180/4190 Idle-Screw Repair* on page 50.

PERFORMANCE STEP

42 Center-Squirter

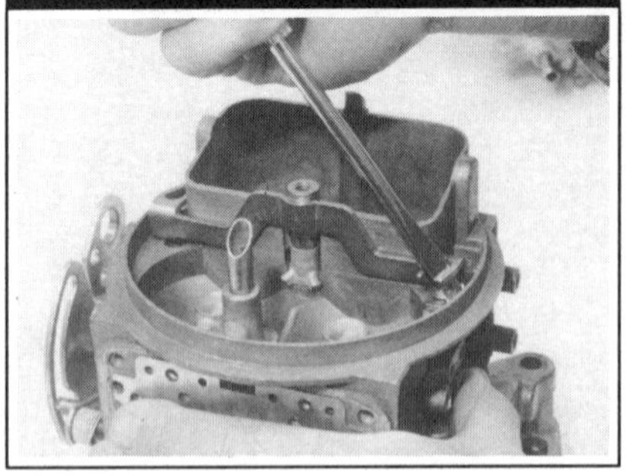

42—Center-squirter racing carburetors have a special air-cleaner mounting bracket that must be removed to access the discharge nozzle screw. First, bend the outside tabs as shown. Then loosen both Phillips-head screws and lift the bracket straight up.

NOTATION REQUIRED

43 Remove Discharge Nozzle

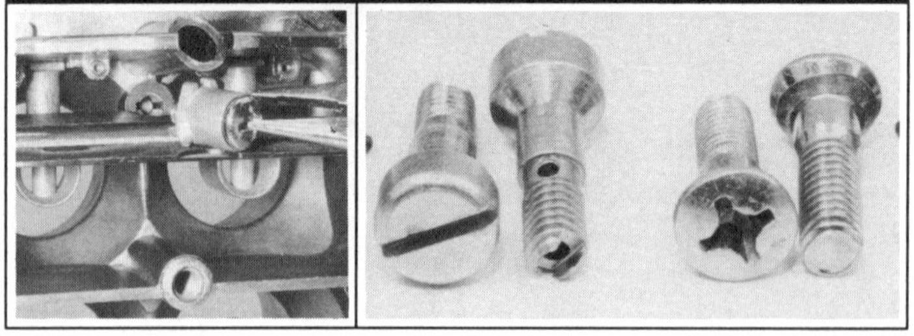

43—Loosen and remove the discharge nozzle and screw. Standard-flow nozzle screws use a Phillips head, while high-flow screws use a straight-slot. Record the screw type, nozzle design and size (two-digits stamped on the nozzle) on your Work-A-Long sheet.

4180/4190 Idle Screw Repair

Model 4180/4190 Holleys are emissions-calibrated carburetors. To prevent casual adjustments to the idle air/fuel ratio, the idle-mixture screws are moved from their usual location in the primary metering block to a "concealed" location behind plugs in the throttle plate.

Model 4180/4190 Holleys are emissions-calibrated carburetors that are often used on 1979 and later Ford trucks and passenger cars. Their most distinguishing characteristic is the lack of idle-mixture adjustment screws in the primary metering block. To prevent the casual "mechanic" from altering the idle air/fuel ratio and adversely affecting emissions, the idle-mixture screws are concealed behind plugs in the throttle plate.

The rebuilding procedures for the 4180 are nearly identical to any 4160 Holley. But because the idle-mixture screws in the throttle plate incorporate O-ring seals, the throttle plate cannot be immersed in carburetor cleaner without first removing the plugs and adjustment screws. When the plugs are removed, *the idle adjusters should not be unscrewed without following a strict procedure that will ensure that each screw is returned to the same location and adjustment originally established at the factory.*

In order to maintain this original factory setting, the adjustment screws should first be turned in until they *lightly* seat. The number of turns required for each screw should be carefully recorded, and as each screw is removed, it should be tagged with its original location in the throttle plate. Each of the adjuster screws has a tension spring that prevents the screws from turning from vibration. These springs do not affect idle mixture and need not be tagged. After the throttle plate and adjusting screws are cleaned and new O-rings are installed, the screws and springs should be assembled, the O-rings lubricated, and each reinstalled in its original location. Screw them in until they *lightly* seat again; then back them out the same number of turns that was recorded during disassembly. Any alteration of this original adjustment should only be attempted with factory service information and specialized test equipment at hand.

1—To remove the idle mixture concealment plugs, drill a 3/16-inch hole about 1/4-inch from the edge of the casting. Use care when drilling to avoid touching the mixture screws with the drill.

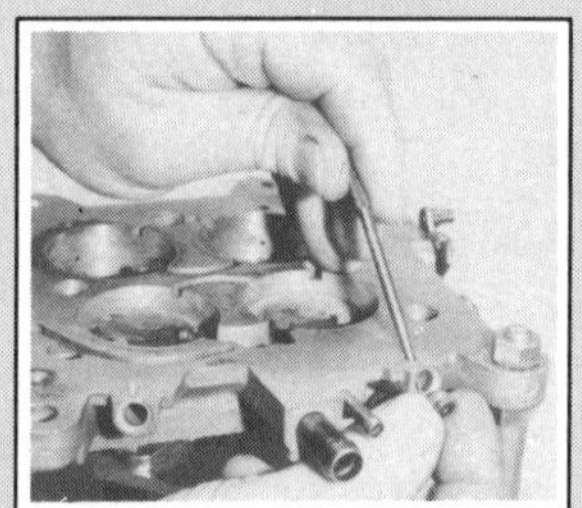

2—Insert a pointed pick or similar tool into the hole and pry out the plug.

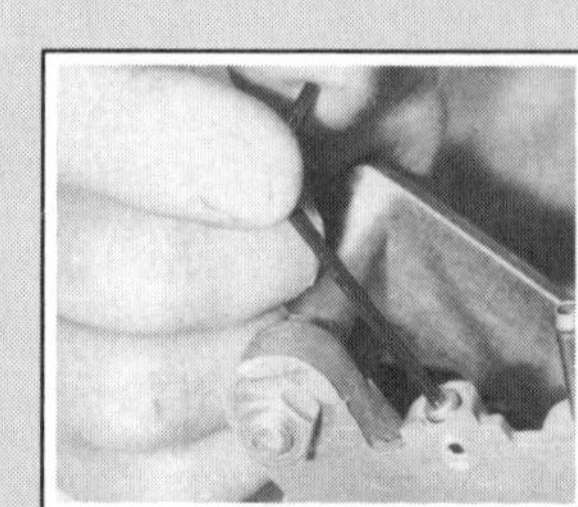

3—Record the number of turns required to lightly seat the adjuster screw and then remove it. Label each screw so that it can be returned to its original location.

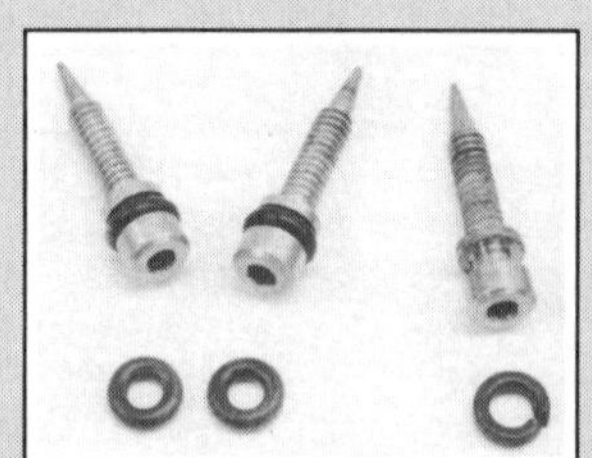

4—Examine the tapered tips of the adjuster for any damage and install new O-rings.

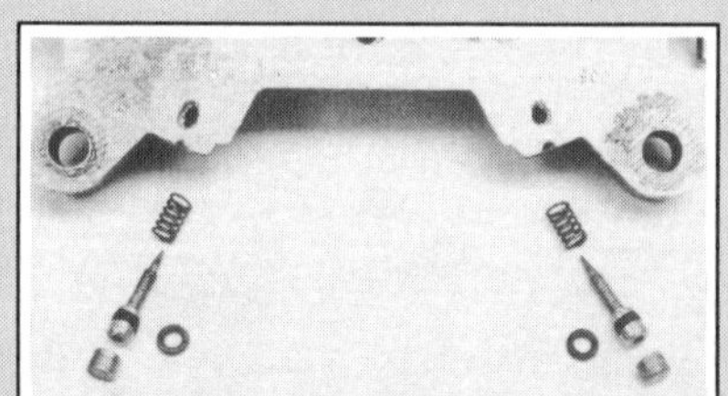

5—After cleaning, new O-rings are installed, the screws and springs are reassembled, the O-rings are lubricated, and each is reinstalled in its original location. Screw them in until they *lightly* seat again; then back them out the same number of turns that was recorded during disassembly.

4165/4175 SPREAD-BORE CARBURETORS

4165

44 4165/4175 Nozzle Tips

44—4165 and 4175 carburetors use wide anti-pullover nozzles. The primary-side nozzle won't clear the choke shaft when it's lifted straight up. Don't bend the choke shaft to remove the nozzle; follow these simple steps. First remove the nozzle screw, then use needle-nose pliers or hemostats to flip the nozzle upside down. Position the nozzle so that the alignment tabs face away from the booster venturi. Once the nozzle is upside down, rotate—or, perhaps, shuffle—the nozzle under the choke plate toward the front of the carburetor. When it is as far forward as possible, the choke plate should open and the nozzle can be lifted free. Before you set it aside, record the nozzle number on your Work-A-Long sheet.

45 Discharge Needle Valve

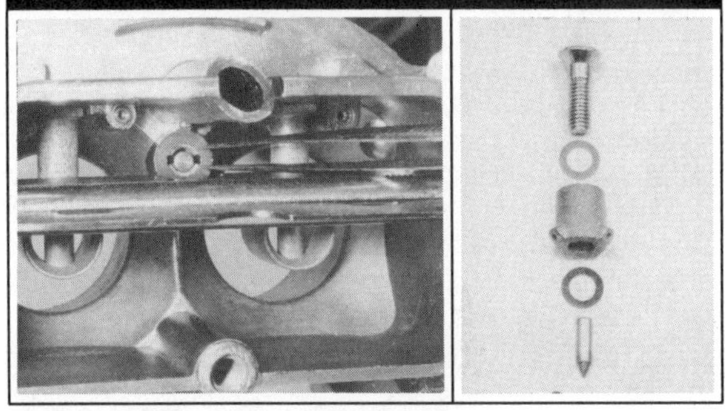

45—Many Holley carburetors have a needle check valve located under the discharge nozzle, oriented as shown in the right photo. The needle usually can be removed by turning the carburetor upside down and letting it fall out onto your workbench (if the needle valve is missing, see the next step). Put the needle valve into bin #1.

IMPORTANT STEP

46 Discharge Needle Missing

46—If there was no needle valve located under the discharge nozzle, make sure your carburetor uses an anti-pullover discharge nozzle, as shown on left. If your Holley is equipped with any one of the standard nozzles (like those on the right), a check valve must be used to prevent fuel from dribbling out of the nozzle when air flows through the venturi. Some carburetor kits include a replacement needle (or check balls); we'll remind you to install it during reassembly. Note the presence or absence of a pump check valve on your Work-A-Long sheet.

PERFORMANCE STEP

47 Secondary Nozzle

47—If your carburetor has mechanically-actuated secondaries, like this Dominator, remove the secondary discharge nozzle and needle valve (see previous step if there was no needle valve installed below the discharge nozzle). Record the secondary nozzle type and number on your Work-A-Long sheet.

4500 DOMINATOR

4500

48 Disassembly Complete

48—For the purpose of rebuilding, the 4500 main body is now disassembled. Do not remove the interior secondary-actuation linkage. If you want to install a different secondary-actuation cam, refer to the *reassembly chapter,* **Step 56** on page 95. Now skip ahead to **Step 52**, *Primary Metering Block Disassembly* on page 53.

49 Remove Hot-Idle Compensator

49—Some Holleys are equipped with a hot-idle compensator valve built into the main body. If your carburetor has this system, remove the cover screws and slide out the bi-metal spring valve. Put the cover and spring valve into bin #2.

50 Diaphragm Disassembly

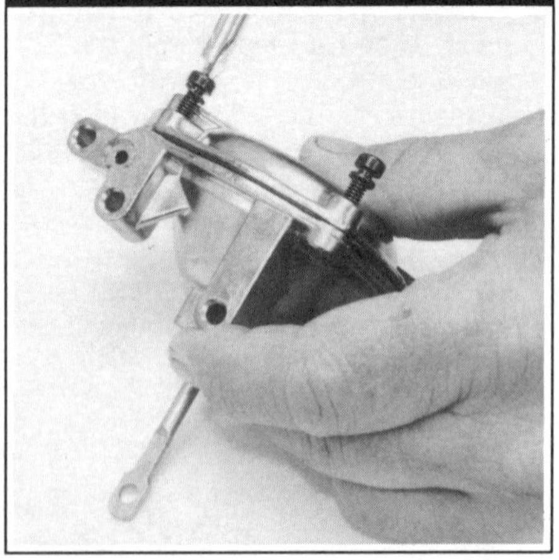

50—If your Holley is equipped with a vacuum secondary diaphragm, perform this step, otherwise skip ahead to **Step 52**, *Primary Metering Block Disassembly*. Remove the cover screws of the vacuum-secondary diaphragm. If you plan to reuse the rubber diaphragm, remove the screws carefully; the diaphragm often sticks to the threads and can be wound up and torn as they are removed. Try spraying silicone lubricant on the screws after they are loosened a turn or so. **Note:** Some Holleys have a manual-choke cable bracket or automatic-choke ground wire attached to one or more screws on the vacuum diaphragm. If your carburetor has either of these attachments, note its location on your Work-A-Long sheet.

NOTATION REQUIRED

51 Remove Vacuum Diaphragm Cover

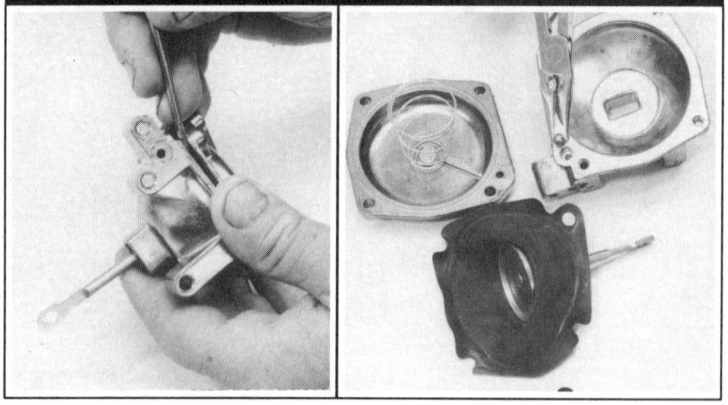

51—Use a small screwdriver to pry the cover loose. As the cover begins to come off, orient the assembly in a vertical position. Remove the cover and diaphragm. Inspect the vacuum feed passage (arrow) for the presence of a check ball. Many Holleys use a check ball to control the secondary opening rate; note its presence or absence on your Work-A-Long sheet. Then pull the diaphragm spring free from the button on the cover and note any color markings (paint or stain) on the spring (indicates stiffness).

Power Valve Basics

Gasoline engines require a lean mixture for fuel economy during cruise and a richer mixture for performance during heavy acceleration. The power valve is a vacuum-actuated valve that performs this essential enrichment function. It senses manifold vacuum and turns on extra fuel flow when vacuum drops to a predetermined level. Despite this straightforward on-off operation, the power valve has been criticized as the Achilles' heel of the Holley carburetor. Many misguided enthusiasts have removed the power valve and installed a power-valve plug to eliminate this "unwanted blight" from their Holley. In most cases, the only thing eliminated is fuel economy and driveability.

The power valve has developed a "bad rap" because of misuse, improper installation, and misunderstanding.

One of the most common beliefs is that power valves rapidly fail. In fact, power valves are very reliable, but there are two principal reasons for *premature* failure. *The first is engine backfiring*, usually due to improper timing, valvetrain malfunctions, ignition problems, etc. Backfiring causes high pressure that can rupture the valve diaphragm. *The second reason for power valve failure is simply low quality*; not in Holley power valves, but in the replacement valves that are included in virtually every non-Holley rebuild kit. Genuine Holley valves are made of top quality materials and are calibrated and stamped with an opening-point number. Non-Holley replacements vary widely in design and quality. Furthermore, they are almost never stamped with their opening point. If that wasn't enough, many replacement valves have limited flow capacity and are not suitable for high-performance applications.

Improper installation is also a major contributor to the power-valve "bad rap." Overtightening often causes outright failure or stretching and subsequent malfunction. Holley window-type power valves (see photo) should be torqued to 100-inch pounds, while those with drilled discharge holes and all aftermarket valves should be torqued to 40- to 50-inch pounds. Using the wrong gasket can also cause malfunctions. There are two types of gaskets used on genuine Holley valves; they are not interchangeable. A third type of gasket is supplied with many aftermarket power valves. Unfortunately, the design of this gasket partially obstructs the feed holes and can prevent full enrichment, resulting in poor performance and surging.

The bottom line is: Always use a genuine Holley power valve of the recommended number, use the correct gasket, and torque the valve to the proper tightness. Follow these simple rules and you'll obtain the all benefits of good fuel economy and driveability with no "bad rap."

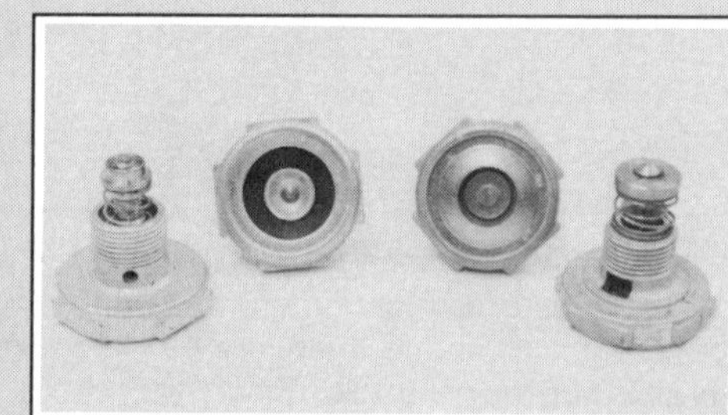

All Holley valves are calibrated and numbered. Most aftermarket valves are not numbered and some are not even calibrated. The two non-Holley valves on the left have no provision for a factory-calibration adjustment. Also notice the restrictive fuel orifices.

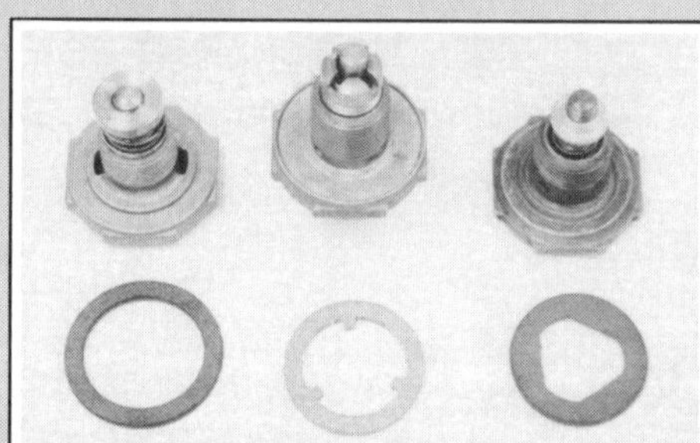

There are three types of power valve gaskets currently available. The plain circular type must be used on valves with an inner stepped locating shoulder (left). The gasket with three locating tabs is used on Holley valves that have no inner shoulder but incorporate a ridge around the outer circumference (center). The third type is often supplied with aftermarket valves. It has a triangular center section that can restrict flow; avoid this gasket.

NOTATION REQUIRED

52 Primary Metering Block Disassembly

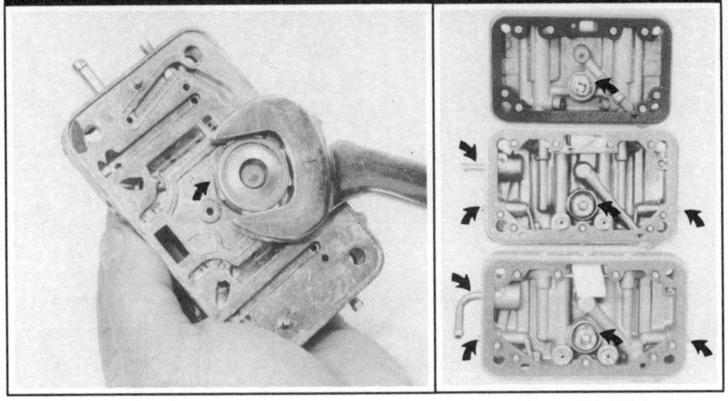

52—Locate the primary metering block. If you did not scribe an "s" on the secondary block, and you cannot determine which block is the primary or secondary, use this rule: On most Holleys, the primary block is the one that has a majority of the following characteristics: 1) a power valve, 2) idle screws, and 3) a vacuum takeoff (arrows). Using a 1-inch open-end wrench, socket, or adjustable wrench, unscrew the power valve. Note the power valve number and type (several small feed holes or two large passages) on your Work-A-Long sheet.

NOTATION REQUIRED

53 Primary Main Jets

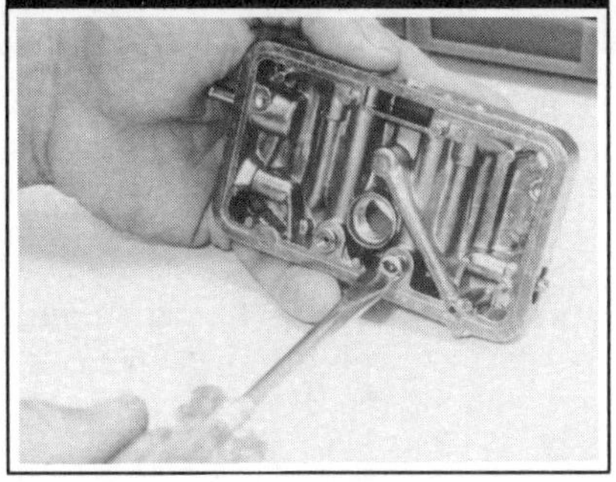

53—If you haven't already, note the primary jet size(s) on your Work-A-Long sheet. Using a straight-blade screwdriver—at least as wide as the diameter of the jets—unscrew both primary jets and put them in bin #1.

NOTATION REQUIRED

54 Remove Idle-Mixture Screws

54—Before you unscrew the idle-mixture screws, gently screw them in and count the number of turns until they seat (just stop turning). Log this information for each idle screw on your Work-A-Long sheet. Now loosen and remove both screws. Use a hooked probe to remove the cork sealing gaskets; they can get pretty "gooey" if you immerse them in carburetor cleaner. This completes the disassembly of the primary metering block.

55 Secondary Metering Block

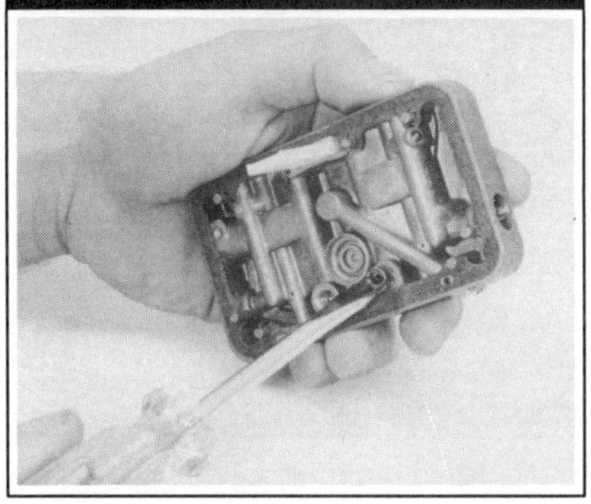

55—Some Holley carburetors use a secondary metering block instead of a metering plate. These carburetors may or may not use a secondary power valve and adjustable idle screws. If your carburetor has a secondary metering block, repeat the same disassembly procedures used in the last three steps on the secondary side. If you haven't already, scribe a small "s" on the side or bottom of the secondary metering block.

56 Float Bowl Disassembly

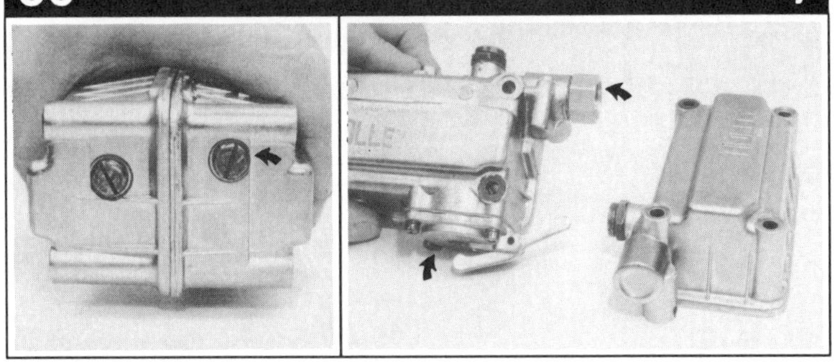

56—Locate the primary float bowl. If you did not scribe an "s" on the secondary bowl, and you cannot determine which bowl is the primary or secondary, use this rule: On most Holleys, the primary bowl has a majority of the following characteristics: 1) a higher sight plug position, 2) an accelerator pump, and 3) a screw-type fuel-inlet fitting (arrows). On a few racing carburetors, like the competition Dominator, the sight-plug positions and other bowl characteristics are identical. In these few cases, the float bowls are interchangeable.

NOTATION REQUIRED

57 Accelerator Pump

57—Before you begin, note the accelerator pump size (30cc or 50cc) on your Work-A-Long sheet. Then, remove the four screws that hold the accelerator pump cover in place, and remove the cover, diaphragm, and spring; light prying will usually remove the most stubborn cover.

NOTATION REQUIRED

58 Pump Inlet Check Valve

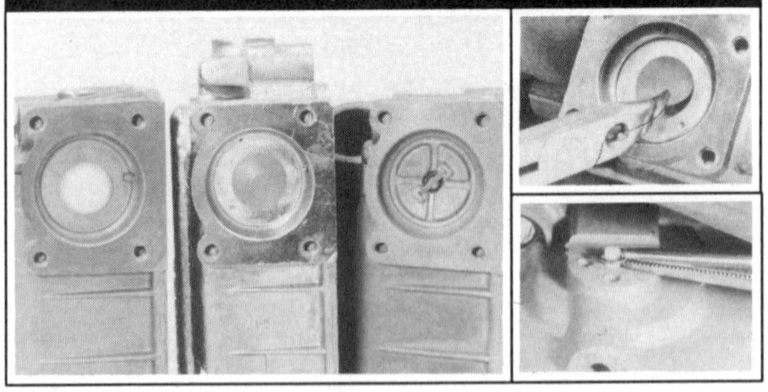

58—Holleys are equipped with three types of accelerator-pump inlet check valves: A steel ball retained by a metal strap, a red-plastic "umbrella," or a white-plastic "cup" valve. Note the type you find on your Work-A-Long sheet. If a red-plastic check valve is used, pull it out with needle-nose pliers. If you find the white-plastic cup valve, remove the small "E" clip on the stem and place the plastic cup, the small metal weight (looks like a washer) and the clip in bin #3.

NOTATION REQUIRED

59 Types Of Holley Needles-And-Seats

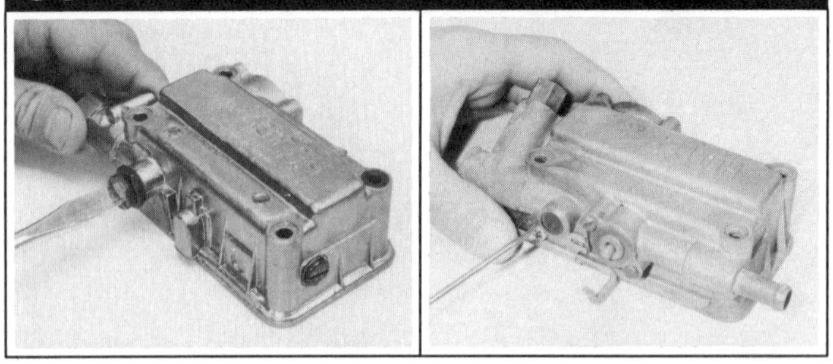

59—Holleys are equipped with two basic types of inlet needles-and-seats and bowl assemblies: 1) externally adjustable, identifiable by a straight-slot screw and hex nut visible on the outside of the bowl (left photo), and 2) internally adjustable with no visible screw-and-nut assembly (right photo). If your carburetor is equipped with internally-adjustable bowls, skip ahead to **Step 64**, *Removing The Bowl Baffle*. NOTE: If the screw and nut on your bowl don't look like those pictured here, STOP, and refer to **Step 61**, *Non-Holley Needles-And-Seats*.

60 Adjustable Needles-And-Seats

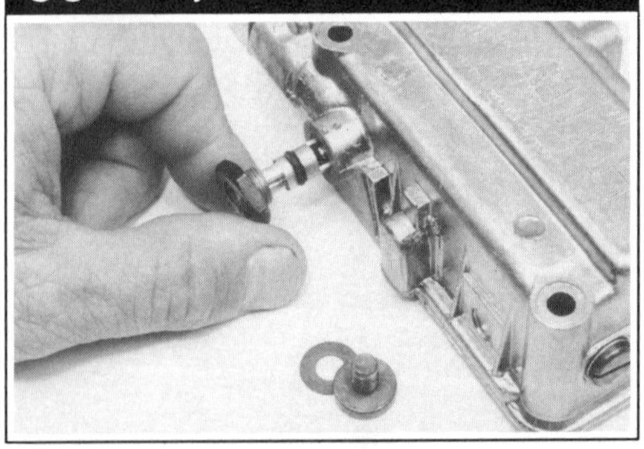

60—Remove the slotted lock screw. Unscrew the needle using the hex nut (you may have to use a wrench to break it loose). If the nut is difficult to turn, spray silicone lubricant on the threads from outside and inside the bowl. Jump to **Step 62**, *Note Needle-And-Seat Type*.

IMPORTANT STEP

61 Non-Holley Needles-And-Seats

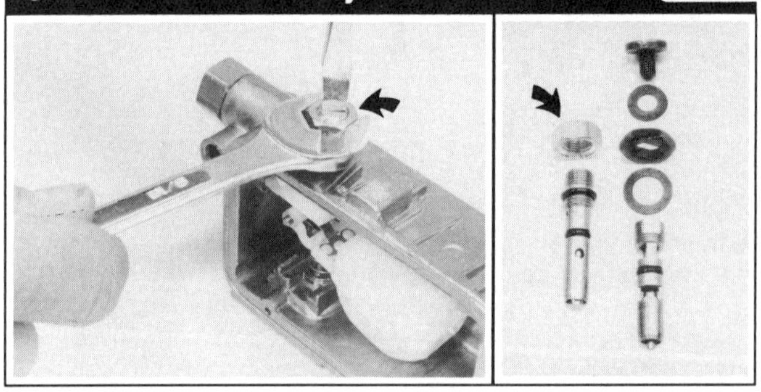

61—If you have externally-adjustable float bowls with hex nuts and screws that look like this (arrows), the needles-and-seats were not manufactured by Holley (Holley type is shown on the far right), and they are not removed by first loosening the screw. Instead, loosen the locknut *then* unscrew the slotted center piece. Refer to *Selecting Carburetor Kits* on page 32 and 33 for more information on needles-and-seats.

NOTATION REQUIRED

62 Note Needle-And-Seat Type & Size

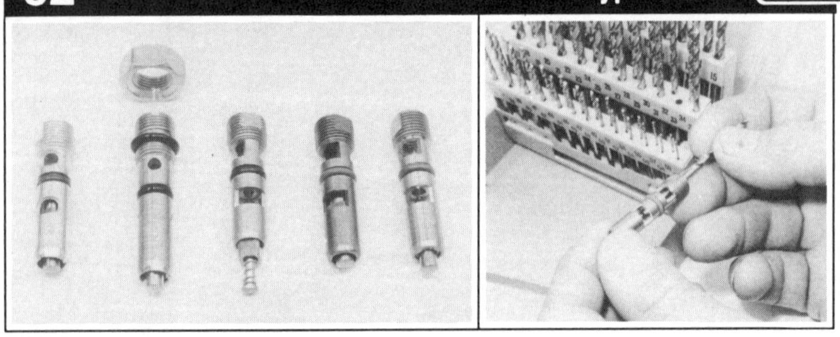

62—When the needle-and-seat is removed, refer to your Work-A-Long sheet and note the overall design and orifice size. Orifice sizing is not possible with the non-Holley type assemblies (arrow). Orifice diameter can be determined by carefully inserting number drills into the seat until one just fits through the orifice (a number-drill to thousandths-of-an-inch conversion table can be found in the Appendix, beginning on page 125).

63 Remove Sight Plug

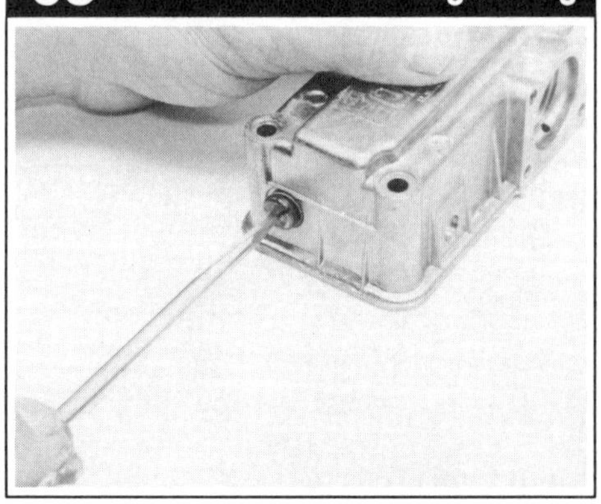

63—Remove the fuel-level adjusting sight plug (and gasket if it comes off easily) on the side of the float bowl.

64 Removing The Bowl Baffle

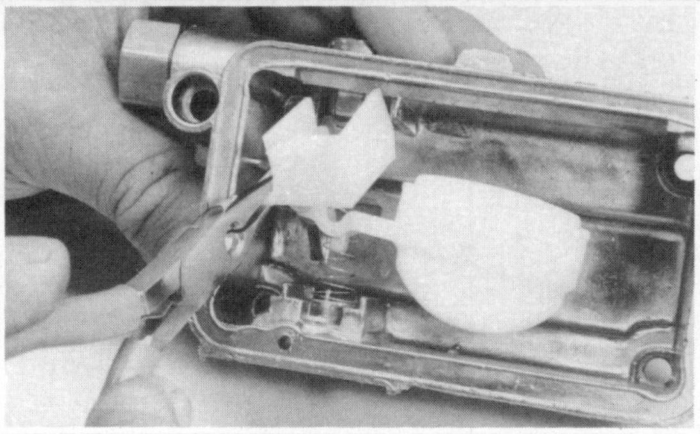

64—If your Holley is equipped with center-pivot bowls, skip ahead to **Step 67**, *Center Pivot Floats.* Most side-pivot bowls have a fuel baffle installed around the needle-and-seat assembly. Remove the baffle with needle-nose pliers by grasping the flat surface as shown. If it does not easily come out, wiggle the pliers as you pull.

NOTATION REQUIRED

65 Removing Primary Float

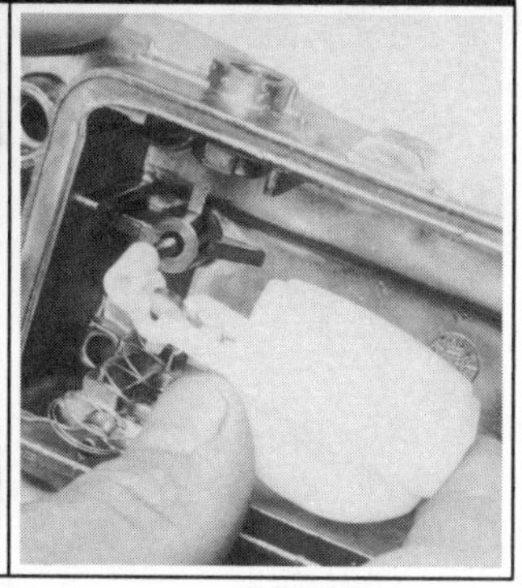

65—Remove the "E" clip with a small screwdriver or probe. Don't lose this clip; some carburetor kits do not include a replacement. Now, pull the float and the counterspring from the locating pin. Note the float type on your Work-A-Long sheet.

Float Bowl Basics

The fuel bowl provides a reservoir of fuel for the various metering systems of the carburetor. In addition, it allows fuel vapors to be separated and vented from the carburetor through tubes located in the airhorn. Without unrestricted venting, a pressure differential in the bowls would prevent accurate fuel metering.

There are two different bowls currently available. The side-pivot (also called "side-hung") and the center-pivot (also called "center-hung" or "cathedral") are named for the location of the float pivot point. Although conversion from one type to the other is possible on almost any Holley, each has specific advantages and disadvantages.

Side-pivot bowls are found on many street, performance, and emissions-type carburetors. The main fuel inlet is often attached to the primary float bowl and fuel is delivered to the secondaries through an aluminum transfer tube, although some models use separate feeds for each bowl. The smaller bowl volume and float size are advantageous for off-road racing since they offer good resistance to sloshing and flooding. Side-pivot bowls are also popular on multiple-carburetor applications, since they are considerably more compact than center-pivot designs. However, fuel-level control under hard cornering is not as stable as with center-pivot bowls. Side-pivot bowls use solid Nitrophyl, hollow brass, or white plastic floats. All side-pivot floats are interchangeable (although each may require different float-heights).

Center-pivot float bowls are commonly fed by individual fuel lines and are called "dual-feed" bowls. The center-pivot design positions the float pivot point in the center of the bowl rather than at one side, making the inlet needle less susceptible to side-to-side fuel movement, a particular advantage in road-racing vehicles. Although not as well suited to straight-line drag racing, an important, often overriding, advantage is increased fuel-flow capacity. Since center-pivot floats displace a larger volume than side-pivot designs, center-pivot bowls can reliably control a larger needle-and-seat diameter, and the greater fuel volume can help maintain a steady air/fuel ratio throughout the rpm range.

Center-pivot bowls are also equipped with solid Nitrophyl, hollow brass, or white plastic floats. NOTE: Brass center-pivot floats have a longer pivot arm and can only be used with metering blocks; *metering plates* protrude into the bowl and will interfere with long-arm brass floats. Nitrophyl and white-plastic center-pivot floats have shorter arms, are fully interchangeable, and can be used with either metering blocks or plates.

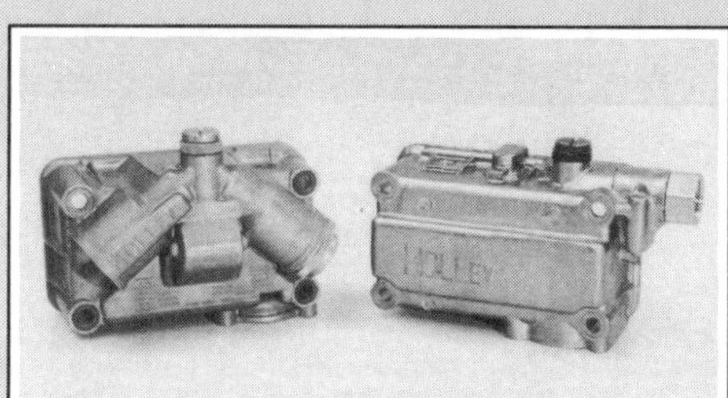

Side-pivot (right) and center-pivot (left) bowls are used on Holley modular carburetors. Each type has been manufactured with numerous minor variations.

This unique LeMans float bowl, originally used on a few Ford high-performance vehicles, is still available from Holley as *part 34-14.*

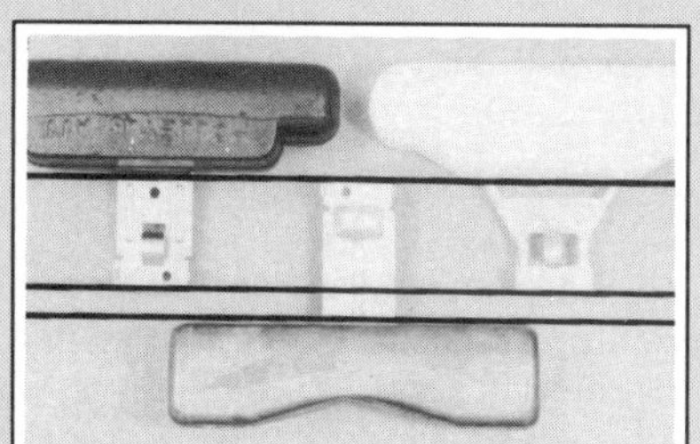

Holley brass center-pivot floats have a longer arm than the Nitrophyl or Duracon plastic floats. The longer brass floats cannot be used with metering plates, see text.

Almost any Holley can be converted from or to side-pivot or center-pivot float bowls. Holley offers several chrome bowl kits that add an impressive custom appearance to any engine compartment.

66 Internally-Adjustable Bowls

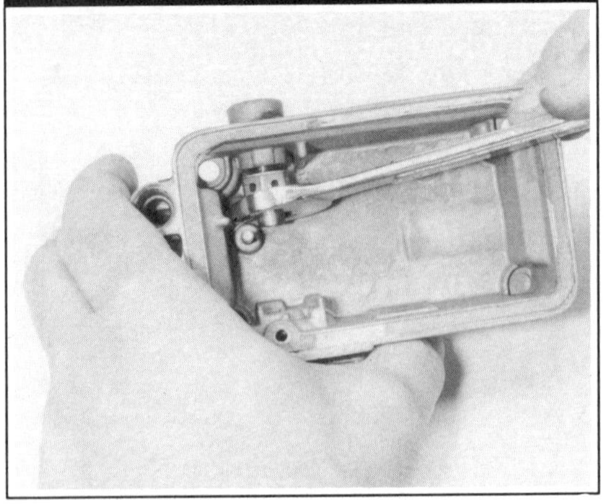

66—If you have internally-adjustable bowls, perform this step, otherwise proceed to **Step 69**, *Remove Bowl Vent Assembly*. Remove the needle-and-seat by sliding the needle out of the seat, if it hasn't already fallen out. Then, unscrew the seat with an open-end wrench and remove its sealing gasket. Now, jump ahead to **Step 69**, *Remove Bowl Vent Assembly*.

NOTATION REQUIRED

67 Center Pivot Floats

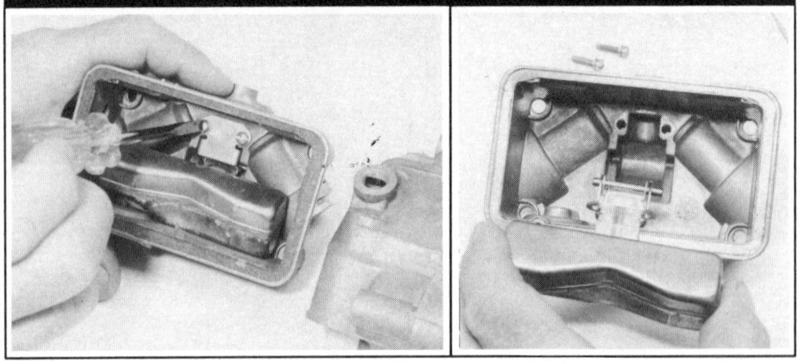

67—If you have center-pivot float bowls (including the LeMans type), remove the float assembly by loosening the two screws that retain the float bracket. As you remove the float assembly, be careful not to lose the pivot, pivot pin, and counterspring (spring is used with metal and Nitrophyl floats). Note the type of float on your Work-A-Long sheet. Metal floats have a longer pivot arm; see the sidebar *Float Bowl Basics*, on page 55 for more information about floats.

PERFORMANCE STEP

68 LeMans Float Bowls

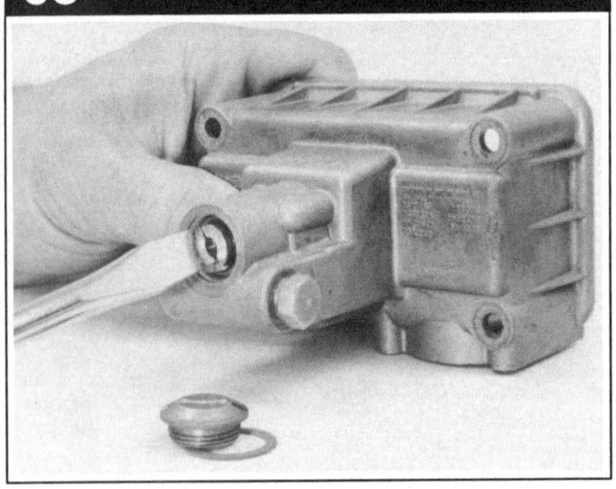

68—The LeMans bowl, originally equipped on the Ford GT-40, Shelby, and Cobra race cars, uses a unique needle-and-seat. Disassembly is easy. Remove the bowl plug (arrow) then unscrew the needle-and-seat assembly. Record the orifice diameter on your Work-A-Long sheet. Remove the fuel inlet fittings and plugs. Jump ahead to **Step 71**, *Disassemble Secondary Bowl*.

69 Remove Bowl Vent Assembly

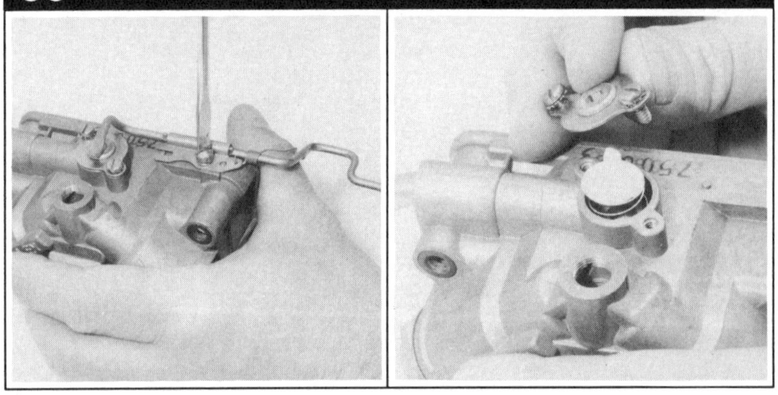

69—Some Holleys have a throttle-arm actuated bowl vent. Several vent designs have been used, so you may want to note the valve configuration on your Work-A-Long sheet to help you during reassembly. The steps for removal generally are: 1) remove the screw holding the arm bracket and remove the arm assembly (some carburetors have a valve retainer clip that must be removed before the arm can be lifted free), 2) remove the two screws holding the valve in the bowl, and 3) lift out the valve and spring (if so equipped).

NOTATION REQUIRED

70 Remove Fuel Inlet Fitting

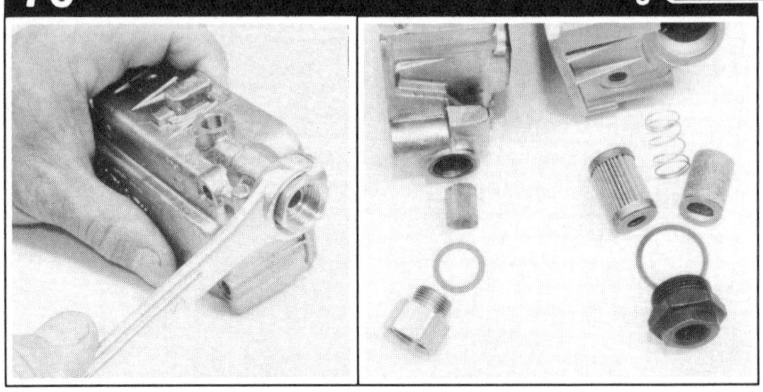

70—Now remove the fuel inlet fittings (some dual-fuel feed Holleys use multiple fittings and/or bowl plugs). Holley has used many different inlet fittings, and you may need open-end or box-type wrenches from 9/16- to 1-inch. When you remove the fitting, you may find a wire screen, sintered bronze element, paper element, or no filter at all. Record what you find (and the location of multiple inlet fittings or plugs) on your Work-A-Long sheet. If your bowl contains a wire screen or sintered-bronze element, make sure it is placed in bin #2 and receives thorough immersion cleaning.

NOTATION REQUIRED

71 Secondary Bowl

71—Jump back to **Step 56**, *Float Bowl Disassembly* and repeat the same process for the secondary float bowl. Don't forget to record the vital statistics for the secondary bowl on your Work-A-Long sheet.

72 Disassembly Is Complete

72—Your Holley carburetor is now disassembled and ready for chemical cleaning. If you haven't done so already, it's helpful to organize component parts into various bins. The method I suggest is: Place plastic parts, floats, electrical components, and very small parts that will not be immersed in carburetor cleaner into bin #1 (if you're not sure, put the part in this bin anyway, additional sorting information is given in the next chapter); place all larger metal components in bin #2; place old gaskets, used rubber parts, etc. in bin #3 (old parts can be helpful during inspection and reassembly); and if you have opened your carburetor rebuilding kit, put the new parts and gaskets in bin #4.

Home/Shop Bead-Blast Cleaning

In the next two chapters I will discuss cleaning, parts inspection, and custom finishing parts for show. A tool that is helpful and time saving in all of these operations is the bead-blast cabinet. Until recently, the cost of a cabinet was well over $500 and outside the reach of most enthusiasts. Now, the Eastwood Company (see appendix) sells a quality blast cabinet *in kit form* that brings this capability to almost any home shop with adequate space. In a matter of a few hours you can assemble a cabinet that rivals many used in professional shops.

Bead blasting is one of the most effective methods of cleaning and prepping metal parts. Corrosion, paint, scale, rust, and other debris are quickly removed by "shooting" millions of small glass beads (or sand or other materials) with compressed air against the part. Metal surfaces are left with a uniform satin finish that may be clear coated, plated, or painted. Glass beading also helps stress-relieve parts, preventing the formation of surface cracks. Many automotive parts are perfect candidates for glass beading; including cylinder heads, valves, intake manifolds, brackets, housings, and many carburetor parts. It's really amazing how fast a rusty choke pulloff or corroded vacuum-secondary housing can be made to look like new. But avoid cleaning assembled throttle plates with a bead blaster. Microscopic particles can become imbedded between the shafts and the throttle plate, requiring disassembly to clean them out.

The major drawback of bead blasting is its demand for large volumes of compressed air. Typically, a 5hp or larger air compressor is required to keep up with steady blasting. To ease the demand for compressed air, the bead-blast gun supplied with the Eastwood cabinet accommodates various size nozzles that allow the user to adjust the airflow to match the capacity of his compressor (additional nozzles in steel and ceramic are available from Eastwood). Even with a small nozzle, blast cabinets still "chew up" lots of air, but if you're patient and allow your compressor to recover between blasts, you can get by with an average home compressor.

Whatever your source of compressed air, make sure you use a pressure regulator, filter, and water trap in the air line. A suction-type blast cabinet, like the Eastwood model, *needs clean, dry air for proper operation.* Air pressure should be adjusted only high enough to maintain effective cleaning. Continued operation at high pressures can quickly "mash" glass beads into dust.

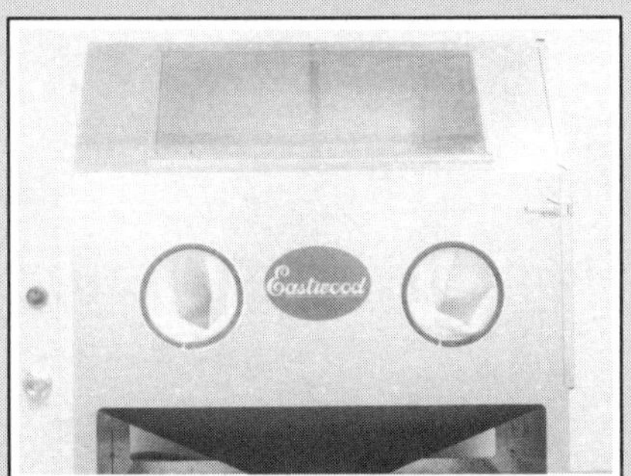

This bead-blast cabinet by the Eastwood Company is large, easy to assemble, and one of the most affordable on the market. It can easily accommodate intake manifolds, cylinder heads, and even wheels. It comes with a blast nozzle and various tips.

Bead blasting requires the use of a filter, water trap, and pressure regulator in order to supply clean, dry air of the correct pressure. Excess pressure will quickly destroy the glass beads.

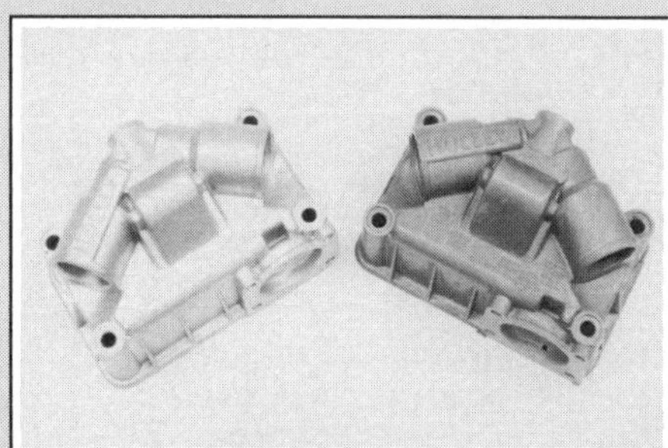

Once you've used a bead-blaster you'll wonder how you got along without it. After a few minutes you can give old parts, like the tarnished float bowl on the right, a like-new appearance. But don't blast thin metals, such as Holley metal floats. Even very low pressure will distort the brass skin.

HOLLEY Rebuilding & Modifying

7—Cleaning & Parts Inspection

Many carburetor maladies are quite obvious—like fuel-logged floats, ruptured power valves, etc.—but others are much more difficult to find. Some carburetor malfunctions are caused by internal fuel leaks, plugged orifices, air leaks, or contamination from sand or other solid particles. Finding and correcting these problems requires knowing what to look for, a keen eye, and patience. These facts make the information in this chapter indispensable. The following pages explain in detail how to clean, inspect, and prepare the component parts of your Holley for successful reassembly and use.

STRIPPING PARTS BEFORE CLEANING

The cleaning and parts-preparation process should begin by carefully inspecting the major component parts of your Holley. Start with the metering blocks. Make sure the idle-mixture needle cork seals have been pulled out of the idle screw holes. (Note: If the idle-mixture needle screws, jets, or other component parts are still in place, STOP, and refer to the previous chapter for tips on removing these items and recording information on your Work-A-Long sheet before you proceed.) If the idle-needle cork seals—or any other rubber parts or O-rings—are left in place, they often will swell up and turn into a gooey mess after a short immersion in carburetor cleaner.

If the surfaces of the metering block, float bowls, and other components are covered with hard, caked-on deposits or sandy particles, or are extremely varnished, you should consider pulling the metering block idle- and main-well plugs before immersion in carburetor cleaner. For further information on this procedure, see the sidebar *Metering Block Cleaning* on page 63.

Check the float bowls to ensure that the sight plugs and other screws have been removed. Make sure that all screws are removed from the main body and other sub-assemblies. If screws are left in place, carburetor cleaner can ooze into threads and may contaminate parts during reassembly (or during surface finishing, if you decide to customize the appearance of your Holley). The only exceptions to this are the linkage attaching screws on the throttle shafts, the butterfly screws, and choke blade screws. These fasteners are difficult to remove and install and can easily be reassembled

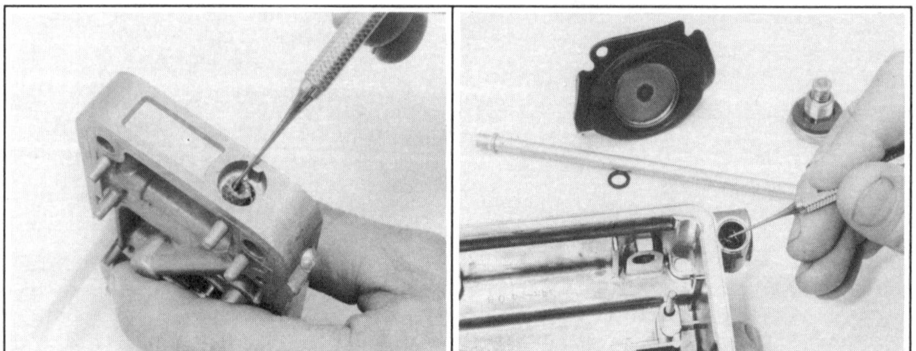

Remove any rubber parts from components before they are immersed in carburetor cleaner. Idle-needle cork seals, O-rings, power valves, and other rubber parts will become a gooey mess if immersed in carburetor cleaner.

If the surfaces of the metering blocks, float bowls, and other components are covered with hard, caked-on deposits or sandy particles, don't take any chances; soak all components in fresh immersion cleaner. If contamination is heavy, you may want to pull the metering block idle- and main-well plugs to ensure complete cleaning.

improperly, so unless you have to rebush the throttle plate (see the sidebar *Repairing Loose Throttle Shafts* on page 91) or replace the choke plate, leave these screws in place.

REMOVING OLD GASKET REMNANTS

Pull off any pieces of stuck-on gaskets that will easily come free and contaminate your carburetor cleaner. Remaining gasket remnants should not be scraped off, since they'll just about fall off after a short time in carburetor cleaner. Furthermore, scraping with a putty knife or other metal tool can easily gouge sealing surfaces that may then develop internal or external leaks. Even more importantly, scraping can slice off or flatten the thin ridges on metering blocks that surround critical fuel and air passages. These narrow—but important—ridges hold the gasket in place and add a back-up seal that can mean the difference between proper function and a "sick" Holley.

If you will not be using an immersion-type chemical cleaner, spray-on gasket removers can remove gasket material without damaging carburetor surfaces. Permatex® *Gasket Remover*, a popular spray-on stripper from the Loctite Corporation, is a powerful combination of chemicals that loosen gaskets in 5 to 10 minutes. (Make sure you read and heed all the safety information on the back of the can.) However, do not use a metal putty knife as recommended. Instead, use a plastic scraper that can be purchased in many paint and hardware stores. The plastic edge is generally sturdy enough to remove softened gaskets, but it won't damage carburetor surfaces. Several applications of spray-on remover may be necessary on stubborn gaskets. For more information on cleaning without immersion cleaners, see the sidebar *Non-Immersion Cleaning Tips*, on the next page.

SORTING PARTS

As I've mentioned earlier, sorting parts into various bins can make your rebuilding job easier and more organized. This also applies to component cleaning. Before you proceed, separate the parts of your Holley into three bins, based on the most appropriate cleaning method for each part. This step is helpful even if you won't be using an immersion cleaner.

Bin #1 should hold all the parts that are not compatible with, or should not be immersed in, carburetor cleaner. These parts consist of any rubber parts—like a choke pulloff, vacuum secondary diaphragm, or power valves—electrical solenoids or sensors, plastic choke covers (many have electric heating elements), and Nitrophyl floats. Nitrophyl is a solid composite, non-metallic material. Carburetor cleaners will attack Nitrophyl and render your floats unusable, so clean composite floats in solvent only. Furthermore, floats will do what they're supposed to do—float—in carburetor cleaner, unless they are held down by something in the parts basket, like the main body. Hollow metal or plastic floats can be submerged using this procedure, but they are easily crushed or damaged if subjected to too much weight. So, just to be on the safe side, you may want to put hollow floats into bin #1. Finally, all parts that are small enough to slip through the holes in the dip basket should be placed into bin #1. This usually includes small screws, pins, balls, light springs, and linkage clips. Don't take any chances here: A small part that finds its way to the bottom of your carburetor cleaner bucket might as well have disappeared into another universe.

Next, locate components that should only be immersed in carburetor cleaner for a short period of time and place them in bin #2. The first candidate is the throttle plate assembly. Many Holley carburetors use Teflon-coated throttle shafts. The green Teflon surface helps prevent air leaks by minimizing throttle plate wear, and the slippery Teflon helps reduce throttle-shaft sticking. Unfortunately, Teflon is not completely resistant to carburetor cleaner. Because of this, the very best way to clean throttle plates with Teflon-coated shafts is to remove the shafts before immersion or only use a solvent wash. Since disassembling the

If you won't be cleaning your Holley with immersion cleaner, you may need to use a spray-on remover for stuck-on gaskets. Permatex® *Gasket Remover* will loosen most gaskets in 5 to 10 minutes.

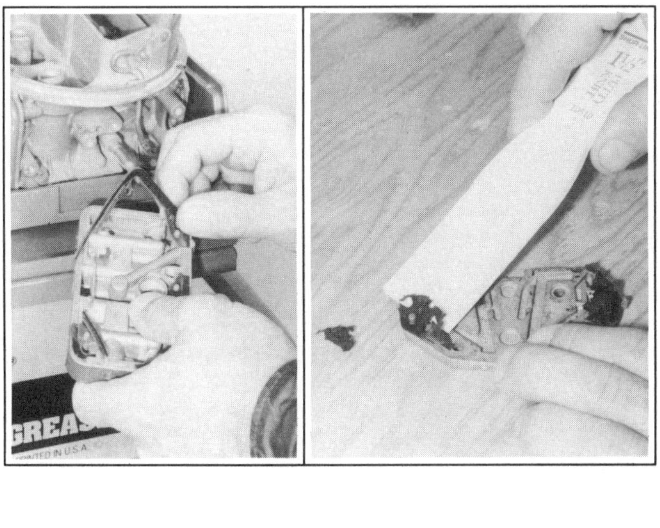

Extend the life of your immersion cleaner by pulling off loose pieces of gaskets before you dip carburetor components. Don't use a metal putty knife to scrape gaskets; it can easily damage sealing surfaces. Let the carburetor cleaner do the work for you. Gasket remnants are easily removed with a plastic scraper after a short time in carburetor cleaner.

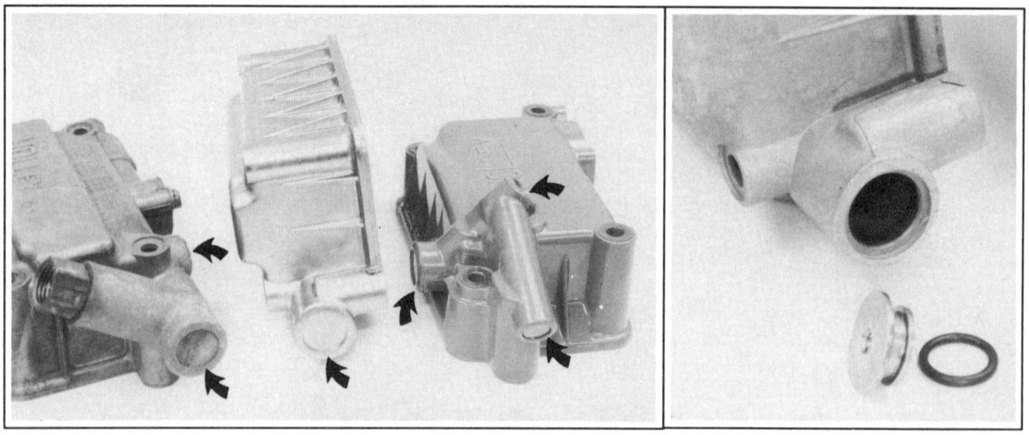

Many side-pivot float bowls have drilled passages that are sealed with non-serviceable internal O-rings (arrows). If these bowls are immersed in carburetor cleaner for extended periods of time, the O-rings will swell up and may not maintain their seal. Place bowls of this type in bin #2 for a short immersion cleaning. And when they are removed from the chemical bath, soak them at least a half-hour in wash water to remove remaining chemical residues.

throttle plates is a difficult job and solvent often won't really clean the plate assembly, a short immersion in carburetor cleaner is the choice of many rebuilders. Immersion of up to an hour or so doesn't seem to have an adverse effect on the Teflon coating. However, if you leave the assembly immersed overnight, the Teflon will almost certainly dissolve into "mush" that may contaminate the throttle-shaft bores. To avoid these problems, either put a lightly-soiled throttle plate in bin #1 for a solvent-only wash, or place the assembly in bin #2 for short immersion cleaning. Regardless of how the throttle plate is cleaned, if you see any signs of Teflon deterioration, don't reuse it until new shafts or bushings have been installed. See the sidebar *Repairing Loose Throttle Shafts* on page 91 for additional information on this procedure.

Here's an important warning about cleaning the throttle plate, automatic choke assembly, or other components with close-fitting parts: Make sure all liquid cleaners (solvent, carburetor cleaner, etc.) are thoroughly free of "grit," sand, or other hard particles. Washing the throttle plate in dirty solvent can force contaminants into the shaft bores, accelerating wear and possibly causing the throttles to stick. And once the bores are contaminated, it's almost impossible to clean them thoroughly without removing the throttle shafts.

Many side-pivot float bowls have drilled

Non-Immersion Cleaning Tips

If you are only fixing a single carburetor, you may not need to purchase an immersion cleaner. Many Holley carburetors do not become heavily varnished or contaminated during use (particularly when the fuel filters are regularly serviced). These carburetors often can be cleaned with a combination of solvent and aerosol carburetor cleaner or lacquer thinner. However, if there are any signs of varnish, sandy, or hard deposits on the interior surfaces of the metering blocks or float bowls (left photo), it is essential that immersion cleaning be used. On these "really dirty" carburetors, the few dollars you spend on immersion cleaner is a wise investment, since without it there is a good chance your Holley won't work properly after rebuilding.

If you're working on a Holley that doesn't require immersion cleaning, here are a few ways to get the "most clean" for the "fewest bucks."

Many aerosol carburetor cleaners are available in parts stores. They consist of powerful solvents that remove grease, dirt, and oil. To a lesser degree, these solvents also remove fuel varnish and dried-on deposits. They work best when they are sprayed on and allowed to soak in for several minutes. All contain toxic chemicals; use only with care.

Lacquer thinner and acetone will also dissolve fuel varnish, oil, and grease. These solvents dry very quickly, so repeated applications are often required. These chemicals are also toxic, so always read and follow safety instructions on the labels.

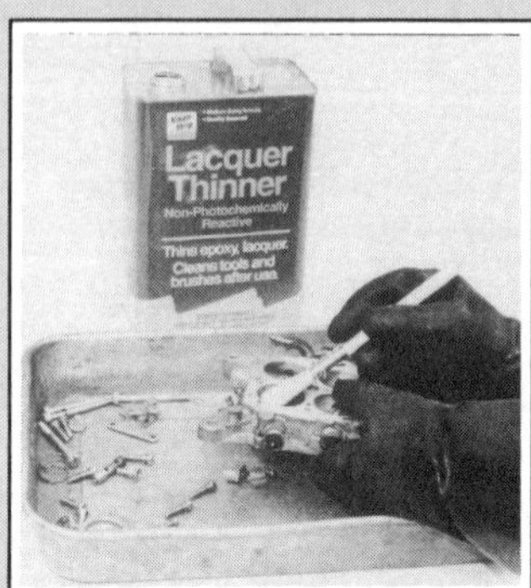

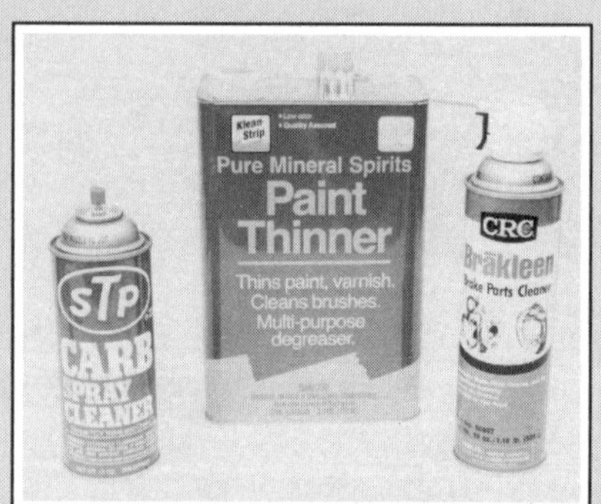

Use solvent (paint thinner works fine) as a final wash. If you can't find solvent, you can use spray-on brake cleaner or some types of carburetor cleaner spray that dry quickly without leaving a residue. But the safest wash is a hot-water rinse bath (see Step 9 on page 64).

Many Holleys use green or dark colored Teflon-coated throttle shafts (left) that help prevent wear and sticking. But Teflon is not completely resistant to carburetor cleaner. Since disassembling the throttle plates is a difficult job and solvent often won't really clean the plate assembly, a *short immersion* in carburetor cleaner is the choice of many rebuilders.

passages that are sealed at one end with a plug and an internal O-ring. Float bowls of this type are easy to identify: They have a passage or passages visible from the exterior that end in a flat surface with a rolled-over or staked edge. Some 4175 carburetors have as many as three plugged passages in each bowl. Float bowls with internal O-ring seals should not be immersed in carburetor cleaner for extended periods of time. Although the O-rings are partly "sealed" inside the groves in the plugs and don't come in full contact with carburetor cleaner, if you immerse them for several hours, the rubber will swell up and may not maintain its seal. Since these plugs and O-rings are not serviceable, any bowl that develops fuel leaks must be replaced. And float bowls aren't cheap! So if your carburetor uses bowls of this type, place them in bin #2 for a short immersion.

I strongly recommend using immersion cleaner, like *TYME-1®*, *Johnsen's Carburetor & Metal Parts Cleaner®*, or others that are available in many auto part stores. If the carburetor has any signs of sandy or white internal deposits, or is particularly dirty or varnished, using a chemical immersion cleaner is essential.

And when they are removed from the cleaning bath, let them soak at least a half-hour in the final wash water to remove as much chemical cleaner residue from the O-ring seals as possible.

One additional note about sorting parts for bin #2: Only put relatively clean metering blocks in bin #2. Metering blocks contain numerous small internal passages and restriction orifices that must all be free of debris for the carburetor to function properly. If the metering block is even slightly suspect as the source of carburetion troubles, give it a full-length immersion cleaning.

The remaining parts should be placed in bin #3 to receive a thorough immersion cleaning in carburetor cleaner. This usually includes the main body, float bowls, metering blocks, automatic choke assembly, accelerator-pump discharge nozzles, sintered-bronze filter elements, and some larger screws and hardware, like bowl screws, fuel-inlet fittings, etc. Throttle plates that do not have green Teflon-coated shafts can also be placed in bin #3 for a full-length immersion.

THE STEP-BY-STEP CLEANING PROCESSES USED IN THIS BOOK

The upcoming step-by-step photos and sidebars will show you how to thoroughly clean the components of your Holley carburetor and prepare them for reassembly. The effectiveness of this cleaning process relies on using immersion-type carburetor cleaner and hot rinse water. If you are only interested in rebuilding a single carburetor (to simply "fix" a problem, for example) and do not wish to purchase immersion cleaner, I have included a sidebar that offers tips on how to get the most "clean" for the least "buck." Even so, I strongly recommend using any of the immersion cleaners found in auto-part stores, and in some cases these chemical cleaners are absolutely essential (see the *Non-Immersion Cleaning Tips* sidebar on the previous page for more information on when you can "get by" without using an immersion cleaner).

The cleaning process presented here works! It is basically the method recommended by Holley, but with a few changes. It can be performed in most small shops, uses no special or expensive equipment, and requires only a minimum amount of workspace. The upcoming steps show parts being immersed in a hot-water wash. After the parts are thoroughly soaked and heated, they are removed and blown dry. This hot-water rinse performs two essential tasks: 1) removes traces of carburetor cleaner from small passages, threads, orifices, etc., and 2) heats the components so that they dry almost instantly when removed from the water. Components washed in this fashion are both very clean and free from water and chemical residues that can lead to corrosion. This technique may not be the "ultimate" in carburetor cleaning, *but it's close!*

Assuming that your carburetor components are sorted as described earlier in this chapter and that you are familiar with the information presented in Chapter 2 on carburetor cleaners and general shop practices, let's begin the most important part of carburetor rebuilding: the cleaning process.

The cleaning process I use requires no special equipment and needs only a minimum amount of workspace. After parts are removed from carburetor cleaner, they are reimmersed in two hot-water rinse baths. This technique removes traces of carburetor cleaner and heats the components so that they air dry almost instantly.

Step-By-Step Carburetor Cleaning

1 What Supplies You'll Need

1—The cleaning process illustrated here requires some basic equipment and supplies. First of all, you'll need immersion-type carburetor cleaner. I have selected *TYME-1*, but most carburetor cleaners work fine. You'll also need a dip basket (not supplied with many carburetor cleaners), preferably with a hook on the side to make draining easier, at least a gallon of solvent, a parts washing tank or wash pan, a couple of plastic or metal buckets large enough in diameter to allow the dip basket to reach the bottom, a source of compressed air, and a supply of clean rags.

2 Locate A Source Of Hot Water

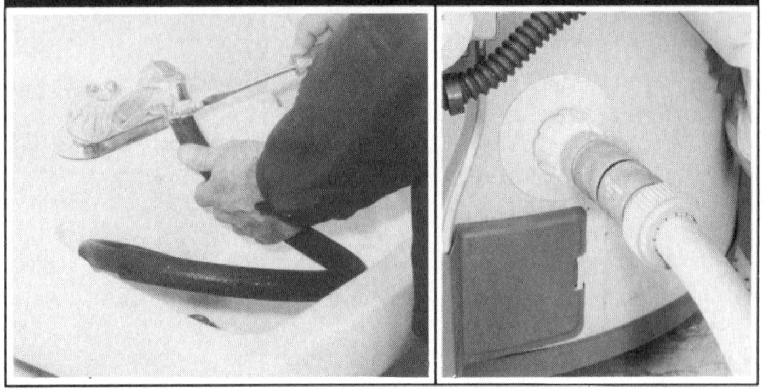

2—You'll also need a ready source of hot water. Many sinks are not deep enough to conveniently fill a bucket. Temporarily attaching a hose to the faucet can make the job easier and safer. 5/8-inch heater hose and a screw clamp will usually get the job done. You may obtain hot rinse water from the drain valve directly on your water heater, but here's a couple of things to consider: 1) old water drain valves may not open easily and, even worse, they may not reseal when closed, and 2) drain valves are located at the bottom of the heater where the water is the least hot.

3 Hotter The Better

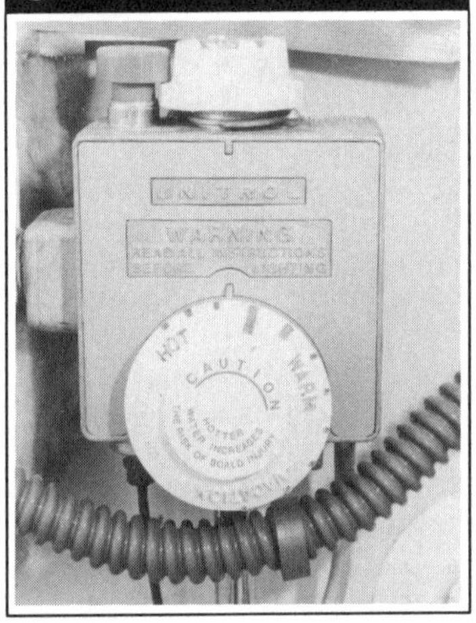

3—The hotter the water, the better it will clean and the faster your parts will dry. You can temporarily turn up the temperature on your water heater, but be sure to turn it back down when you're through. Very hot tap water can be dangerous to unsuspecting family members, especially young children.

4 Protect Yourself

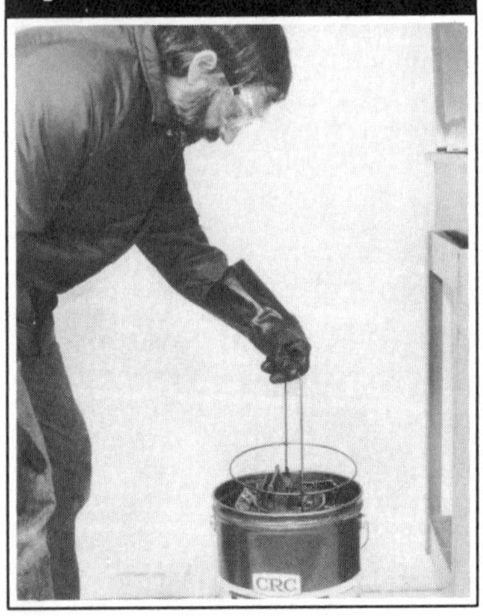

4—For just a few extra dollars you can protect your eyes and hands from hot water and the harsh chemicals used in carburetor cleaners. Among many other hazards, it is easy to accidentally drop a full dip basket into carburetor cleaner. The resulting splash can cover your face with caustic chemicals. No carburetor, regardless of its monetary value—or project, regardless of its immediacy—is worth compromising your eyesight or personal safety.

MASTER MECHANIC TIP PRO TIP

5 Pre-wash Parts

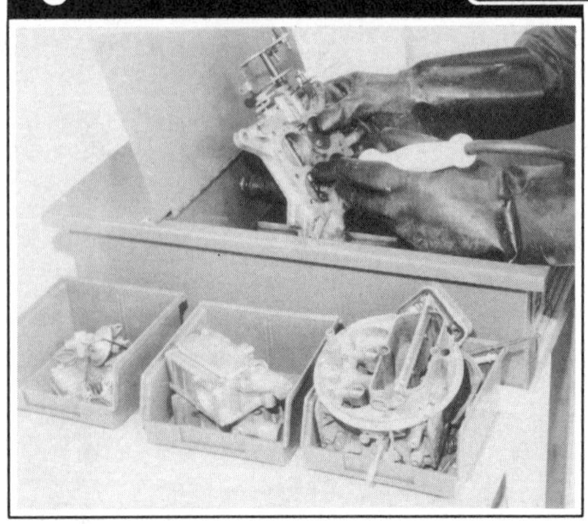

5—As discussed earlier in this chapter, carburetor parts have been sorted into bin #1 (don't immerse), bin #2 (short immersion), and bin #3 (full-length immersion). Although you may have washed the entire carburetor with solvent before you began disassembly, giving the individual components in bins #2 and #3 a final solvent prewash before immersion will keep as much dirt as possible out of your carburetor cleaner, maximizing its strength and extending its useful life.

Step-By-Step Cleaning Continued On Page 64

Removing the idle-mixture screws and power valves (if so equipped) before immersion cleaning is all that's required to prepare Holley metering blocks during most carburetor rebuilds. However, if there is evidence of sediment, sand, or other hard-material contamination in the float bowls—or the carburetor produced an erratic idle or main system malfunction either before or after rebuilding—there is a possibility that the idle and main-well passages and small orifices inside the metering block need special attention.

The only effective way to remove some types of internal contamination from metering blocks is to remove the cup or flat plugs located over the main and idle wells and expose the interior passages. Many main wells (inner pair) have flat, rather than cupped plugs. These may be simple plugs or they may have emulsion or metering tubes attached to them. These tubes also contain small orifices that should be given a careful inspection, since almost invisible particles trapped in these tubes can dramatically affect engine performance.

The following steps for metering-block plug and/or emulsification tube removal require careful and precise drilling and other delicate procedures. If, after reading these steps, you are unsure of your ability to safely perform these operations, consider having a professional carburetor shop rebuild the metering blocks for you. Or you can purchase new metering blocks from Holley, although they are expensive.

Drilling operations are best performed on a drill press; but can be accomplished by hand if you are *very careful.* Cupped plugs are removed by drilling a 1/8-inch hole through the center. *Don't let the drill slip down when it breaks through the plug;* there may be precision orifices just below the plugs that are easily damaged. Insert a #8 sheet metal screw (with about .040-inch of the tip ground off) into the hole and use two small crows-foot prybars, pull out the plug. The procedure for flat-plug removal requires a slightly different approach. If your metering block has plugs that are recessed, measure and record their depth before you begin. Drill a 1/8-inch hole in the center of the plug no more than 1/8-inch deep (any further and you risk damaging integral passages that may exist below the solid part of the plug). Insert a sheet metal screw and pry the plugs loose in the same manner as for the cupped plugs.

New cup-type well plugs are available from Holley in packages of ten, **part 1007-107**, but flat plugs, either with or without emulsion tubes, are not available. However, cups and plugs can be reused if you: 1) deburr the drilled hole with a fine file, 2) solder the hole closed (being careful not to let solder run down emulsion tubes into critical orifices, 3) thoroughly clean all passages and orifices, 4) apply a small amount of fuel-resistant epoxy (such as *JB Weld®*) to the plug wall before you press them back in place, and 5) return plugs that are recessed below the surface of the metering block to the same depth you recorded earlier.

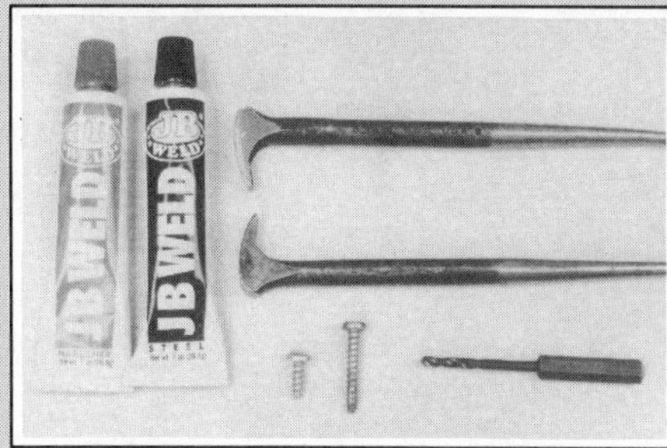

1—To remove and install plugs, you will need a fuel-resistant epoxy, two crows-foot prybars, #8 sheet metal screws with .040-inch of the tips ground off, an 1/8-inch drill bit, and a thin ruler or depth micrometer for plugs not installed level with the surface of the metering block. A drill press makes the job easier, but it can be accomplished by hand if suitable care is used.

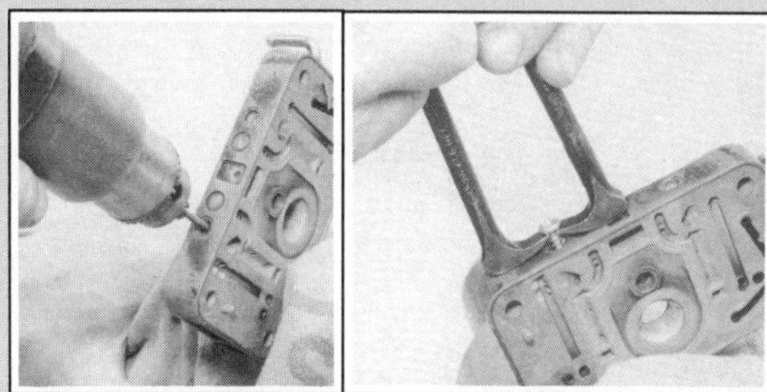

2—Carefully drill an 1/8-inch hole in the center of the cupped type plug. If your metering block has plugs that are recessed, measure and record their depth before you begin. Use the same 1/8-inch drill, but do not drill over 1/8-inch deep. Insert the #8 sheet metal screw and pry the plug loose with two crows-foot prybars.

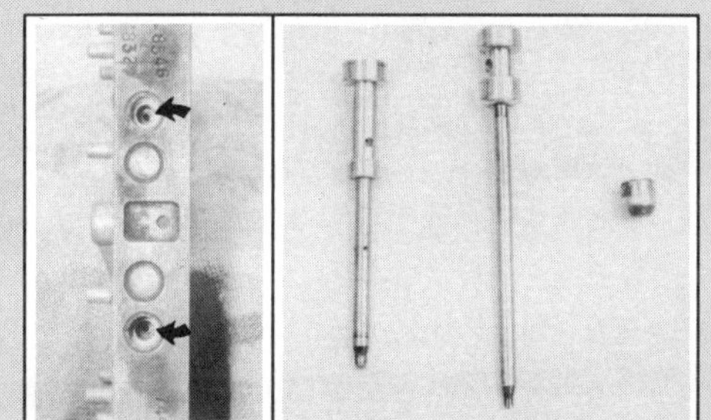

3—Idle-feed restrictions are sometimes located directly under the idle well plugs (arrows). Drilling too far can easily damage these restrictions and ruin the metering block. The main-well plugs may also have emulsion tubes attached and can be damaged by drilling over 1/8-inch deep.

4—If new plugs are not available, the old ones can be soldered. Be very careful not to allow any solder to run down into exposed passages on emulsion tubes.

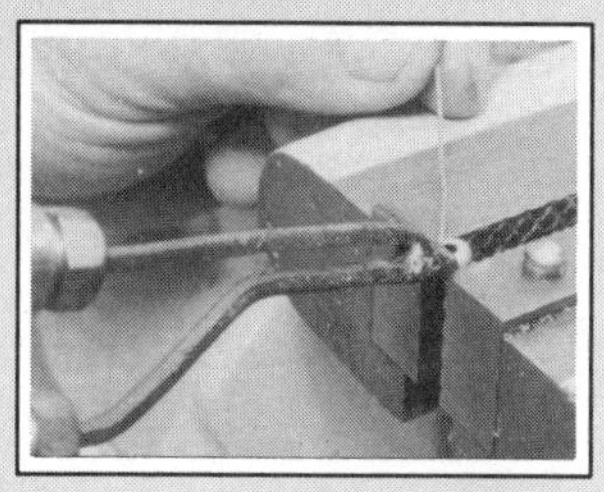

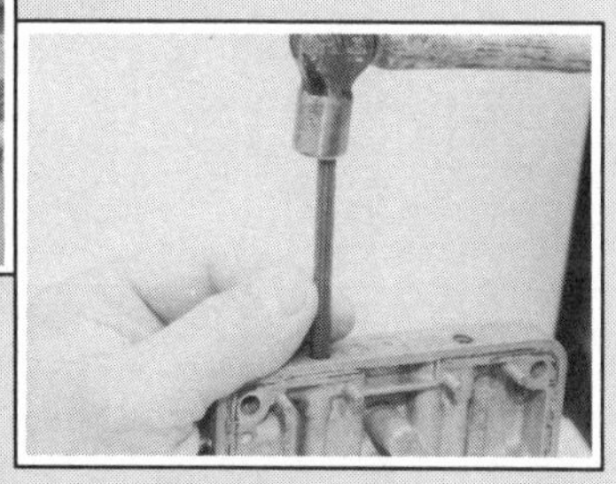

5—Lightly coat the outer edges of the cups or plugs with epoxy and drive them into the metering block with a small drift and hammer. Return recessed plugs to the same depth you recorded earlier. IMPORTANT: Use only a minimum amount of epoxy; excess cement can run down inside the metering block and close off passages/orifices.

6 Prepare Parts From Bin #2

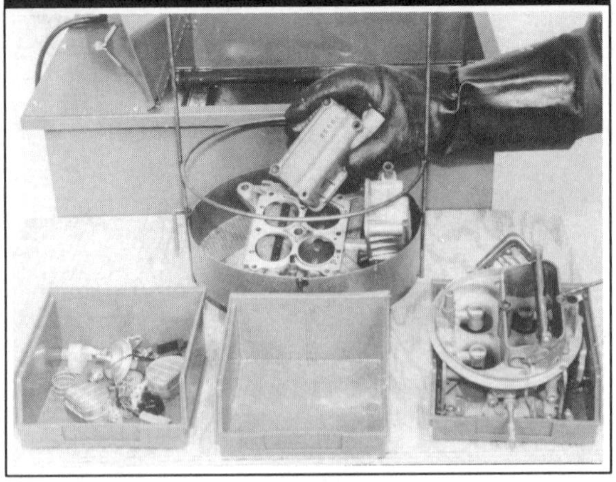

6—Since the parts in bin #2 should only be soaked for a short period of time, let's begin by transferring these parts into the dip basket. If you don't have any parts that require this special handling, proceed to **Step 14**, *Clean Parts In Bin #3*.

MASTER MECHANIC TIP PRO TIP

7 Float Cleaning Tips

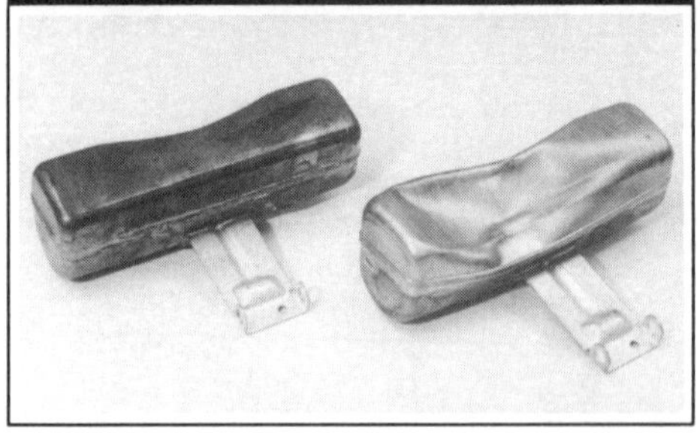

7—Cleaning Holley metal floats in immersion cleaner (never immerse Nitrophyl or Duracon plastic floats) is often not worth the trouble or risk. Since floats will *float* in chemical cleaner, they must be held down in the dip basket. And even if you are very careful, they can be crushed if other parts shift. Play it safe and clean the floats in your solvent tank. If you want to remove fuel stains, use spray-on carburetor cleaner and follow with a solvent rinse.

8 Lower Parts Into Cleaner

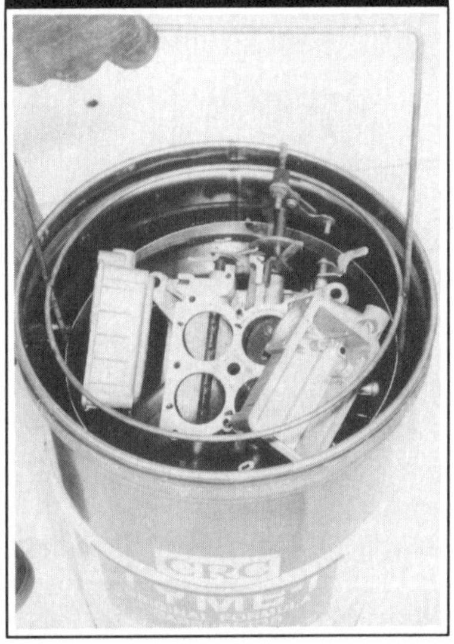

8—Make sure the parts are oriented so that liquid will freely drain from them when the dip basket is raised (float bowls facing down, etc.). Then slowly lower the basket in the cleaner and replace the lid on the bucket to minimize evaporation and breathing chemical vapors. Check the clock and when 25 minutes have elapsed, proceed to the next step.

9 Prepare Hot-Water Rinse

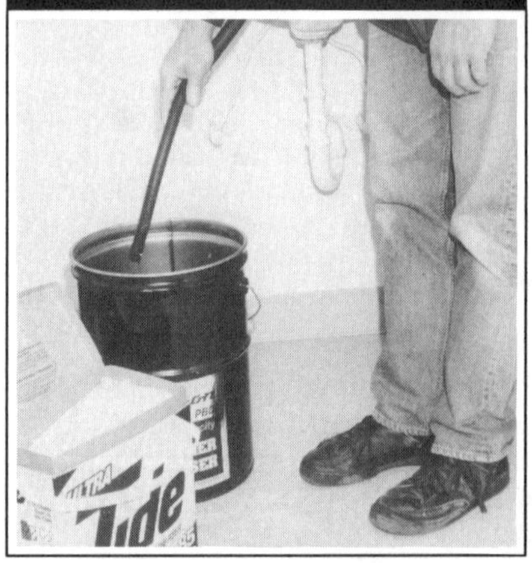

9—Now fill both buckets about 2/3 with hot water. You can optimize the cleaning process by adding about a cup of detergent (laundry or dishwasher detergents work fine) into the first bucket. Then move both buckets near the carburetor cleaner; position the soap-wash closest to the immersion cleaner.

10 Rinse Off Chemical Cleaner

10—Pull the dip basket from the carburetor cleaner and hook it on the side of the bucket to let the parts drain. You can use a small brush to determine if the surface varnish or dirt has been loosened; if the parts need more time in the cleaner, reimmerse them for another 15 minutes or so. I don't recommend immersing Teflon-coated throttle shafts more than about one hour, maximum. After the parts have drained for a minute or so, carefully lower the basket into the first hot-water (optionally, hot *soapy* water) bucket.

11 Hot Water Soak

11—Continually raise and lower the basket in the wash for about two minutes, then lift the basket and let it drain for a few seconds. Now, transfer the basket to the second hot-water wash. Raise and lower the basket a few times then allow the parts to soak about 5 minutes. This thoroughly heats the parts and removes residues of immersion chemicals (and soap) from pores in the castings.

12 Dry Parts With Compressed Air

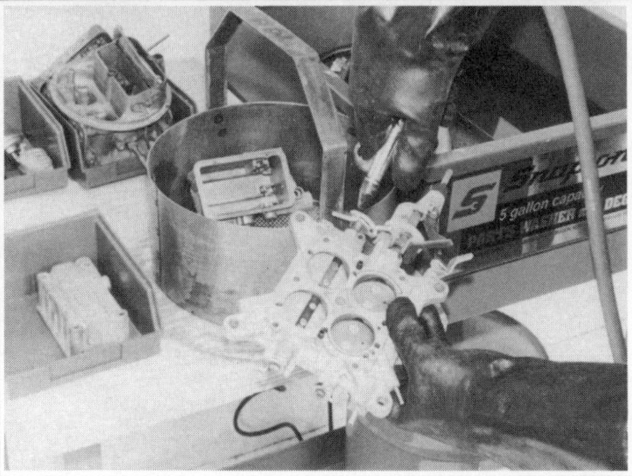

12—After about 5 minutes, remove the parts from the hot-water wash. If the water was very hot (near boiling), the parts will dry almost immediately. Use compressed air to remove any remaining water and residue from passages and surfaces (make sure you use a safety nozzle). Remnants of stuck-on gaskets can be removed with a plastic scraper. If gasket remnants won't easily scrape off, put the part into bin #3 for a second immersion cleaning. After you blow each component dry, set it aside in a clean parts bin.

SAVE MONEY $

13 Using Canned-Air

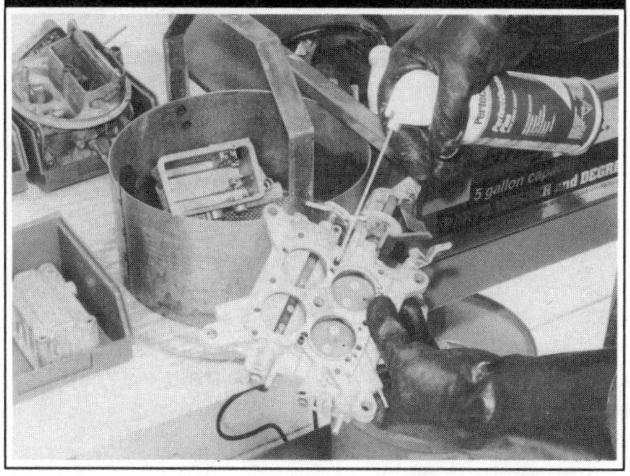

13—If you don't have an air compressor, try drying parts with "canned-air." This method works particularly well when the rinse water is very hot, since most components dry with only a short blast from the can. Make sure to direct the nozzle into passages where residues collect. You can maximize the "air" pressure by immersing the can in warm water, but don't heat it beyond the maximum safe temperature (indicated on the label); over pressure can cause the can to explode!

14 Cleaning The Parts In Bin #3

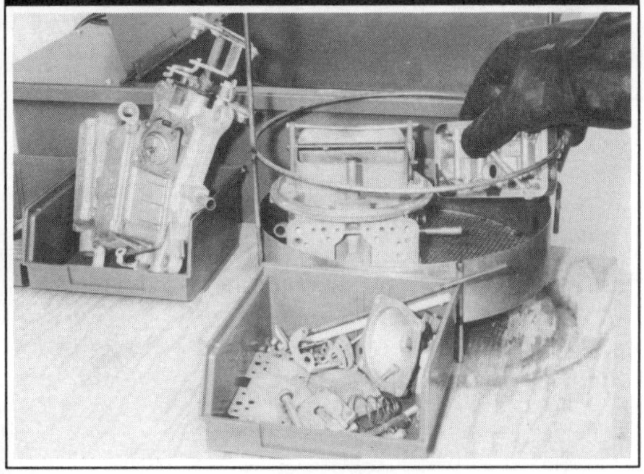

14—Now move the parts that need extended soaking, including those with stuck-on gaskets, from bin #3 into the dip basket. Then go back to *step 8, Lower Parts Into Cleaner* and perform the last six steps again, except extend the immersion time to several hours or overnight.

MASTER MECHANIC TIP PRO TIP

15 Cleaning Small Parts

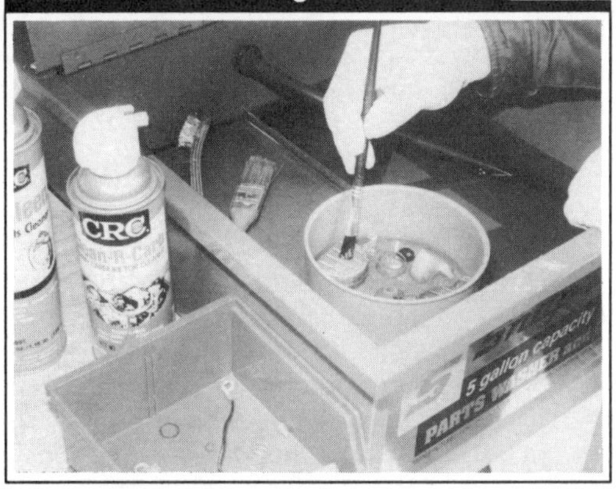

15—Most parts in bin #1—clips, screws, springs, plastic parts, etc.—require little or no cleaning. However, if you decide to wash them, place them in a shallow container of clean solvent (dirty solvent can be nearly opaque, making it very difficult to find small parts on the bottom of the container). Spray especially dirty parts with carburetor cleaner. Let the spray cleaner loosen the dirt a few minutes before you rinse the parts in solvent. A small hobby brush can be helpful for removing dirt and chemical residue. Don't blow small parts dry with compressed air; you'll blast them into the far corners of your shop! Let them air dry on a clean shop rag.

16 Organize Parts For Reassembly

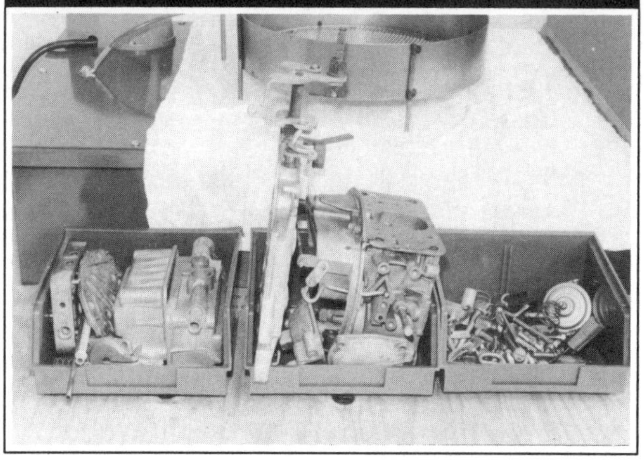

16—After you have completed the cleaning process, reorganize the parts into bins in preparation for final inspection and reassembly. A useful method is: place *fuel-bowl related parts* (float bowls, floats, metering blocks, metering plates, accelerator pump covers and arms, fuel-inlet fittings, and transfer tubes) in bin #1; *place main-body related components* (throttle plate, choke mechanism/linkage, vacuum diaphragm housing, and accelerator pump nozzles) in bin #2; finally, *all screws and remaining small parts* should be placed in bin #3.

PARTS INSPECTION BEFORE ASSEMBLY

Now that the component parts of your Holley have been thoroughly cleaned, the bare metal surfaces can reveal problem sources that may have been masked by dirt, grease, or gasket material. In fact, a detailed parts inspection may locate the principal reason for the original carburetor malfunction (like a damaged O-ring sealing surface or a thoroughly plugged orifice), or it may help you locate a "problem-in-the-making" that could shorten the useful life of your rebuilt Holley (such as warped sealing surfaces that often allow fuel or air leaks after a few weeks or months in operation). Despite the obvious fact that a carburetor can only perform as well as its component parts, a purposeful parts inspection is often overlooked by many mechanics. The reason may be the rush to complete the rebuilding job, or a belief that all maladies are cured by a long soak in carburetor cleaner, but don't let yourself make that mistake! Without a careful inspection of every component part, the success of your carburetor rebuild or repair work is questionable.

The most visible signs of carburetor malfunctions are external fuel leaks. *While minor external leaks may not be the cause of the malfunctions, they can point to trouble spots.* The locations of fuel puddles or stains on the intake manifold that you made on your Work-A-Long Sheet during carburetor removal (see Chapter 5) can now help you locate defective or damaged components. If you noted the appearance of slight fuel leaks around the throttle shafts, this could indicate flooding, worn throttle shafts, or indicate no problem at all. Since this is commonly observed on Holley carburetor installations, let's begin the inspection process with the throttle-plate assembly.

Never assemble a Holley without a detailed inspection of the component parts. Here a close inspection of the metering blocks revealed a plugged emulsion tube, only discovered when the main-well plugs were removed. Although many carburetor malfunctions are not this subtle, without a careful inspection of every component part, the success of your carburetor rebuild or repair work is questionable.

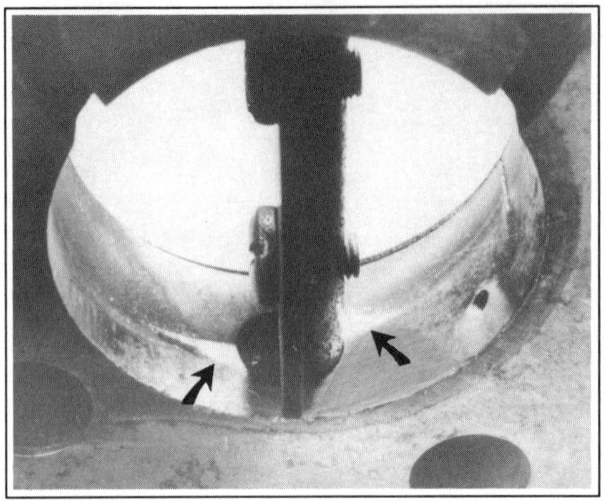

If your Holley suffered from inconsistent idle speeds, whistling or other abnormal sounds at idle, the throttle shafts or plate may need disassembly and repair or even replacement. The first thing to look for is signs of visible wear, particularly erosion of the primary throttle bores (arrows) from lateral (side-to-side) movement of the butterflies.

Throttle-Plate Assembly Inspection

Despite the sophisticated fuel and air metering control that a Holley carburetor provides, it contains very few parts that are subject to any measurable wear. A notable exception to this is the throttle-plate assembly. It contains one or two throttle shafts that are constantly moving in response to driver input. Furthermore, the shafts must seal-in engine vacuum and atomized fuel.

If your Holley suffered from inconsistent idle speeds, whistling or other abnormal sounds at idle, or you were unable to set a stable idle air/fuel ratio, the throttle shafts or plate may need disassembly and repair or even replacement. First, look for signs of visible wear, particularly erosion of the primary throttle bores from lateral (side-to-side) movement of the butterflies. Lateral throttle-bore wear is often the result of improperly designed throttle linkage; see the sidebar *Carburetor Linkage Basics* on page 101 for tips on correcting this problem. If you find any lateral wear in the bores—more than a few thousandths of an inch—excessive air leakage can occur, producing an unstable idle. In these cases you must replace the throttle plate, primary butterflies, and (probably) throttle shaft(s) and Teflon seals. If you're faced with this prospect, it's often easier to obtain a replacement throttle-plate assembly.

Some rebuilders inspect throttle-shaft fit by wiggling them up and down in their bores. But because of design variations, visible clearance is not a reliable indicator of wear or possible vacuum leaks. In fact, the only definitive test of throttle-shaft condition is to perform on-car testing (see text on this page and the sidebar *Repairing Loose Throttle Shafts* on page 91 for more information on diagnosing and repair).

Some rebuilders inspect the fit of the throttle shafts by wiggling the shafts in their bores. Because of design variations, throttle shafts that seal fine can seem to fit too loosely, while others—of a different design—may seem snug, but can cause excessive vacuum leaks. Unfortunately, there are no definitive off-car measurements of throttle-shaft wear. If you suspect worn or leaking throttle shafts, refer to the sidebar *Repairing Loose Throttle Shafts* on page 91 for more information on diagnosing and repair.

Another very common throttle-plate problem is distortion and cracking from overtightening or using too-thick or too-compressible base gaskets. When the

Many Holley carburetors have booster venturis that are cast as one piece with the main body, as those on the left. Others have separate boosters that can be replaced (identifiable by the rolled metal radius—arrow), but installing booster venturis is only performed by a few custom carburetor shops (see text).

throttle plate is forced to bend down at the bolt flanges, it often cracks through the throttle-shaft or butterfly bores. These cracks can leak air and adversely affect idle, or occasionally, cracks run through an idle-fuel passage and may cause internal or external fuel leaks. It is very difficult and costly to repair cracked throttle plates. If you find any cracks, obtain a replacement from Holley or a professional carburetor shop (sometimes you can find a used plate for a bargain) or, if the cost is too high, you may want to obtain a replacement carburetor (see the sidebar, *When Not To Rebuild* on page 43).

Mike Jones Carburetion uses this specially fabricated fixture (above) to tighten booster venturis. It uses the same principle of burnishing and rolling the booster arm metal firmly into the main body as the tooling used at Holley (right) during the original manufacturing process.

Main Body Inspection

The main body can suffer from many maladies, some of which are very visible (when you know what you're looking for), while others can be overlooked by even the most experienced rebuilder.

First, let's look for the obvious.

Engine backfires can melt or loosen booster venturis. Inspect the boosters for missing chunks or a distorted shape, and if they are not cast into the main body, check them for looseness (a slight, barely detectable movement is OK). When the damaged boosters are cast as one piece with the main body, you'll almost certainly have to obtain another carburetor, since main bodies are not available as replacement parts. But if your main body is very rare and valuable, you can have the old boosters machined out and replacements installed, however, this procedure requires special machine work and is expensive. If loose or damaged boosters are not cast into the main body, they can be tightened or replaced at a moderate cost. Only a few custom shops, like Mike Jones Carburetion, The Carburetor Shop, and Holley's Custom Shop, perform these services. Give one of these companies a call to determine if any of these special techniques are appropriate for your situation.

Now let's look for slightly less-obvious maladies.

The main body can suffer from many maladies, but one of the most obvious is melted booster venturis, often caused by engine backfires. Always inspect the boosters for missing chunks or a distorted shape.

Inspect all the threaded holes in the main body, especially the eight holes for the fuel-bowl screws. Make sure the threads are clean and intact. If any threads look damaged, you can usually correct the problem by installing a Heli-coil insert (for thread repair tips see the sidebar *Repairing Damaged Threads* on page 68). Check any suspect or repaired threads by inserting a screw (and the assembly it attaches for proper spacing). The fastener should solidly tighten to its full-rated torque and screw in and out easily without jamming.

If your carburetor uses a brass tube with O-rings to transfer accelerator pump fuel from the metering block to the main body, carefully inspect the counterbore for damage. Gouges, deep scratches or casting flaws can allow fuel to leak from the accelerator pump circuit directly into the intake manifold, resulting in idle and part-throttle stumbling and roughness. A damaged transfer tube bore will often necessitate scrapping the main body and starting over with a replacement carburetor.

Now let's search for nearly hidden flaws in the main body.

Many Holleys have vacuum passages that require multiple drillings during manufacture. The ends of unused intersecting passages are plugged with staked-in lead balls. The secondary-diaphragm feed passage is a good example. These balls can occasionally come loose and fall out. If the secondaries open very late or don't open at all, the problem could be a leaking or missing lead ball. The secondary diaphragm passage is the only passage sealed with a lead ball on most Holley main bodies. But Holley has introduced hundreds of variations in their basic modular carburetor design, so it's almost impossible to predict what unique feature you'll find.

Repairing Damaged Threads

Threaded holes in critical carburetor components must be in top shape to prevent fuel or vacuum leaks. Partially or completely stripped threads can cause elusive fuel-metering problems and/or external leaks that could pose a fire hazard. Fortunately, many stripped threads, except for those in safety-critical locations—like fuel-inlet fittings—can be easily and inexpensively restored.

If the threads are merely corroded or slightly damaged, repair can often be accomplished by carefully retapping the threaded hole with a cutting tap of the proper size and thread pitch. However, a *cutting tap* has a substantial taper at the tip and will not cleanup the last few threads deep in the hole. To repair threads to their full depth, follow with a *bottoming tap* of the same size.

For threads that are stripped or require more than retapping, there are two repair methods commonly available to the do-it-yourselfer. The first technique involves drilling out the remaining threads, retapping with a special tap, and installing a stainless-steel threaded insert that returns the hole to original thread size. Helicoil® manufactures a well-known brand of threaded inserts. A second common method of repair, developed by the Loctite® Corporation, uses a two-part chemical epoxy mix that is first inserted in the damaged bolt hole. While the epoxy is still a liquid, the fastener is coated with a "releasing" compound and screwed in the damaged hole. Extra epoxy is forced out between the threads, and once it hardens and the bolt is unscrewed, the hardened epoxy becomes a rock-hard substitute for the missing threads.

Each of these repair methods has advantages and disadvantages. Both allow fasteners to be tightened to their full recommended torque, but for pure thread strength, Helicoil inserts are superior. In fact, when carefully installed, they are often stronger than the original threads. Plus, Helicoil repairs are ready to use immediately while epoxy requires a minimum of 30 minutes to harden. However, the epoxy method is less expensive, since a special tap, installation tool, and stainless-steel thread inserts are not required. Furthermore, a drill motor, the correct drill bit, and some degree of skill is required for successful Helicoil repairs, while the Loctite epoxy technique requires no special tools and can be performed almost anywhere.

Whichever technique you choose, if the repaired threads will not allow the fastener to reach full recommended torque, the damaged component should be replaced.

The most popular brand of threaded inserts are manufactured by Helicoil. This kit from Holley uses Helicoil inserts to repair float-bowl and other 12-24 fasteners.

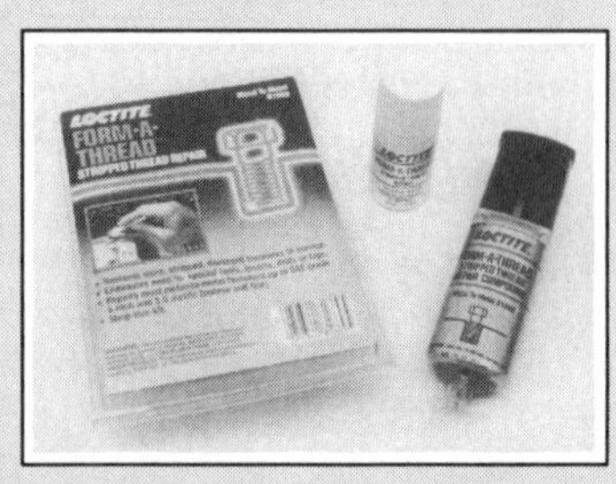

The Loctite *Form-A-Thread®* repair kit is a chemical method of thread repair. While not as strong as some other methods, it restores most carburetor threads; however, it should not be used for repairing fuel-inlet threads, needle-and-seat, or other safety-critical threads.

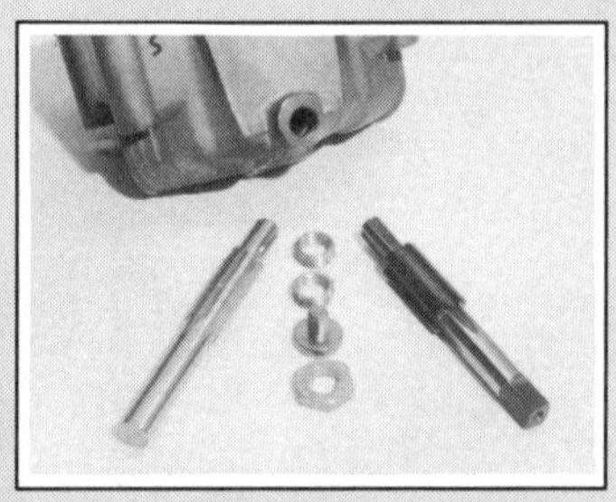

If the needle-and-seat threads in the float bowl are damaged, the safest method of repair is simply to replace the bowl. However, if you need to salvage the bowl (to preserve the original appearance of a restored car, etc.), the threads can be repaired with a special tap and thread insert developed by *The Carburetor Shop.*

The main body also contains many other small passages. Some of these make right-angle turns, have exits that are completely hidden, or incorporate small-diameter orifices pressed or drilled in one end. These passages occasionally become restricted with hidden debris. Compressed air can be a valuable tool to help you determine the condition of these passages, but make sure you use low-pressure (regulated) air of no more than about 25 psi. Forcing high-pressure air into some carburetor passages can blow-out lead plugs, air bleeds, and other orifices, often rendering a main body useless. Furthermore, a lead ball or brass plug is a formidable projectile when blasted free with 100psi or more air pressure. Using a rubber-tipped blow gun (and eye protection), force regulated air into the main and idle air-bleed passages and make sure an air stream emerges from the small orifices in the brass air bleeds (the brass bleeds *are* there, aren't they?). Then check air flow through the booster venturi, idle-fuel, and secondary-diaphragm passages. If any passage is clogged, try forcing air in from the other direction, and if that's not possible, insert a short length of wire, but be careful not to touch any orifices. Working the wire around inside the passage will often dislodge foreign material. If you run into a solid end, make sure the passage was open originally; many Holley main bodies have partially drilled, unused passages. When the contamination is removed and air seems to flow freely through all passages and/or orifices, reimmerse the main body in carburetor cleaner for a few more hours to loosen any remaining contamination. Then rinse, blow dry, and make a final confirmation that all passages are clear using low-pressure air.

If your Holley uses a brass tube with O-rings to transfer accelerator pump fuel, carefully inspect the counterbores for damage. Gouges (as visible here), deep scratches, or casting flaws can allow fuel to leak directly into the intake manifold and may require scrapping the main body and starting over with a replacement carburetor.

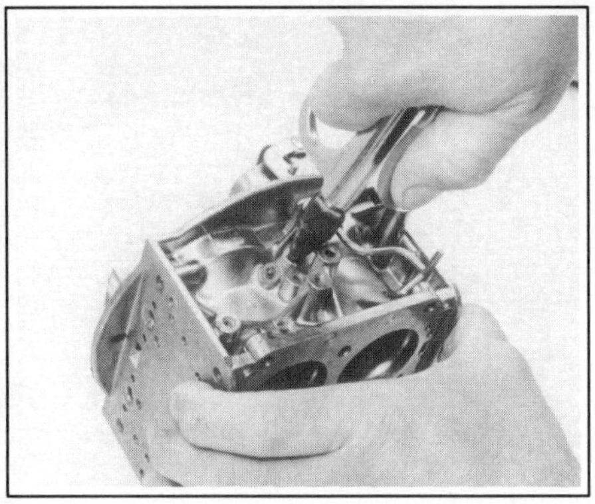

Use only low-pressure (25psi maximum) compressed air to help you determine the condition of passages in the main body. Forcing high-pressure air into some carburetor passages can blow-out lead plugs, air bleeds and other pressed-in orifices, often rendering a main body useless.

Perhaps the most hidden flaw in Holley main bodies isn't even visible with the unaided eye, but nevertheless is responsible for some malfunctions. The three main-body surfaces that mate with the metering blocks and/or plate and the throttle-plate assembly must be flat to allow the gaskets to properly seal the interconnecting air and fuel passages. But sometimes they are warped or bowed and don't uniformly compress the gaskets, leading to leaks that have unpredictable effects on idle and part-throttle performance. Fortunately, it's a simple procedure to check these surfaces; all that's required is a straightedge and feeler gauge. If you measure surfaces that are warped more than .008-inch, you should file or machine these surfaces flat. Refer to the sidebar *Refinishing Warped Surfaces* on the next page for detailed information on this procedure.

The inspection of the main body is just about complete. Before you set it aside for reassembly, give it a final "once over" for casting flaws or gouges on sealing surfaces, cracks, and any debris in interior passages. If you've filed or deburred any surfaces, make sure you blow off any metal particles with compressed air. When everything looks OK, let's move on to the metering blocks and plates.

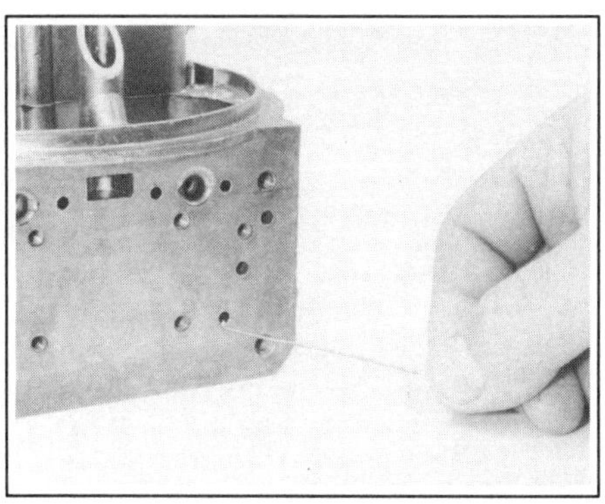

If any passages are clogged, try forcing air in from the other direction, and if that's not successful, insert a short length of stiff wire, but be careful not to damage any orifices. Working wire around inside the passage combined with short blasts of compressed air will often dislodge foreign material.

Inspecting Metering Blocks/Plates

The metering blocks are complex castings designed to combine fuel and air and deliver this mixture to the booster venturis. Like the inspection process for the main body, it is essential to scrutinize the metering block(s) for both easy-to-see maladies, like corrosion damage or surface gouges, and subtle problems, like interior contamination and plugged orifices.

If the carburetor you are working on was contaminated with sandy particles or hard scaly deposits, these materials may have found their way into the metering block and could have plugged interior orifices and connecting passages. Even a thorough soaking in carburetor cleaner may not unplug some orifices that are thoroughly packed with debris. In these special situations, it may be necessary to remove the idle- and/or main-well plugs and clean the passages directly. See the sidebar *Metering Block Cleaning* on page 63 for tips on when and how to do this procedure.

And while we're on the subject, if you don't need to pull the well plugs—and the vast majority of carburetor overhauls fall into this category—make sure that the well plugs are tight in their bores. The outer set of plugs seal the idle wells, and the inner set seal the main wells. The idle-well plugs are usually brass cap plugs (look like little freeze plugs) while the main-well plugs vary widely in configuration. Many main-well plugs are deeply recessed, others are solid, some have numbers stamped on them, and others look exactly like the idle-well plugs. Regardless of their configuration, they should be present and tight in their bores. Give all the plugs a poke with a probe and make sure they don't move. If you find a loose one, remove the plug and either replace it with a new plug (they are hard to find; your best bet is to call Holley or The Carburetor Shop) or apply a thin coating of epoxy adhesive (like JB Weld) to the edge of the plug and reinstall it. Once the epoxy hardens, it will form an airtight seal. For additional tips on working with well plugs, refer again to the sidebar *Metering Block Cleaning* on page 63.

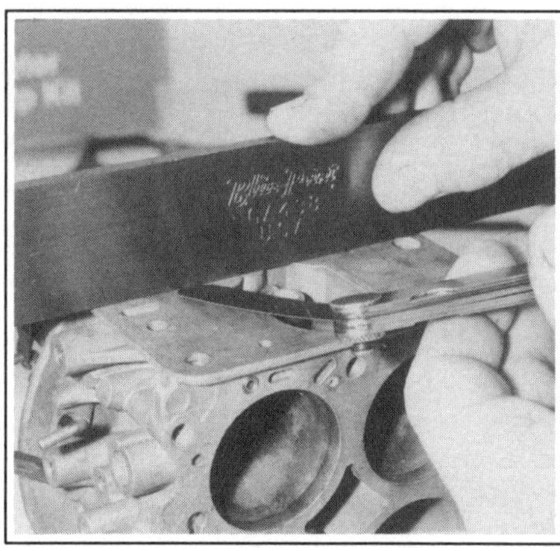

A few sources of potential trouble are almost impossible to directly observe. Main-body gasket surfaces can become warped or bowed, causing internal air or fuel leaks. Mating surfaces should be checked with a straightedge and feeler gauge (see sidebar *Refinishing Warped Surfaces* on page 70). Interior metering block passages can be another trouble spot. Even a thorough soaking may not unplug some orifices. In these situations, it may be necessary to remove the well plugs and clean the passages/orifices directly.

If the interior passages look clean, there are only a few additional areas on the metering block to inspect. First, most metering blocks, like main bodies, use a ball or plug to seal at least one drilled passage (accelerator-pump fuel delivery). Make sure the plug is securely staked in place. Next, the metering block (and metering plate) gasket surfaces have thin ridges around some or all of the passages that prevent the gasket from moving and help it seal. If these ridges are flattened, it doesn't mean that you have to throw out the metering block, but it's not a good sign. The carburetor may work fine when first assembled, then, when the gasket loses some of its compression, an interior leak may develop, degrading carburetor performance. If you have a metering block (or metering plate) with some of the ridges flattened out, make a note of this on your Work-A-Long sheet and put an extra emphasis

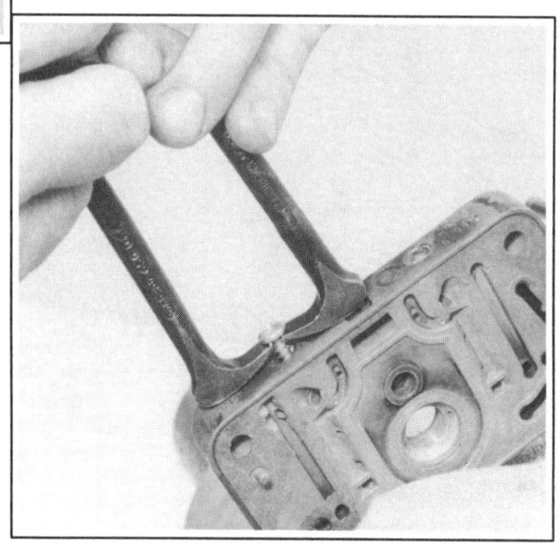

Refinishing Warped Surfaces

Most Holley carburetor problems are caused by internal or external fuel leaks. And these leaks can sometimes be traced to warpage of "flat" mating surfaces that no longer uniformly compress gaskets. Minor internal leaks are difficult to find and can cause a wide variety of performance ailments. Major leaks can easily lead to flooding or even underhood fires. Checking and, if necessary, correcting mating surface flatness will prevent these malfunctions and should be an essential part of every Holley rebuild.

The main-body to metering-block (or metering plate) surfaces on the main body are the most frequent locations for warpage and potential problems on a Holley carburetor. Other, less common, areas of warpage that should be checked are: 1) the top and bottom surfaces of the throttle plate, 2) the bottom of the main body, and 3) the accelerator pump cover and it's mounting surface on the float bowl. Check these surfaces by placing a straightedge diagonally, horizontally, and vertically on the surface and verify that a .008-inch feeler gauge cannot be inserted between it and the component part. If there is more than .008-inch warpage, the surface should be repaired.

A vertical milling machine equipped with the proper size (usually 4-inch or less) flycutter is the best choice for repairing any warped carburetor surfaces. However, if you are patient and very careful, using a top-quality medium or fine flat file can produce adequate results. Before you begin filing, make sure the part is securely clamped in a vise and cannot move, even slightly. Use long, slow, straight strokes. Apply even, level pressure and regularly check your work using the same straightedge and feeler-gauge technique. Tipping the file will result in a bowed surface that may leak worse than ever. Applying machinist's dye to the surface between periods of filing (let it dry completely before filing) can be a helpful aid in identifying high and low spots and in pointing out the start of surface bowing from accidentally tipping the file. When a .005-inch feeler gauge will no longer fit anywhere between the straightedge and the filed surface, STOP. There is no need to completely "cleanup" the low spots to ensure leak-free gasket sealing. And, finally, metal particles from filing will work their way into every small passage and crevice. Completely clean all components before you reassemble your Holley using solvent and compressed air (and, of course, eye protection).

The photos in this sidebar show the general steps involved in using a flat file to correct surface warpage on a main body and other parts. If you decide to perform this procedure keep two things in mind as you proceed: 1) take your time, and 2) constantly check your work for the desired results. If you follow these two simple guidelines, you should be able to restore the sealing surfaces on your Holley to like-new condition.

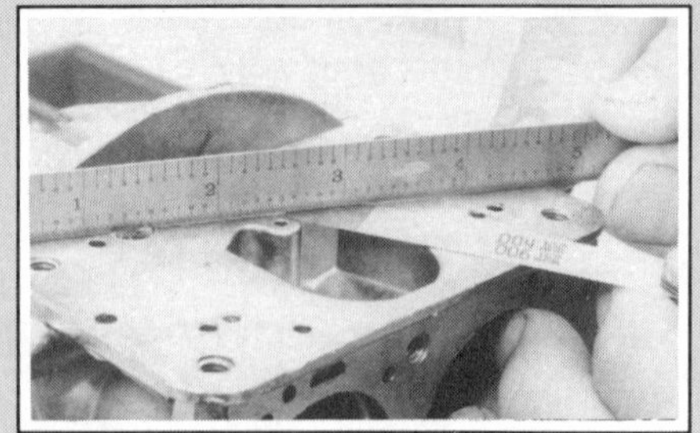

The metering block to main body surface is the most common trouble spot on Holley carburetors. Check with a straightedge and a feeler gauge. Check these surfaces by placing a straightedge diagonally, horizontally, and vertically on the surface and verify that a .008-inch feeler gauge cannot be inserted between it and the component part.

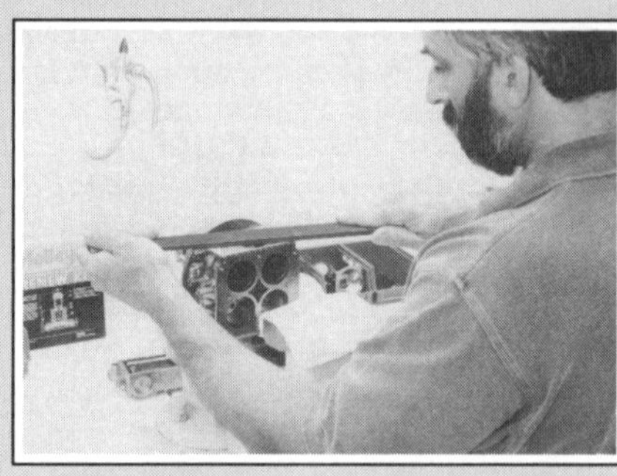

Use one hand on each end of the file to balance the pressure and to keep the file as flat as possible. Use long strokes. Apply even, level pressure and regularly check your work using the straightedge. Tipping the file will result in a bowed surface that may leak worse than ever.

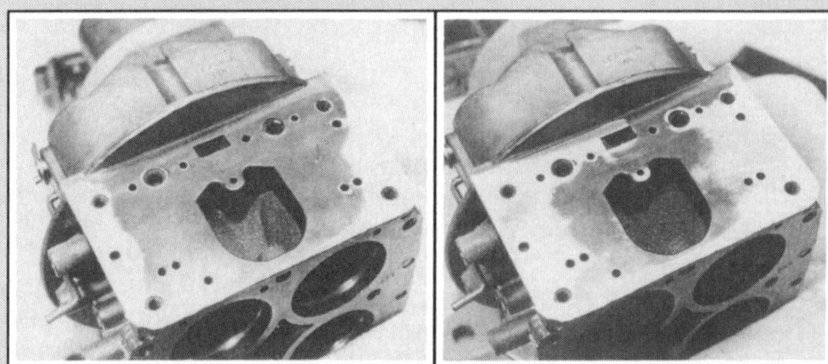

As you continue to file, regularly check your work. Machinist's dye applied to the surface between periods of filing (let it dry completely before filing) can be a helpful aid in identifying high and low spots. Most hardware stores carry, or can order, this fast-drying metal stain.

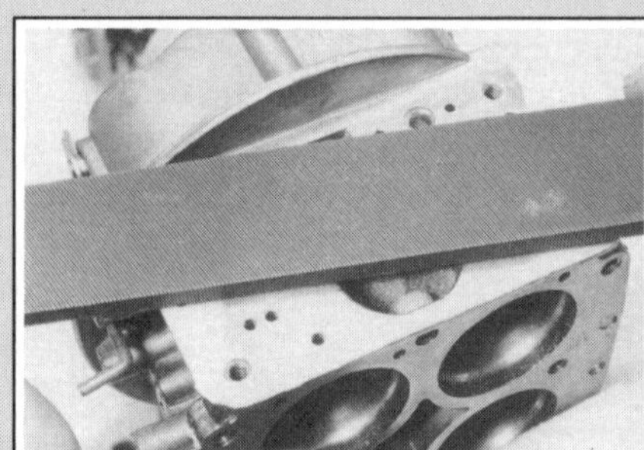

Also, regularly check the file grooves for "loading up" with "blobs" of metal that can cause scratches in carburetor surfaces. Clean the grooves with a metal pick or probe. A clean file will cut quickly and produce a smooth surface.

There is no need to completely "cleanup" the low spots. As long as a .005-inch feeler gauge will no longer fit between the part and the straightedge, the surface gasket will properly seal. The accelerator pump fuel passage was one of the last surfaces to come in contact with the file.

Metering block and plate surfaces have thin ridges around some or all of the passages that hold the gasket and help it seal. If these ridges are flattened (arrows) an interior leak may develop, degrading carburetor performance. When this problem exists, put extra emphasis on torquing the bowls and metering plate (and retorque them after the carburetor has been in operation for two weeks).

on torquing (and retorquing after the carburetor has been in operation for a few weeks) the bowls and metering plate. Next, if your carburetor uses a brass fuel-transfer tube for accelerator pump fuel, check the counterbore just as you did in the main body. The interior surface should be smooth and contain no gouges. Finally, give the passages one final close look. Make sure there is no debris in the idle-feed restrictions (when they are visible), PVCRs, air-bleed orifices, or the accelerator pump feed channel. If all checks out OK, put the metering block(s) back into your parts bin.

The cast-aluminum metering plate is considerably easier to clean since all of the passages are clearly visible. In fact, the only things to look for here are junk in the main and idle feed orifices, damage to the gasket sealing surface, and the rare crack in the casting. When you have completed your inspection of the metering plate, return it to the parts bin.

Inspecting Float Bowls And Other Components

The inspection of major components is nearly over, but a few critical pieces remain. Let's continue with the float bowls.

The float bowls are simple looking but deceptively complex castings. They contain from six to ten threaded holes, three or more machined surfaces, and several interior passages. The best place to start your inspection is with all the threaded holes, including the needle-and-seat hole,

The secondary-vacuum diaphragm cover has ridges that hold the diaphragm in place when tightening the cover. In addition, the lower half of the housing has four raised points (arrows) that permit the rubber diaphragm to be compressed without damaging it or squeezing it out from between the top and bottom halves. If either of these sealing aids are missing, assembling the secondary diaphragm will be considerably more difficult.

the fuel-inlet passages, the accelerator-pump screw holes, the sight-plug holes, and any interior float-mounting holes (found on center-pivot and other performance bowls). Make sure all the threads are clean and undamaged. Although virtually any threaded hole can be repaired with special tools and inserts (including the needle-and-seat threads), a replacement bowl is usually the most cost-effective choice if you find any damaged threads. The only exception to this may be when you are restoring a rare carburetor and you wish to keep all the original parts.

If your float bowl(s) uses a steel ball as a check valve in the accelerator pump cavity on the underside of the bowl, make sure the ball is present and its seat is free of corrosion.

Now check the flatness of the accelerator-pump mounting flange with the same technique used on the main body surfaces. This surface is almost always flat enough to prevent leaks when the pump cover is properly tightened, but occasionally a bowl will have warped enough to require filing this surface. And while you're at it, check the flatness of the pump cover, too.

Your parts bins contain many additional parts that should all get at least a moment of your attention before you set them aside for reassembly. Some general things to look for are:

- Make sure there are no residues of carburetor cleaner or dirt in the automatic choke mechanism.
- Check the secondary-bowl fuel feed tube for cracks, deep scratches, or bending at the ends.
- Make sure all remnants of gasket material have been removed from sealing surfaces on the metering plate, blocks, support plate, throttle plate, and main body.
- Check the floats for damage or leaks and refer to the sidebar *How To Weigh And Check Floats* on page 84 before you proceed.
- Check the accelerator-pump discharge nozzles for contamination or plugged orifices.
- Inspect the secondary-vacuum diaphragm cover ridges (they help hold the diaphragm in place and improve sealing) and *don't file the raised points off the lower housing mounting surface*; they allow the rubber diaphragm to be compressed without cutting it or squeezing out from between the top and bottom halves (see Reassembly Chapter, **Step 35** on page 89).
- Give all the castings a final inspection for casting flaws; places where the casting is incomplete, porous, or cracked.

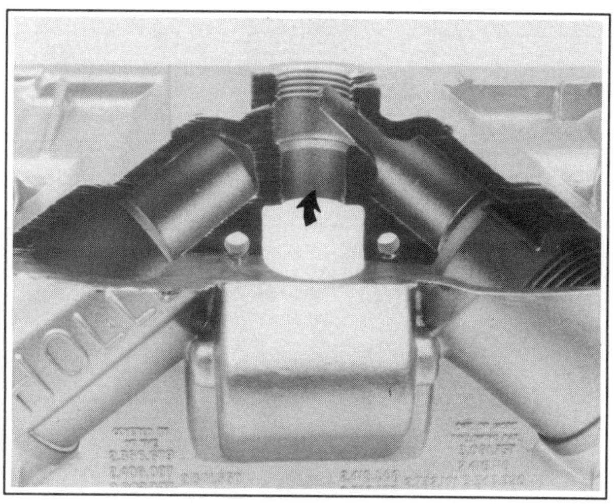

Look over all the threaded holes in the float bowls, especially noting the condition of the needle-and-seat threads and smooth bore below them. If the threads or sealing bore are damaged, proper float adjustment will be impossible and fuel leaks and flooding may occur. Although virtually any threaded hole can be repaired (including the needle-and-seat area), a replacement bowl is usually the most cost-effective choice.

HOLLEY Rebuilding & Modifying

8—Custom Finishing

CUSTOM FINISHING YOUR HOLLEY

Many automotive enthusiasts have spent several hours and great care rebuilding their carburetor only to be disappointed at the final result. Their disappointment is rarely caused by a carburetor malfunction, rather it is due to the unspectacular appearance of their rebuilt Holley. If you are interested in having your Holley look as good as it performs, the information in this chapter is for you.

There are several options open to you. The most straightforward one is to simply bolt on any of the chrome accessories available from Holley. Just the simple addition of chrome float bowls adds an impressive custom look. If you really have the do-it-yourself bug, I'll show how to paint your Holley with some of today's best fuel-resistant epoxy paints. But if you won't rest until you turn your Holley into a real eye-popper, then you may find the information on using the latest fluorescent paints just your cup of tea. If all this do-it-yourself stuff is a bit too much, then you may find what you need in the sidebar on Holley's Custom Shop. This special service from Holley Replacement Parts provides virtually everything from simple replating to complete chromed and custom-finished (and even completely rebuilt) award-winning carburetors, like the 850 double-pumper on the cover of this book.

If your carburetor rebuilding project is primarily one of a practical and functional nature, then you can skip ahead to the next chapter where we begin step-by-step reassembly. On the other hand, if you want your Holley to be a work of art as much as you want it to work correctly, then read on, because here's where the fun begins!

WHAT ABOUT REPLATING?

Holley carburetors are treated with a dichromate "chemical film" plating that gives them their unique yellowish-gold appearance. This very thin plating effectively protects the interior passages and other surfaces from corrosion due to the myriad of chemicals—and even small amounts of water—that are present in gasoline. Holley recommends against using any carburetor that no longer has a visible dichromate coating on the interior surfaces.

When the dichromate protection has become thin, your replating options are somewhat limited. A few years ago, it would have been possible to obtain redichromating services at many local

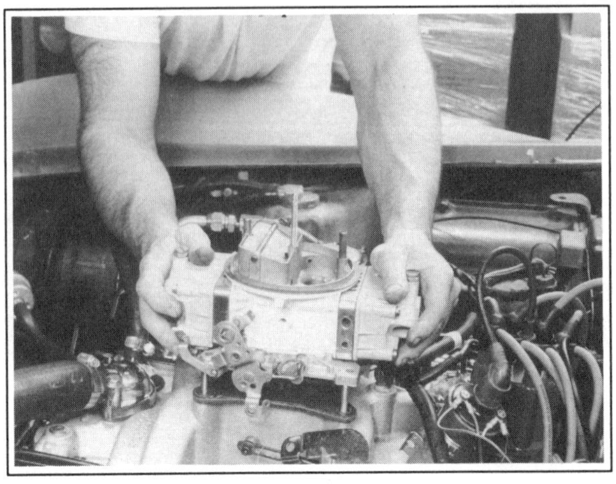

This chapter covers some of modifications that can add an impressive, custom look to your Holley. After painting with white and fluorescent red paints and buffing the metering blocks to a chrome-like shine, the author reinstalls this rebuilt 780cfm Holley on a 1976, 413cid smallblock Corvette. This Holley looks terrific, dresses up the engine compartment, and it works great, too!

Owners of street rods, show cars, restored muscle cars, race cars, or just automotive enthusiasts often desire more than just a functionally perfect carburetor. The Holley Custom Shop® offers a variety of services for those who would like a carburetor that stands out in appearance as well as function. These special services include custom rebuilding and modifying for special applications, redichromating, powder coating in a variety of striking colors, and chrome or even gold plating. All of these services are performed by Holley factory technicians with years of experience in custom applications.

Redichromating operations restore the original factory finish to the carburetor main body, float bowls, metering blocks and other small components. The dichromate chemical film provides protection against external and internal corrosion as well as giving a "brand-new" look to the carburetor. Each piece is individually dipped in dichromate solution to ensure complete and uniform refinishing. This is an invaluable service for the enthusiast who is restoring a special show vehicle and wants an out-of-the-box appearance for a rare or obsolete Holley carburetor.

Custom Plating is offered in traditional chrome, black chrome, or 24-karat gold! They will plate one bowl or every part on the carburetor. All parts receive a thorough buffing and polishing prior to plating, which ensures a deep, mirror-like finish. The finished surfaces are highly resistant to heat, gasoline, and other chemicals found in the engine compartment. Custom Shop plating is probably the best-looking finish available anywhere for a Holley carburetor.

Powder Coating is similar in appearance to paint but is much more durable and resistant to heat and chemicals. Many colors are offered including: Ford Blue, Crimson Red, Ebony Gloss Black, Canary Yellow, Chevy Orange, Gold, and Candy Red, Blue and Gold. The Custom Shop can also powder coat other engine components—like valve covers—to match or contrast with your Holley. And you can combine powder coating with plating treatments. For example, a powder-coated main body with chrome or gold-plated float bowls can add a "trophy winning" look to your vehicle.

Custom Rebuilding And Modifying services are also available from Holley's Custom Shop. They include assembled and flow-tested custom carburetors in a variety of custom finishes. The Custom Shop can even completely rebuild, restore, and/or custom finish your carburetor! For more information write them at: Holley Replacement Parts Division, Custom Shop, 601 Space Park North, South Cartwright Street, Goodlettsville, TN 37072 or call (615) 859-3124.

Most very large companies are not interested in working directly with their customers. However, Holley's Custom Shop is available to you for just that purpose. If you would like to "hire" Holley to work on your next carburetor, give the Custom Shop a call.

This is just one example of what Holley's Custom Shop can accomplish. This 850cfm racing carburetor (pictured on the cover of this book) was disassembled and carefully inspected, the bowls and metering blocks were buffed and polished, the main body was powder coated, and the bowls were plated with a deep chrome surface. After surface prep was completed, Holley experts modified the carburetor for optimum performance and reassembled it with the best gaskets and most durable components.

Expert technician Jack Nelson at the Custom Shop assembles another prize-winning Holley carburetor. Jack spent years in the Holley Race Trailer helping racers tune and modify Holley carburetors. This experience sets the Custom Shop apart from other carburetor shops.

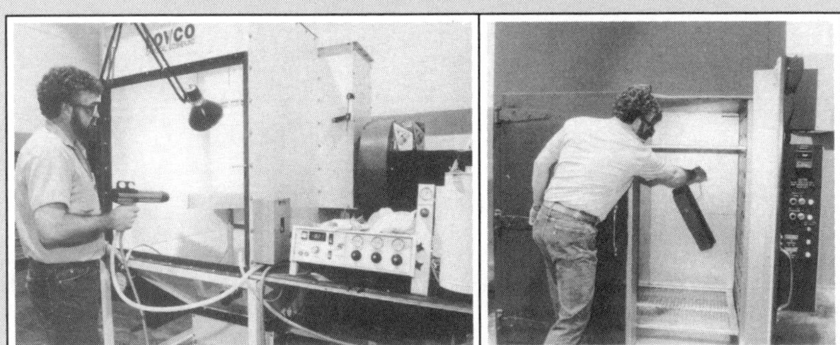

Another expert at the Custom Shop, Terry Carter, applies powder coating to a customer's valve cover. The plastic power is completely dry and wouldn't ordinarily stick to the surface, but it is held in place by a high-voltage charge. After the part has received a uniform coating, it is placed in a high-temperature baking oven. This melts the plastic into a smooth, shiny coating. When cool, the plastic coating is extremely durable and very attractive.

Holley also packages several chrome parts for general sale in speed shops and parts stores. You can combine these chrome accessories with Custom Shop powder coatings or, if you're a die-hard do-it-yourselfer, try painting your Holley, yourself.

plating shops, or you could even do it yourself with a few cents worth of chemicals dissolved in water. But dichromate, like many other plating processes, involves the use of "heavy metals," like chrome, that are now considered hazardous waste by the EPA. Restrictions applied to the use of these materials throughout the U.S. have curtailed their widespread use, and finding local plating shops that will apply chemical films to small batches of carburetor components is quite difficult and/or expensive. Luckily, if you cannot find a local source, Holley's Custom Shop will redichromate your carburetor for a minimal charge. If you have the time to ship your components, dichromate replating is an inexpensive way to add corrosion protection and improve the appearance of your Holley.

PAINTING CARBURETOR COMPONENTS

A fresh dichromate surface can make your carburetor look like new. But if you would like to improve or customize the appearance of your Holley beyond the "new look," then for a small investment in aerosol paints, supplies, and a few tools you can apply a durable, eye-catching finish that will brighten any engine compartment.

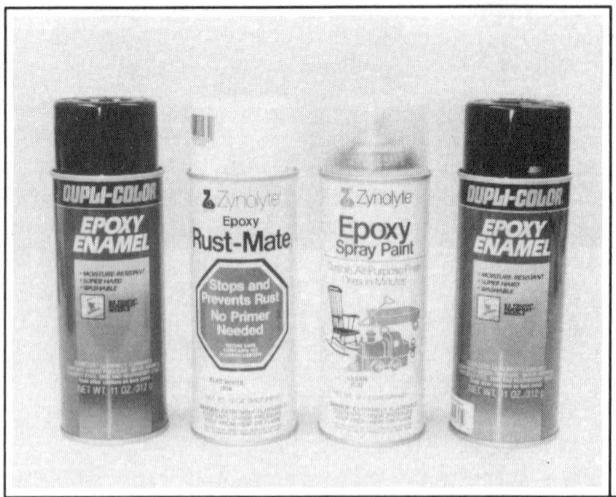

Although the newest engine enamels and epoxy-based paints offer good resistance to chemicals, fuel, and heat, you should carefully consider the type of use and abuse your carburetor will receive. If you only occasionally modify or tune your Holley, painting is probably a practical choice. If your Holley will be used on a race car that requires frequent modifications and jet changes, painting is not as practical.

While painting carburetor components is not a new idea, only recently has it become a practical one. For example, Holley's Custom Shop can paint your carburetor and other engine components with an extremely durable "powder paint." Once applied, this plastic-like, dry powder is baked in an oven to form a hard, chemical- and fuel-resistant surface. This is undoubtedly the ultimate in carburetor paint finishing. Because the powder application and baking equipment is both very specialized and expensive, you must send your carburetor to Holley for this procedure. However, this does not rule

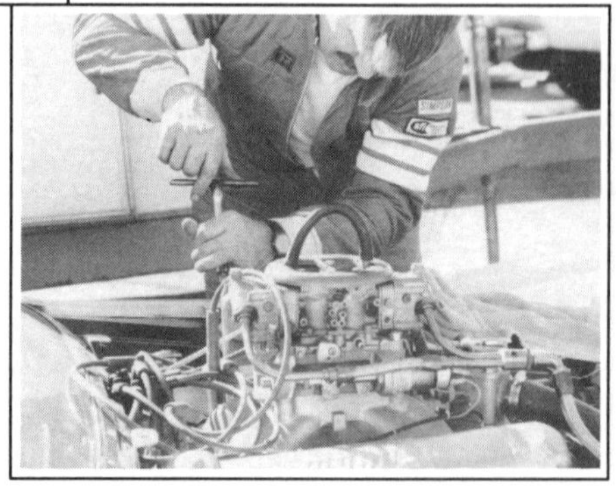

The Carburetor Shop

The Carburetor Shop®, located in Southern California, specializes in the restoration, rebuilding, and modification of all Holley carburetors (and most other brands). The owner, Brad Urban, is widely recognized as one of the country's leading authorities on modifying and race prepping the Holley carburetor. His shop provides a number of services including: complete rebuilding, race modifications, individual component repair, and carburetor colorizing. The Carburetor Shop is also a Holley authorized parts distributor, and they stock a full line of Holley carburetor kits, accessories, and replacement parts.

Brad offers a number of unique services and modification kits that can save a damaged carburetor or make it easier to tune. For example: Loose main body booster venturis are aligned, tightened and/or modified for better flow; float bowls with damaged needle-and-seat adjuster threads can be restored; and Brad developed a kit to modify Holley metering blocks to accept removeable and tunable PVCR restrictions.

A few of the special racing parts offered are: high-flow needle-and-seat assemblies, specially modified floats for Nascar and circle track, and reusable float bowl and metering block gaskets. Brad also provides a color-restoration service, called *ColorRestore®*, that includes bead blasting and a chemical treatment that results in a very attractive, durable surface finish. For more information write or call The Carburetor Shop at: 1457 E. Philadelphia, Unit #24, Ontario, CA 91761, 714-947-9722.

Brad Urban's Carburetor Shop stocks a full line of Holley carburetors and replacement parts. If you need a small part to complete your rebuild, he probably has it!

Specially modified floats are available for various types of racing. Note how modified portions of Nitrophyl floats are resealed with epoxy, essential to prevent the float from becoming laden with fuel.

After bead-blasting, carburetor components are chemically colored with a *ColorRestore®*, a durable and attractive finish.

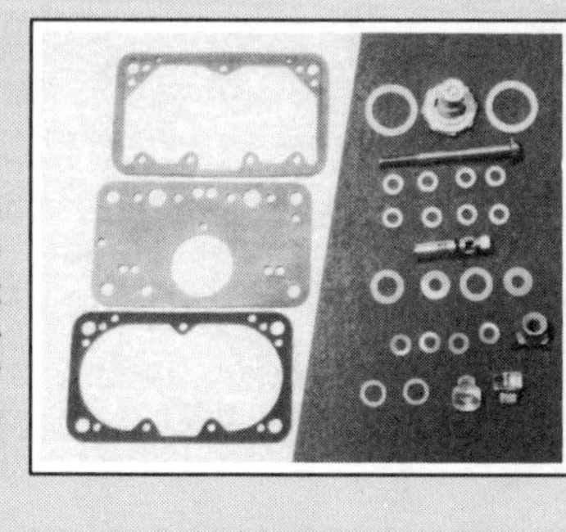

Reusable gaskets and seals make on-track tuning faster. Gaskets are made of a special material that doesn't stick. Brad also offers high-flow needles-and-seats and other components for Nascar and drag racing teams.

Holley produces chrome components and accessories that can add a rich custom appearance to your carburetor. Side and center pivot float bowls, vacuum diaphragm housings, accelerator-pump covers, choke linkage, and even air cleaners can be combined with custom painted components. This two-barrel was painted with fluorescent yellow and dressed with chrome bowls and other Holley accessories.

out the do-it-yourself approach. You can obtain good results and durability in your own garage with some of today's epoxy-based engine enamels. These new-technology paints offer very good, albeit not complete, resistance to chemicals, fuel, and heat, and are available in several attractive colors. The sidebar *Selecting Paint* on the next page will help you buy the right paint for the job.

Before you stock up on spray cans, you should consider the type of use <u>and abuse</u> your carburetor will receive. If your Holley will be used on a street-driven or off-road vehicle that rarely requires carburetor disassembly or removal, do-it-yourself painting is probably a practical choice. On the other hand, if your Holley will be used on a race car that requires frequent jet or tuning changes, spray painting is not as practical, although you may still want to consider Holley's Custom Shop powder-painting process, since it will withstand a great deal of abuse. The bottom line is that as good as the new epoxy paints are (these are the same paints that are recommended for use on engines, transmissions, etc.), they should not be used in areas of continued exposure to hot gasoline; an unavoidable consequence of frequent tuning. Repeated contact with fuel can dull the surface finish and eventually spot dissolve the paint, resulting in a haggard look.

Another consideration is the lack of colors available in fuel-resistant paints. Although there are several colors produced, only a few—most notably white, black, some reds, and grey—are considered attractive by many enthusiasts. Although tastes vary widely, and the choice is certainly up to you, I feel the remaining colors are somewhat subdued and lack punch. This is not a major problem, however, since you can combine the most attractive colors or use them with Holley's chrome accessories. A black main body with grey bowls has an industrial, but appealing appearance. Adding chrome bowls to a black main body and metering blocks looks very impressive. A white or red main body with chrome bowls, or even white bowls, really gets noticed.

If you're looking for the ultimate in do-it-yourself finishes, you may want to consider applying the latest fluorescent finishes, like bright yellow, hot pink, lime green, and super red. These remarkably bright colors are inexpensive and available in many auto parts and paint stores. But there are drawbacks. The most important one is lack of durability. In fact, fluorescent finish is easily dissolved by gasoline, and without some way to protect the paint, it would be completely impractical. Fortunately, you can spray on a clear epoxy overcoat that will add a reasonable degree of protection. In addition, fluorescent colors require an undercoat; white paint is often used. The several layers of paint and the thin protective surface make the use of fluorescent colors a bit limited. But if you're a dyed-in-the-wool do-it-yourselfer looking for a real eye-appealing paint project, refer to the sidebar *Using Fluorescent Paints* on page 79 for the tips you'll need to get the best results, with the least frustration.

PAINTING TOOLS AND ACCESSORIES

Very few specialized tools are required to paint your Holley, however, one requirement is particularly important: patience! As in any precision painting project, careful masking is more than half of the job, and masking all the parts of your carburetor can easily take an hour or more. Furthermore, sloppy work will really show in the finished piece. Patience is also required when spraying paint. You will not be happy with the appearance of one thick coat. Many paints require several thin coats to produce a "deep," shiny look. Fortunately, most paints are run resistant and fast drying,

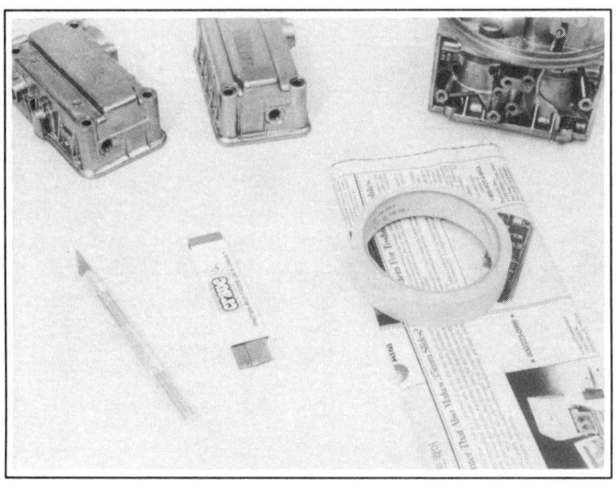

Very few specialized tools are required to paint your Holley. In fact, the most important ingredient, other than paint, is patience! Take your time when masking components (it can take well over an hour), as sloppy work will really show in the finished piece.

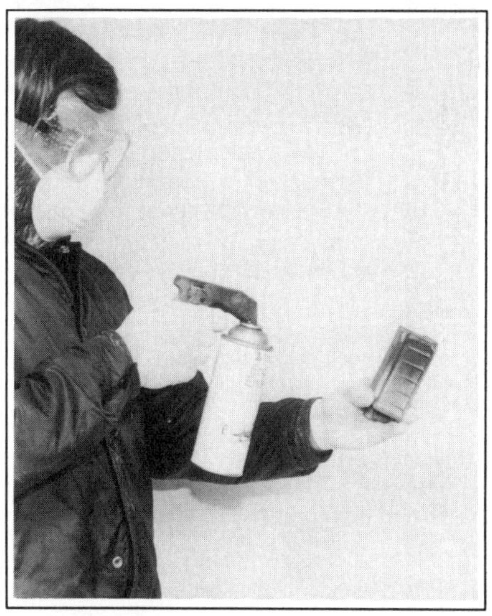

Patience is also required when spraying paint. You will not be happy with the appearance of one thick coat! Many paints require several thin coats to produce a "deep," shiny look. Fortunately, most paints are run resistant and fast drying, so you need only wait a short time between coats.

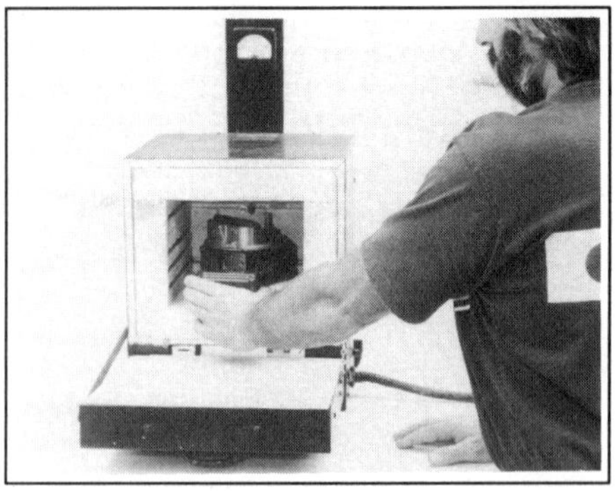

Paint is less susceptible to runs when the components you're spraying are warm. Also, heat drying can improve paint durability and resistance to fuel. Although not essential, a low-temperature kiln or oven with a box large enough to accept a main body can make your painting job easier and more successful.

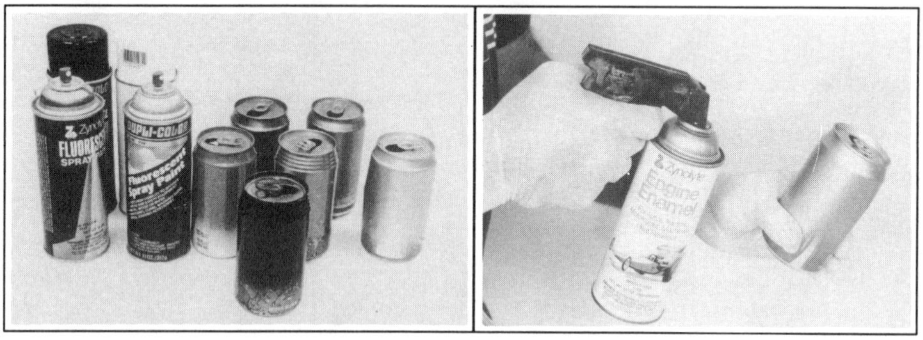

Before you *even consider* painting any component of your Holley, practice and experiment with the paint you've chosen using old soda-pop or beer cans. When you can apply thin, even, run-free layers of paint on the extremely smooth surfaces of pop cans, you'll have little trouble with carburetor parts.

so only a short period of time is required between coats.

The temperature of the air in your shop, and the temperature of the part you are painting will also affect your work. If the components (and the paint) are too cold, the paint will easily run and take much longer to set up. If the parts are warm when painted, runs won't happen as easily, and the paint will dry very quickly. This makes applying multiple coats much easier. Finally, I have found that heat drying can improve paint durability and resistance to fuel. One way to accomplish this is to warm parts before painting and to bake them afterwards in a temperature-controlled kiln having a box large enough to accept a main body. Unfortunately, kilns are expensive. While using one is not essential, having access to a small oven can make your job easier.

Most new epoxy paints are made from really nasty chemicals. They are very irritating to your skin and nasal tissues and just about impossible to get off your hands. The recommendation here is simple: Never paint anything without eye protection, a respirator, and latex rubber gloves. All these items are widely available, inexpensive, and absolutely essential to do a top-quality job. You must closely examine the parts for uniform paint application throughout the painting process. If you don't have adequate eye

Selecting Paint

There are many types, brands, and colors of spray paints available in automotive, hardware, and paint stores. Some stand up quite well to the heat and chemical torture that a carburetor's exterior must endure, while others can be destroyed by the first touch of gasoline or solvent.

The most durable paints for carburetor customizing are brand-name epoxies and heat-resistant engine enamels. These high-quality paints, like Dupli-Color® and Zynolyte®, are widely available in automotive and hardware stores. Avoid off-brand bargain paints; the money saved doesn't begin to cover the hassle involved in disassembling, stripping, and repainting carburetor components.

Also avoid lacquer or fluorescent paints, as they are particularly susceptible to gasoline and other solvents. However, if you find a "must-have" fluorescent or lacquer color, an overcoat of clear epoxy paint can greatly improve durability. Always test paint compatibility before spraying any parts of your carburetor, and allow base coats to cure for at least 24 hours before applying clear overcoats. Refer to the sidebar, *Using Fluorescent Paints*, on page 79 for more information on using clear overcoats. In all cases, read and carefully follow the instructions on spray cans; by sticking to manufacturer's recommendations, you can obtain a remarkably high-quality finish.

Regardless of what paint is used, baking painted components for a minimum of one hour at between 150- and 200-degrees(F) will improve durability and allow the part to be put in service much sooner. And make certain you are satisfied with the color and finish quality before baking the part. Heat-cured paint is extremely difficult to remove. In fact, bead blasting is often the only effective means of removal. Uncured, fresh paint, however, can be removed with spray- or brush-on paint remover, or by a quick immersion in lacquer thinner.

Zynolyte and Dupli-Color offer top quality engine enamels and epoxy enamels. Fluorescent colors should always be top coated with a fuel-resistant clear.

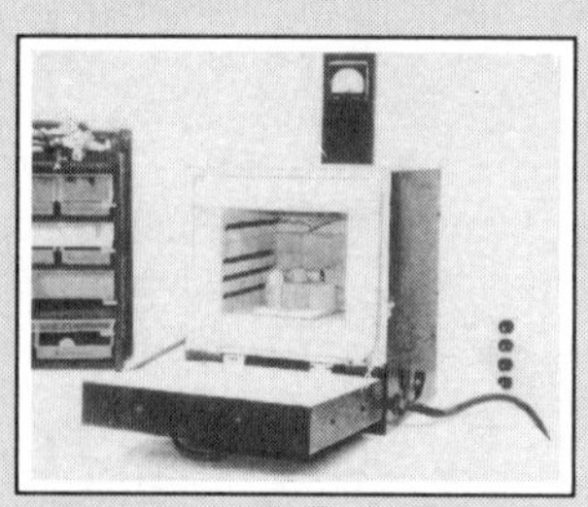

A small temperature-controlled kiln is excellent for baking painted parts. Baking improves durability and speeds the painting process.

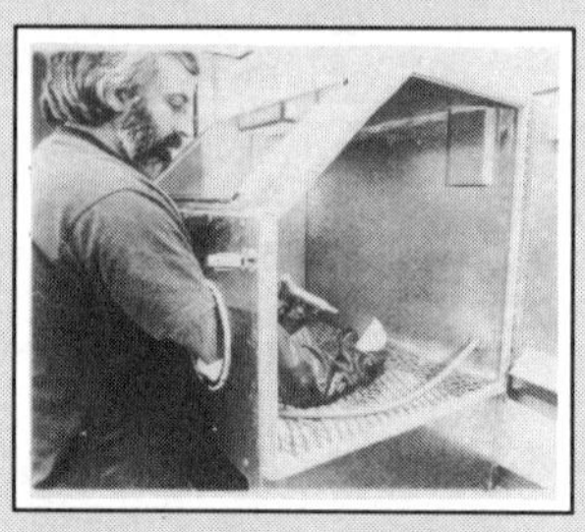

Bead-blasting is often the only effective way to remove heat-cured paint.

Spray- or brush-on paint strippers easily remove fresh, uncured paint. Soaking the part in lacquer thinner also works well.

protection and a respirator, your eyes will water and your nose will drive you crazy. *As much for your own health and comfort as to obtain the highest quality work, don't paint anything without eye protection, a respirator, and rubber gloves.*

The only remaining supplies and tools you'll need are masking tape and a sharp stencil (X-Acto®) knife, a small amount of lacquer thinner, and some clean rags. Optionally, you may want to buy a pistol-grip handle for your aerosol spray-cans. The usually fit a wide variety of cans and make spraying a bit easier.

PRACTICE FIRST

A final word of caution before you begin painting. As resistant to fuel and heat as the new epoxy paints are, they are equally resistant to paint removers. If you are dissatisfied with the appearance of your carburetor once it is painted, about the only way to completely remove the paint and start over is to have the components bead blasted. Luckily, this service is widely available in most cities at machine shops and automotive-part rebuilders. Surfaces that come in contact with fuel must be masked before they are blasted, since bead blasting will also strip dichromate plating. Bead blasting will leave a slightly rough surface, perfect for subsequent paint adhesion, but if you give up on paint and choose to have them redichromated, you may be dissatisfied with their appearance, since dichromate often looks best when applied to a smooth surface. So before you make the decision to apply paint, you should be aware that, in some ways, it is a non-reversible choice.

Three of the best tips that I can give you to help avoid disappointing results are: practice, practice, and practice. And ideal practice surfaces are easy to find: old soda-pop or beer cans. When you are able to apply thin, even, run-free layers of paint on the smooth surfaces of pop cans, you'll have little trouble with carburetor parts. Practicing at different temperatures will also help you become familiar with the coverage, flow, and drying characteristics of the paint you've selected. Only after you're satisfied with the results on pop cans, should you consider applying paint to your carburetor.

1 Remove Any Oil & Dirt

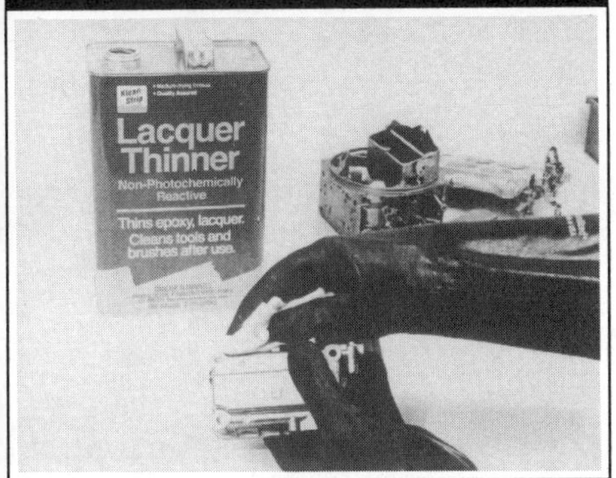

1—Before you begin masking parts, make sure any remaining gasket remnants, oil, and dirt are removed. A rag moistened with lacquer thinner, or spraying carburetor cleaner then wiping with a clean rag, works well.

2 Masking Supplies

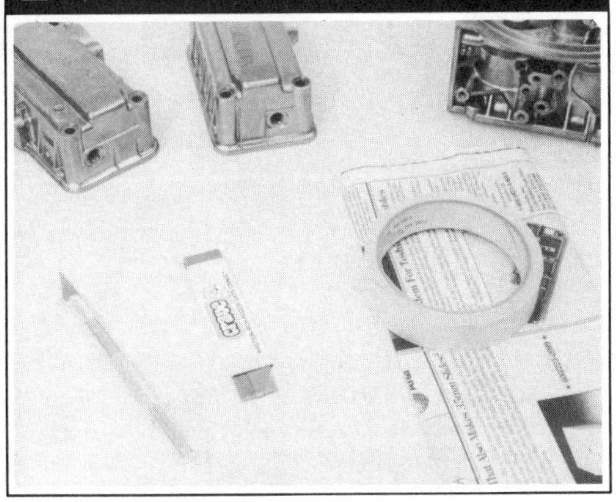

2—A roll of masking tape (the low-adhesion type is best), an X-Acto knife (a single-sided razor blade is not as easy to control but will work in a pinch), and a couple of sheets of newspaper are all the tools required to mask Holley component parts. The only other "supply" you'll need is patience.

MASTER MECHANIC TIP — PRO TIP

3 Bead-Blast Surface Prep

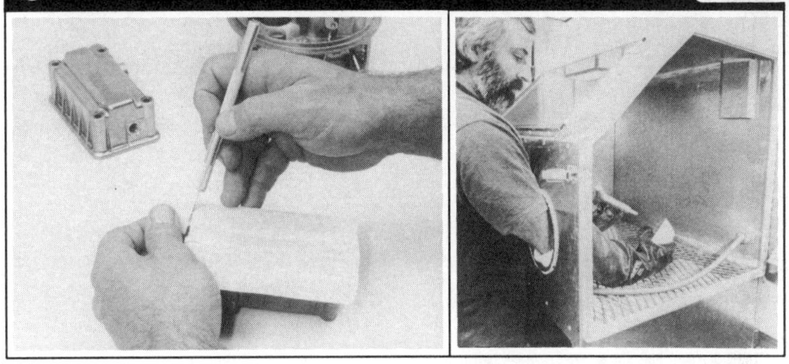

3—Bead blasting will remove contamination and provide the perfect surface for paint adhesion. To prevent blasting the dichromate finish on interior surfaces, mask components before blasting. Then carefully remask before final painting (as illustrated in the following steps); don't use the same mask for both jobs since portions of the tape mask will be eroded by the bead blaster, and dust, forced into the tape seams, will contaminate the paint.

4 Begin By Masking The Float Bowls

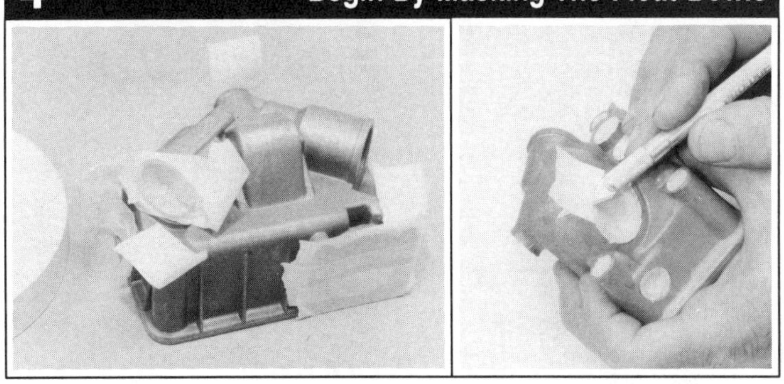

4—Apply pieces of masking tape over the fuel inlet, accelerator-pump cavity, bowl screw holes, sight plug and needle-and-seat openings. Use the X-Acto knife to cut the tape at the edges. Try to use the bowl edge and the knife blade as a sort of "scissors." This technique is fast and leaves a clean edge. Don't mask the interior, you'll need to use the opening to hold the bowl during painting.

5 Mask Remaining Components

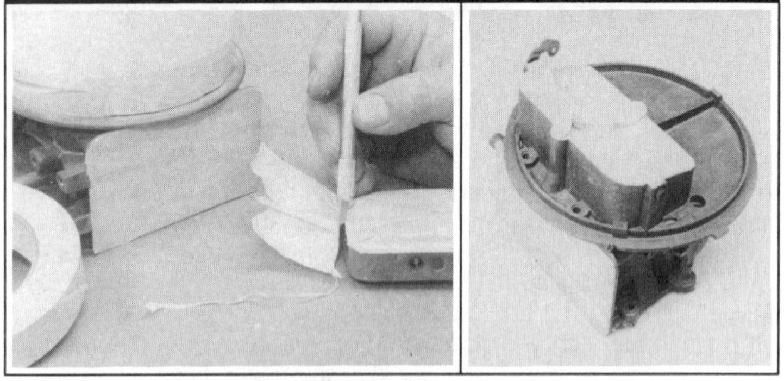

5—Continue masking the remaining parts. Metering blocks are relatively easy. The main body will take a while, however, since you have to mask the metering block/plate surfaces, the bottom surface, and the airhorn area. The easy way to mask the airhorn is to cover everything that is enclosed in the air cleaner. A bit "fancier" approach is to mask only the airhorn opening (this works best on two-barrels) as shown on the right.

6 Leave A Place To Hold

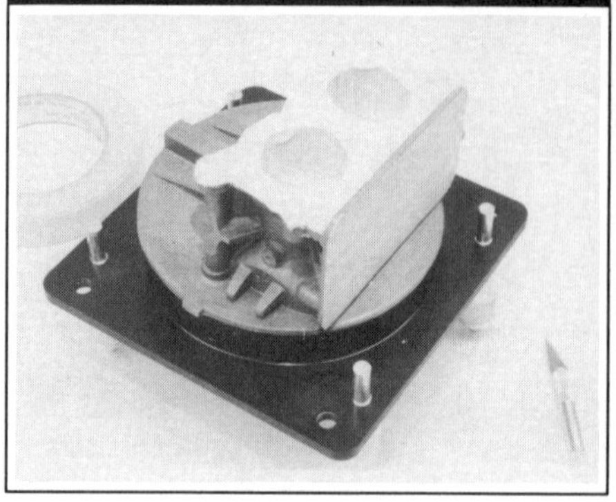

6—Components must be held in many positions during painting, so don't forget to leave a place where you can firmly grasp the main body. Masking about 3/4-inch down the venturi (from the bottom) will provide a good place to stick your fingers.

7 Masking The Throttle Plate

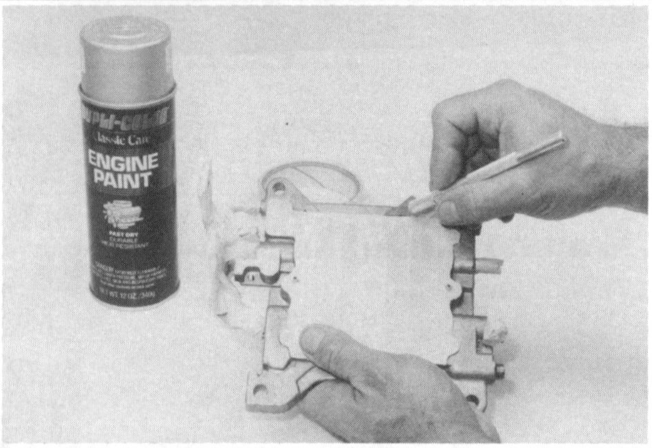

7—Masking the throttle plate is the most difficult job, but since Dupli-Color® makes a durable, natural-looking aluminum paint that makes old, discolored plates look like new, it may be worth the effort. If you decide to give it a try, carefully mask linkage, springs, vacuum fittings, and both upper and lower surfaces.

8 Final Prep Before Painting

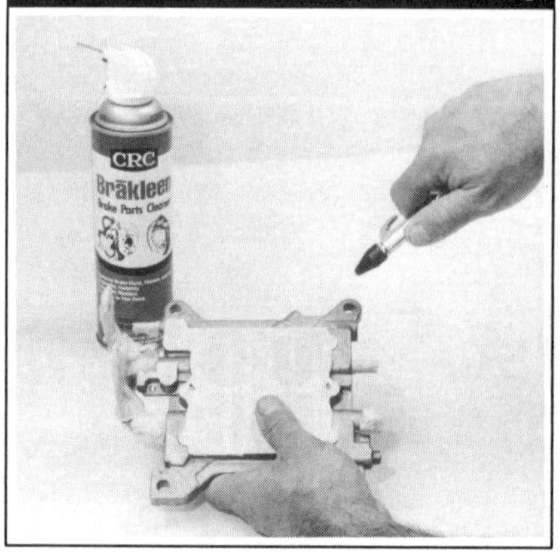

8—When all parts have been masked, give them a quick once over with compressed air to remove dust or dirt. If you find any oil or fingerprints, a quick wipe with lacquer-thinner (or other quick-drying solvent) should do the trick. Also, prepare a place in your shop for painting by spreading newspaper over the floor and/or benchtop. Overspray from just one aerosol can will spread over a 6- to 10-foot radius.

9 Begin Applying Paint

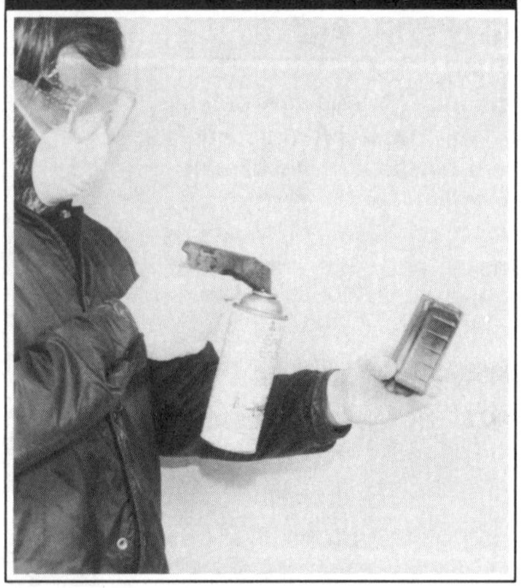

9—Assuming you have selected the right paint, practiced as described in the text, are familiar with the paint manufacturer's instructions, and are using eye, nose, and hand protection, you can begin applying paint. Make sure that the temperature of the parts and your shop are within the recommended range. Cold paint is much more likely to run. As you spray all sides, move the part as much as you move the spray can. Keep your eye, the spray can, and the part in a straight line for detail work.

MASTER MECHANIC TIP

10 Use Extra Light

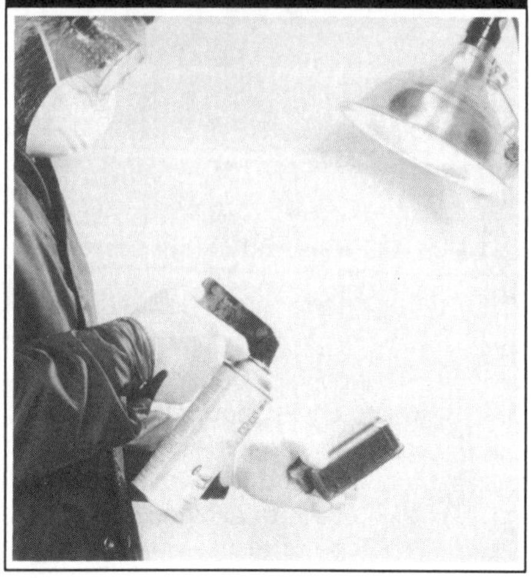

10—You'll get the best results if you follow the paint manufacturer's recommendation of applying several thin coats. Allow the paint to set up for a short time between coats, usually 5 to 10 minutes. And here's another tip that's worth it's weight in gold: *The more light you have in your shop, the better quality your work will be.* This is especially important when you apply multiple coats, since it becomes difficult to see the paint you're applying over previous layers.

11 Baking For Durability

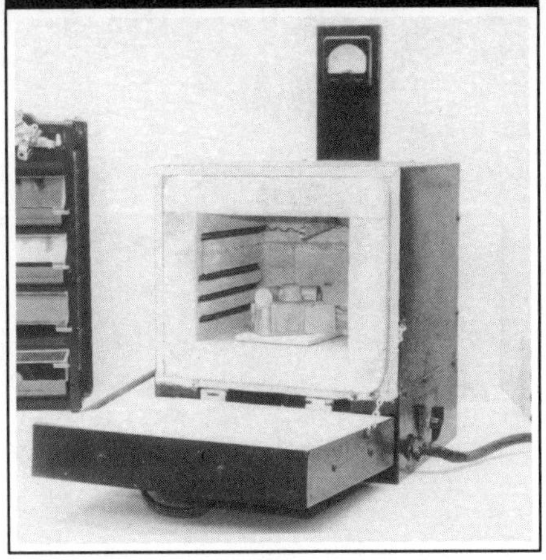

11—A small kiln or oven can make your painting job easier and the results last longer. Warming parts to the high end of the paint manufacturer's recommended temperature will minimize runs. And baking parts for several hours at 150 to 200 degrees(F) after applying the final coat will harden the paint and make it more resistant to fuel and handling during reassembly and reinstallation. Don't heat painted parts above 200 degrees(F) or the paint may oxidize, evident when the parts smoke during baking.

12 Remove Overspray

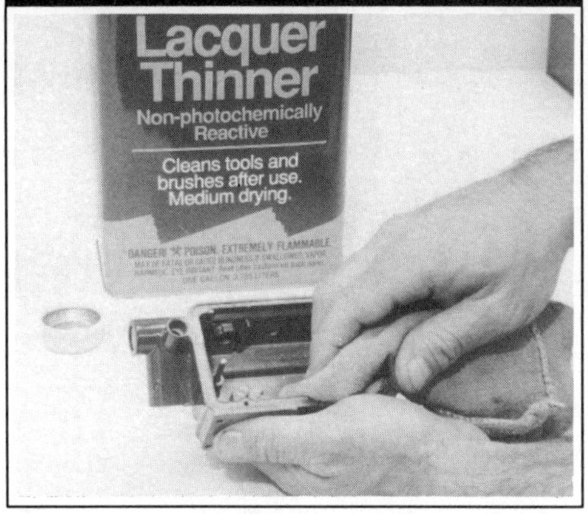

12—Remove the masking tape when the paint is dry. If you have baked the parts in an oven, the masking tape may be a bit more difficult to remove, but as long as the temperature was kept below 200 degrees(F), it should come off without much fuss. Make sure all paint overspray has been removed from the inside and gasket surfaces of float bowls and metering blocks. Finally, blow off all major components with compressed air before reassembly.

Using Fluorescent Paints

Fluorescent paints are available in a number of striking colors that can give a prize-winning appearance to your Holley carburetor. Unfortunately, these paints are very susceptible to damage from gasoline and other underhood chemicals. However, by using an epoxy clear overcoat you can produce a fluorescent finish with adequate durability for a show car or a light-use street machine. Unfortunately, fluorescent paint coatings, even when using the technique described here, are not durable enough for carburetor parts used on daily drivers or race cars that require frequent disassembly.

Most fluorescent paints are very transparent. The metal surface under the paint will clearly show through the paint without an opaque undercoat. The color of the undercoat should be chosen to optimize the brightness of the fluorescent color. Flat white is usually a good choice because it adds high reflectivity and flat paints are generally easier to apply than their gloss counterparts, however, you can experiment with other colors. Because multiple coats of paint are required with fluorescent colors—the base coat, the fluorescent color, and a clear top coat—each coat must be sprayed on as lightly as possible or the results can be disastrous. And after each type of paint has been applied, the part should be baked for at least one hour (at under 200 degrees[F]) and/or air dried for a minimum of 24 hours.

The fluorescent color is applied after the basecoat. Repeated coats give the paint a darker, less brilliant, hue, since less of the white undercoat shows through. Sometimes the desired result can be obtained with a very light coat of fluorescent paint. After the fluorescent color has been applied, you must again bake the part for one hour and/or let it air dry for at least 24 hours. If this is not done, the clear overcoat may attack the fluorescent color and destroy the finish.

Applying a uniform clear epoxy enamel finish is essential for optimum final appearance and durability. Apply no more than two or three very light coats, allowing several minutes drying time between each coat. A clear coat that is too heavy will dull and yellow and may eventually flake or chip off.

Practice the complete painting sequence on several soda pop cans before you actually paint any carburetor parts. Experiment to determine the *minimum* amount of paint that can be applied and still achieve satisfactory results. After practice, you can produce real eye-popping results, but if you need additional durability, powder painting is probably your only alternative. See the sidebar *Holley Custom Shop*, on page 73 for more information on special coatings.

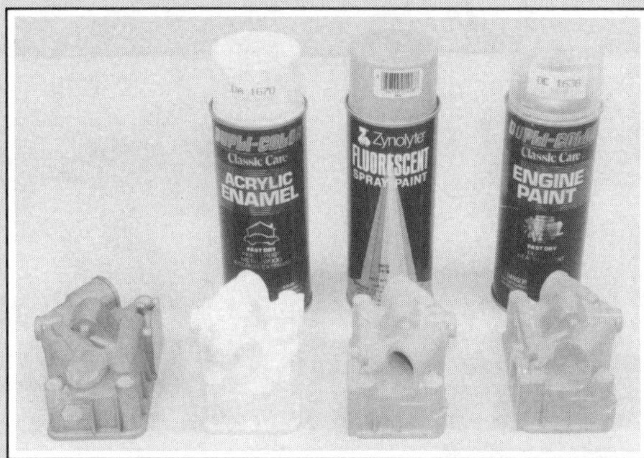

Here are the key steps to using fluorescent colors. The bowl must be thoroughly cleaned (bead-blasting works great) and masked (left). Then a flat-white undercoat is applied and baked (second from left). After it has cured for 24 hours, a very light coat of fluorescent color is applied, baked, and allowed to dry (second from right). Finally, a clear epoxy overcoat is applied, like this Dupli-Color® fast-dry clear. The finished piece (right) will resist only light contact with fuel or other corrosive chemicals. For more durability powder coatings must be used.

Fluorescent paints (left) are available in some really wild colors, including Pasadena Pink, Gold, Lime Green, Bright Red, Blue, and others. Select only a quality clear overcoat (right); it's the only thing protecting the color from damage.

HOLLEY Rebuilding & Modifying

9—Carburetor Reassembly

This chapter provides detailed workbench procedures that will guide you through the step-by-step reassembly of virtually any Holley modular carburetor. In order to properly apply the information in this chapter, you should be familiar with the tools needed to work on Holley carburetors—see Chapter 2. You should also be able to identify the individual component parts of your Holley—see Chapter 3.

WORK-A-LONG SHEET

Many of the steps in this reassembly chapter refer to information recorded on your Work-A-Long sheet; a blank copy is provided on pages 123 and 124. If you disassembled your Holley following the steps in Chapter 6, you should have recorded vital information that you will find helpful during reassembly. On the other hand, if your carburetor was already disassembled when you began your project, consider going over the steps in Chapter 6, particularly those marked with a ***NOTATION REQUIRED*** icon, and record as much information on your Work-A-Long sheet as possible before you begin reassembly. As you proceed through reassembly in this chapter, use your Work-A-Long sheet to record exactly what jets, power valves, check valves, etc., were installed in your Holley if they are different from those logged during removal. Finally, if you do any of the tuning or other modifications discussed in Chapter 11, make sure to keep your Work-A-Long sheet up to date. An accurate record of the components in your Holley will make future troubleshooting or tuning much easier, and it can even add value to your carburetor when you sell it.

STEP-BY-STEP PROCEDURES

Not all of the step-by-step procedures in this chapter apply to all Holley models. For example, if you're rebuilding a 2-barrel, steps related to secondary reassembly are not applicable and should be skipped. In many cases, instructions are provided in the step-by-step captions that direct you to "jump ahead to..." when upcoming steps do not apply to a specific carburetor model. However, all steps are clearly titled and in the proper sequence, so if you come upon a step that does not relate to your Holley, review the subsequent steps until you come to the next relevant procedure.

Several steps in this chapter are labeled with icons—symbols—that represent a unique Holley model, like 4500 or 4165. Unless otherwise instructed, only perform these steps if you are working on that model carburetor. Some steps are labeled with icons that represent important information or special help, such as ***IMPORTANT STEP***, ***MASTER MECHANIC TIP***, or ***SAFETY STEP***. Carefully read these steps unless you are specifically directed to skip over them. These steps contain important information that can often make the difference between a successful rebuild and disappointment.

FINDING ADDITIONAL INFORMATION

If you would like additional information on the following subjects before you begin reassembly, refer to the indicated chapters.

1 Preparing For Reassembly

1—Organize both new and cleaned parts before you begin reassembly. Here is one method: Place *fuel-bowl related parts* (float bowls, floats, metering blocks, accelerator pump covers and arms, and fuel-inlet fittings) in bin #1; place *main-body related components* (throttle plate, metering plates, choke mechanism/linkage, vacuum diaphragm housing, accelerator pump nozzles, and transfer tubes) in bin #2; *all screws and remaining small parts* should be placed in bin #3. In addition, keep *old gaskets, rubber parts, etc.* that you accumulated during disassembly in bin #4, and place *new parts from your carburetor rebuilding kit* into bin #5. For information on selecting carburetor kits, refer to pages 32 and 33.

2 Begin Float Bowl Assembly

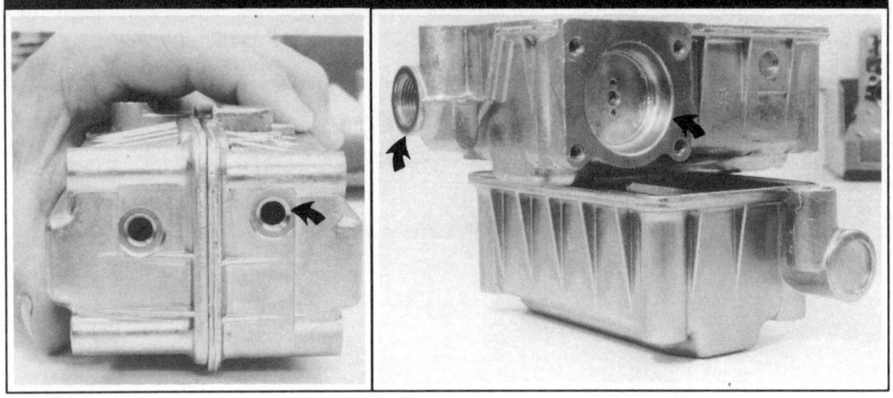

2—Locate the primary float bowl. If you marked the secondary bowl with a small "s" during disassembly, finding the primary bowl will be easy. If you aren't sure, the primary bowl has a majority of the following characteristics (arrows): 1) a higher sight plug position, 2) an accelerator pump, and 3) a screw-type fuel-inlet fitting. On a few racing carburetors, like the Dominator, the sight-plug positions and other bowl characteristics are identical. In these few cases, the float bowls are interchangeable.

3 Fuel-Inlet Filters

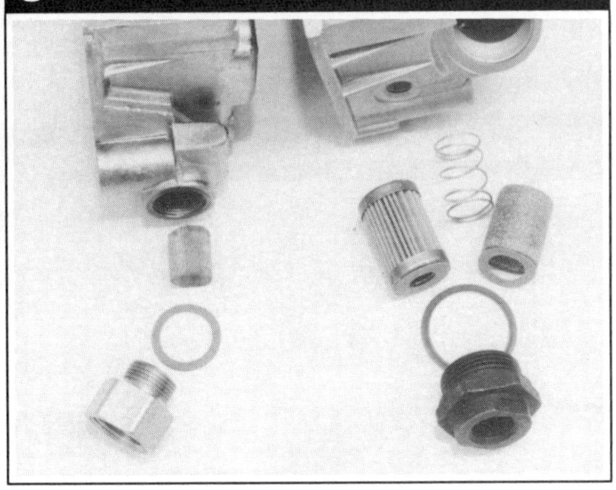

3—Holleys are equipped with several types of fuel-inlet filters, or they may have no filter at all. The inlet filter should only be considered a "last resort" catch of contaminants; a backup for a quality, high-volume in-line filter. If your carburetor was equipped with a wire inlet screen—shown on the left—it offers little restriction and should be reinstalled. If your Holley has a sintered-bronze or paper element, it provides better filtration but offers substantially greater restriction. Use either of these elements only if: 1) the element is thoroughly clean, and 2) the carburetor will not be used on a large-displacement, high-performance engine, especially when the carburetor has a single fuel inlet.

IMPORTANT STEP !

4 Thread And O-Ring Lubrication

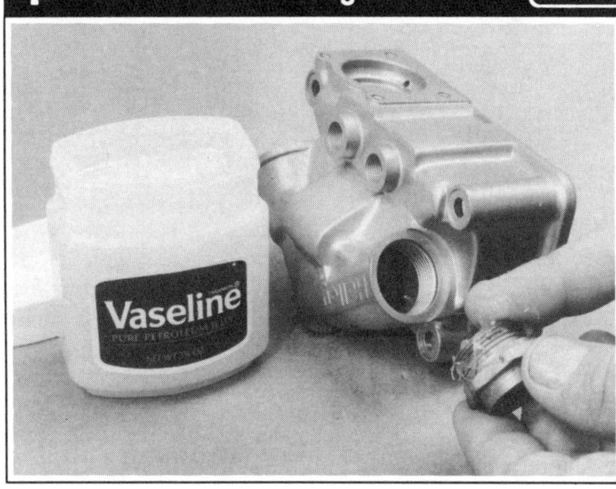

4—O-rings and other rubber parts can be damaged if they are assembled dry. Also, steel fuel-inlet fittings and other fasteners can be difficult to smoothly install since the threads tend to stick to aluminum and other soft-metal castings. *These problems can be eliminated if you apply a little petroleum jelly before assembly* (WARNING: Other lubricants may not be compatible with the parts in your carburetor; refer to the lubricant chart on page 15 for information on lubricant compatibility). When you use a lubricant on threads, you can reduce the torque recommended in the upcoming steps by about 10- to 15%, although this is often not necessary on the small fasteners used on Holley carburetors. Refer to the sidebar *Fastener/Bolt Torquing Tips* on page 83 for important information on torquing fasteners before you proceed.

5 Install Inlet Screen

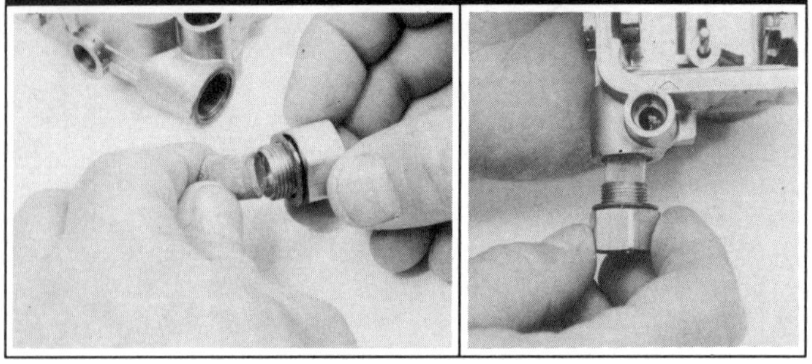

5—Refer to the *Common Holley Gaskets And O-Rings* illustration on page 126 for help in locating the gaskets you will install in the upcoming steps. If your Holley does not use a wire inlet filter screen, proceed to the next step. First, locate the correct inlet fitting gasket from your carburetor kit and install it over the threads. Next, install the wire screen fully into the recess in the inlet fitting. Then install the fitting in the bowl while holding the bowl and fitting upright. This technique prevents the screen from falling out of position.

6 Install Bronze/Paper Element

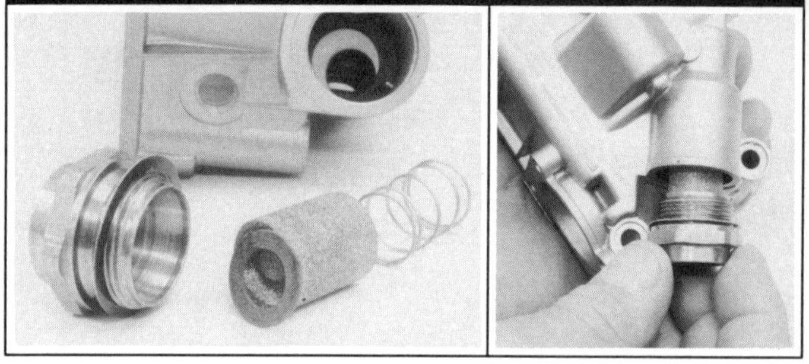

6—If your Holley does not use a bronze or paper filter element, proceed to the next step. Locate the black metal gasket for the inlet fitting and the thin paper gasket for the inlet filter from your carburetor kit. Install the black gasket on the threaded area of the fitting. Then install the paper gasket (arrow) into the recess in the fitting (note: some paper filters have this gaskets already attached to them). Set in the bronze element with the pointed interior cone toward the fitting, or install the paper element with the hole toward the fitting, stack on the spring, and install the components in the bowl (don't forget to apply a little petroleum jelly to the threads on the fitting).

TORQUING FASTENERS

7 Tighten Inlet Fitting

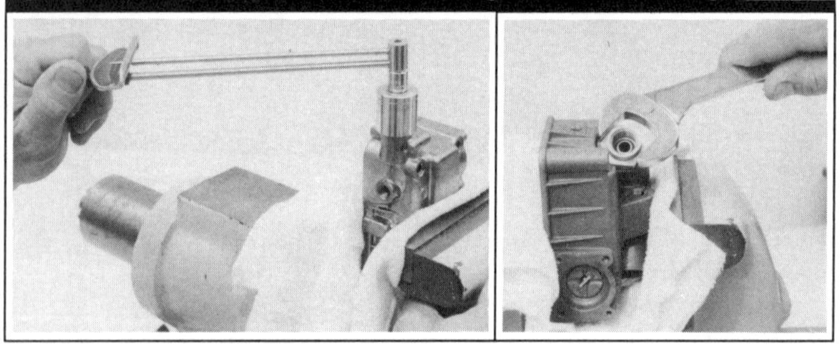

7—If you haven't already, locate the gaskets for the fuel-inlet fittings from your carburetor kit; lightly lubricate and install the fittings in the float bowl. Clamp the bowl in a bench vise with brass jaws (or use a rag) and tighten 9/16-24 fittings on side-pivot bowls to 175- to 200-inch pounds and 7/8-20 fittings and plugs on center-pivot bowls to 200- to 250-inch pounds. If you don't have the special adapters to tighten large inlet fittings and plugs with a torque wrench, use an open-end wrench, but make sure the fittings are tight.

PERFORMANCE STEP

8 Center-Pivot Inlet Plugs

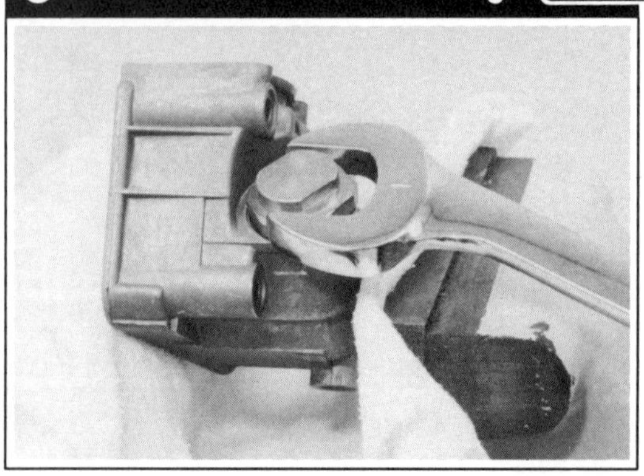

8—Some center-pivot bowls have both sides drilled and tapped for fuel-inlet fittings. These carburetors use an inlet plug on one side. Install a black metal gasket on the plug and screw the inlet plug into the side of the bowl opposite the sight plug hole (left side), since virtually all applications route the fuel lines on the sight-plug (right) side, away from the throttle linkage. Torque the plug to 200- to 250-inch pounds.

PRECISION MEASUREMENT

9 Accelerator Pump Check Valves

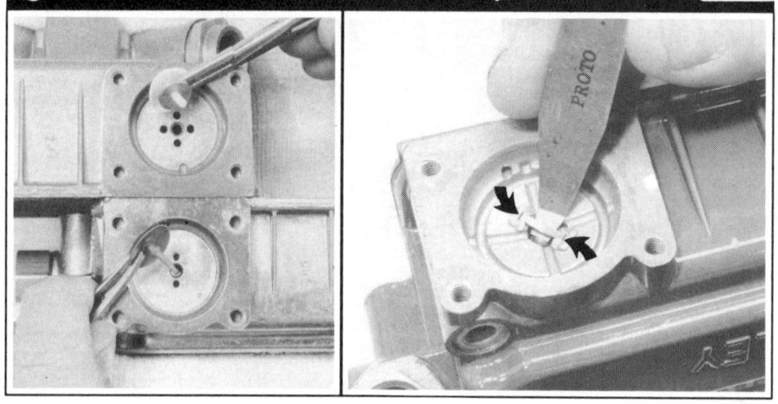

9—Holley carburetors are equipped with three types of accelerator pump check valves: a steel ball retained with a "strap" (a wire clip is used on LeMans bowls), a red-plastic "umbrella," and a white-plastic "cup" valve. If your Holley has either of the plastic valves (left), skip to the next step. If you have LeMans bowls, skip ahead to **Step 12**, *Installing The Pump Assembly.* Insert a thin feeler gauge between the ball and the retaining strap to ensure that .011- to .015-inch clearance is provide for ball movement. Carefully bend the strap with a probe or small screwdriver to adjust this clearance. Finally, make sure the strap is solidly attached to the bowl. If it's loose, peen the saddles (arrows) then recheck the clearance. Skip ahead to **Step 12**, *Installing The Pump Assembly.*

10 Installing Red "Umbrella" Valve

10—If your Holley uses the white-plastic "cup" valve, skip to the next step. The red-plastic "umbrella" check valve is easy to install. First, locate the new plastic diaphragm in your carburetor kit; it has an extended stem (arrow). Apply a little petroleum jelly to the stem and insert it into the hole in the center of the pump cavity. While applying light pressure to the "umbrella" with your thumb, use needlenose pliers or hemostats to pull the neck until the collar "snaps" through the hole. Then cut off the excess neck with diagonal cutters. Skip to **Step 12**, *Installing The Pump Assembly.*

11 Installing White "Cup" Valve

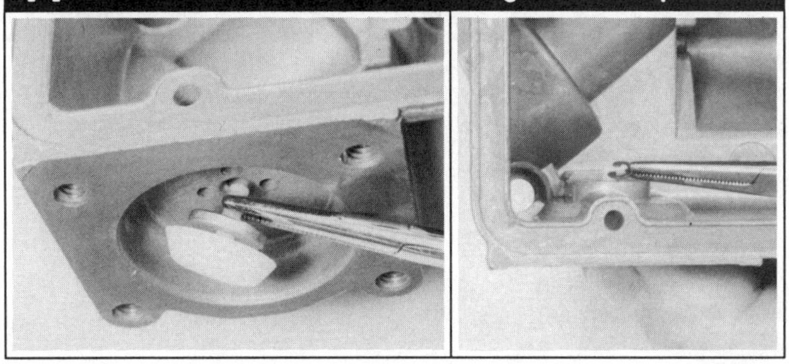

11—Locate the white plastic "cup" valve, metal washer, and small "E" clip from your small parts bin. The metal washer is a weight that holds the plastic valve open, ensures rapid filling, and helps prevents fuel pullover in performance applications. The valve will not work properly without this special metal washer. Install the washer on the stem of the valve and insert the stem into the center hole in the pump cavity. While holding the valve in place with your thumb, use needlenose pliers or hemostats to install the small "E" clip on the stem that protrudes into the bowl. Make sure the "E" clip is thoroughly seated in position.

Fastener/Bolt Torquing Tips

Properly torqued fasteners are essential to a secure mechanical assembly, and carburetors are no exception to this rule. Apply too-little torque and there will be insufficient clamping force to prevent leakage, or fasteners may further loosen and back out completely. Over torquing can warp components, pull threads, or damage fasteners.

Accurate torque readings are only possible when the threads are clean and undamaged. And although the torque specs given in the table on page 125 and in the steps in this reassembly chapter are for dry (non-lubricated) fasteners, I suggest that you lubricate all threads with Vaseline. Lubrication allows fasteners to screw in easier and prevents thread seizure. To compensate for lubrication, you can reduce torque values by about 10- to 15%, although this reduction is often not required for the small, high-quality fasteners used on Holley carburetors.

Torque specifications are often listed as a range rather than a single number. Always attempt to torque to the upper limit of the specification (but never above it). *If the torque reading levels off, do not continue turning the fastener!* Going by the book, the fastener is considered to be of acceptable tightness if the reading leveled off above the minimum torque value. However, torque values that level off indicate that the threads and/or the fastener are about to fail. I suggest that you remove the fastener, examine both the male and female threads, and if the female threads appear undamaged, install a new fastener and try again. If the female threads are damaged, they must be repaired by one of the methods discussed in the sidebar *Repairing Damaged Threads*, on page 68.

When more than one fastener holds a single part, such as the four screws that retain a float bowl, the torque should be applied in several increasing steps and in a crisscross tightening sequence.

Spending a little extra time to properly torque fasteners is the marque of the professional. This "extra" step will help ensure that your rebuilt carburetor functions properly and reliably. Since proper bolt torquing techniques eliminate damage from overtightening fasteners, it can also save you time and money.

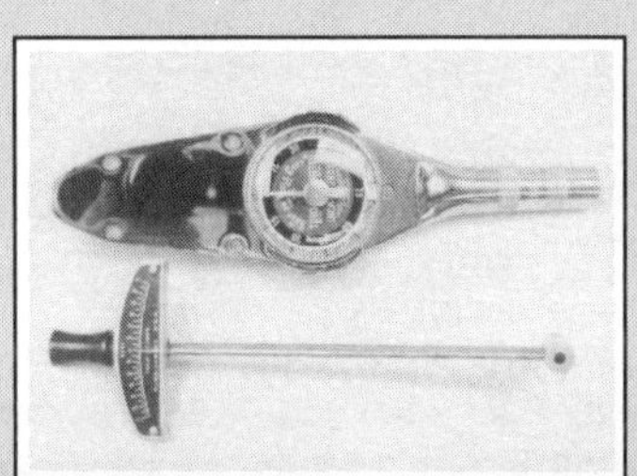

There are many types of torque wrenches available in a wide price range. A model calibrated in inch pounds with a maximum rating of about 200 is best for work on Holley carburetors.

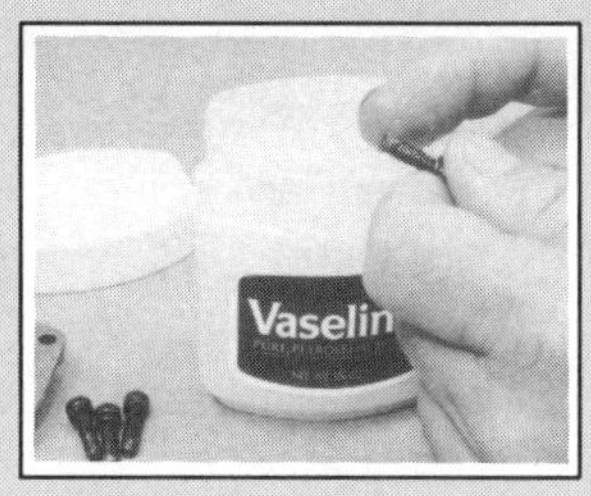

Although most torque specs are for dry fasteners, I suggest that you lubricate all threads on Holley carburetors with Vaseline (see text).

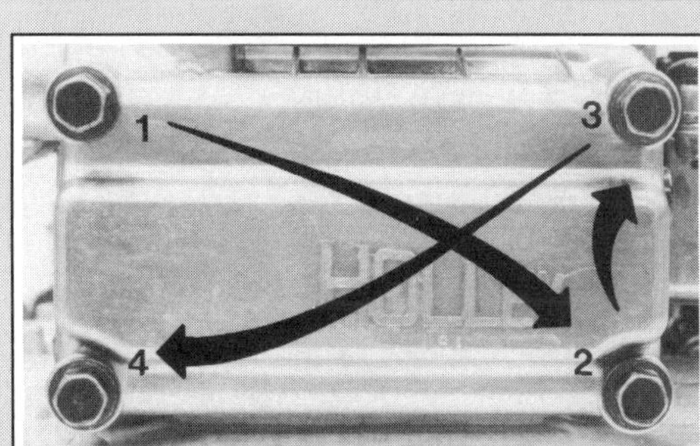

When more than one fastener holds a single part, such as the four screws that retain a float bowl, the torque should be applied in several increasing steps and in a crisscross pattern.

TORQUING FASTENERS

12 Installing The Pump Assembly

12—Refer to your Work-A-Long sheet to determine if your Holley uses the standard 30cc or the taller 50cc pump on the primary bowl. With the bowl inverted, place the pump spring (1/2-inch tall for the 30cc pump and 1-inch for the 50cc pump) into the bowl cavity. Locate the new diaphragm, and notice that the rivet in the center has two distinct ends. The thick head (arrow) should face up, away from the bowl. Position the diaphragm on the spring and set the pump cover over the diaphragm. Finally, install the four screws (8-24x3/8-inch for the 30cc cover and 8-24x5/8-inch for 50cc pump cover); most screws have integral lock washers. Make sure that the diaphragm is centered under the cover; tighten the screws to 12- to 18-inch pounds.

13 Types Of Floats

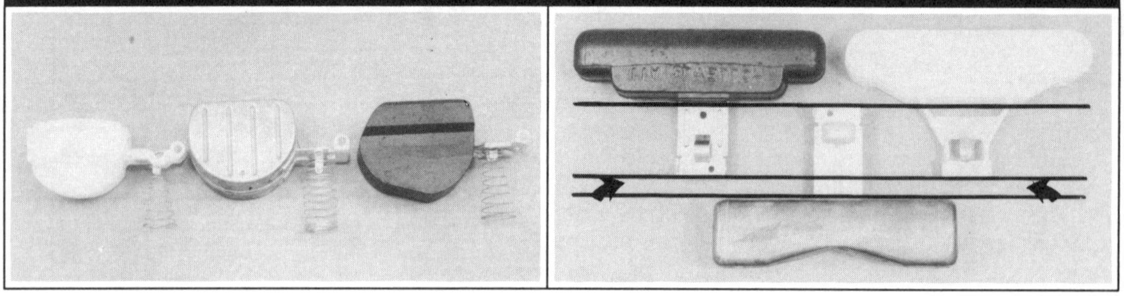

13—Both side-pivot (left photo) and center-pivot (right photo) floats come in hollow metal, solid Nitrophyl, and white-Duracon plastic. All floats are interchangeable within their groups (although different compositions may require different float height settings), except *brass-type, center-pivot floats. These floats have longer float arms (indicated by the arrows) and will interfere with secondary metering plates; they must be used with metering blocks only*. All other center-pivot floats use shorter float arms and can be used with either metering blocks or plates. All floats should be checked for proper buoyancy and/or leakage before reassembly; refer to the sidebar *How To Weigh Floats,* below. For more information on floats and float bowls see the sidebar *Float Bowl Basics*, on page 55.

How To Weigh Floats

The floats in the bowls of all Holleys rise and fall in reaction to small changes in fuel level; rising to shut off incoming fuel; lowering to allow additional fuel to enter the bowl. If all is well, the level of fuel in the bowls will remain remarkably constant. However, if a float loses any of its buoyancy, the fuel level will rise, producing a rich mixture, stalling, or even flooding. All carburetor floats should receive a close inspection to verify that no fuel has leaked (or been absorbed) into the float.

There are three types of Holley floats—hollow brass, hollow white plastic (Duracon), and a solid black Nitrophyl. The hollow floats are just "cans" that can be easily inspected for leakage. However, Nitrophyl floats have a thin liquid-proof exterior. If this surface layer is damaged, fuel will soak into the interior material, increasing float weight.

Regardless of the type of float in your Holley, first check for visual damage, such as dings, dents, or any visible cracks. If the float passes this first inspection, the next step is to perform a leak test or, in the case of Nitrophyl floats, a saturation test. If the floats have been dry for more than a few hours, they should be completely immersed in clean solvent for at least twenty minutes, preferably overnight. After immersion, examine brass or plastic floats by shaking them. If any liquid has accumulated inside the floats, replace them. Nitrophyl solid floats should be weighed after immersion. Floats should not exceed the weights indicated in the photo on the right. If you have any doubt about the integrity of a Nitrophyl float, replace it.

The minimal investment in both money and time you spend in inspecting and replacing questionable floats will be more than offset by the predictable, reliable operation of your rebuilt Holley.

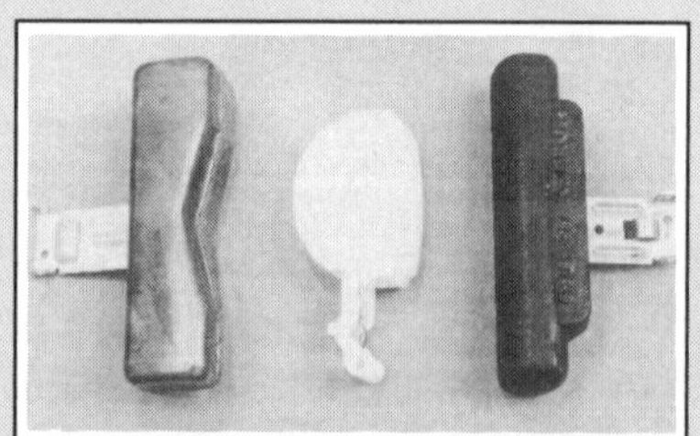

Holley floats are either hollow brass or plastic, or solid Nitrophyl. Hollow floats can be shook to test for fuel leakage; Nitrophyl floats must be weighed (see text).

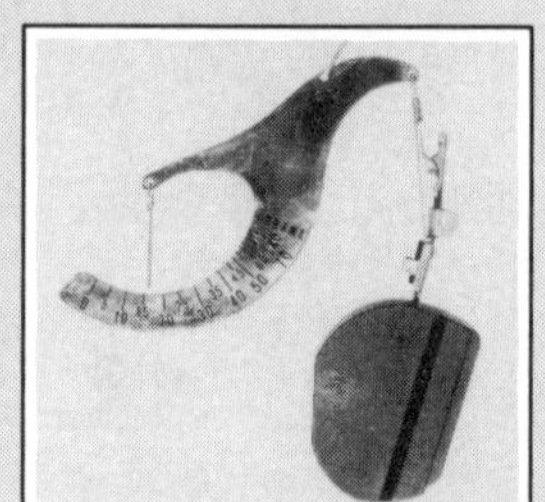

Inexpensive float scales are available to weigh Nitrophyl type floats. Excessive weight indicates that the float is saturated with fuel and must be replaced.

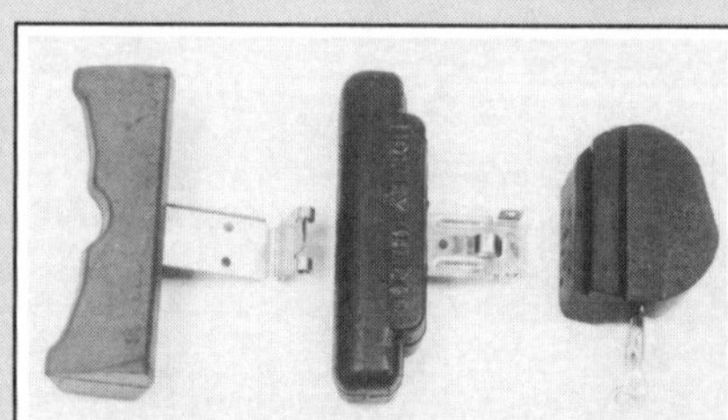

Nitrophyl solid floats should be weighed after immersion in solvent for at least 20 minutes. Center-pivot floats should not weigh more than 12 grams and side-pivot floats should not exceed 16 grams. If you have any doubt about the integrity of a Nitrophyl float, replace it.

14 Types Of Needles-And-Seats

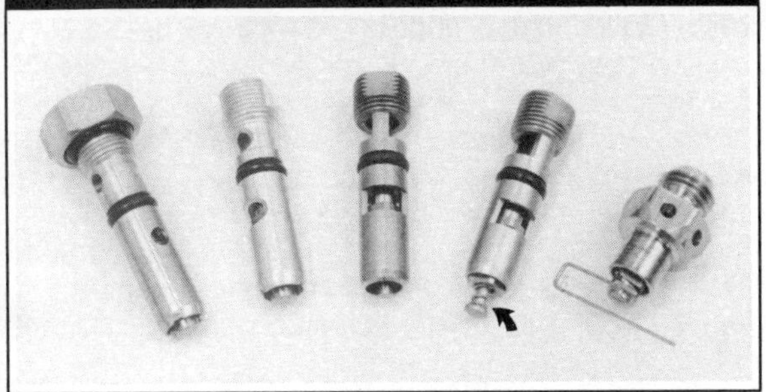

14—Several types of needles-and-seats are available from Holley and other manufacturers. The two on the left are non-Holley replacements and, because of their limited flow capacity, are not recommended for performance applications. The next part is a standard Holley externally-adjustable assembly. The fourth needle-and-seat—also available from Holley, **part 6-513**—uses an extra spring (arrow) that helps damp float bounce in off-road applications. Finally, the rightmost unit is used in side-pivot bowls that do not have an external float-height adjustments; if your carburetor is equipped with this last type of needle-and-seat, skip ahead to **Step 23**, *Internally-Adjustable Bowl.*

TORQUING FASTENERS

15 Install Center-Pivot Float

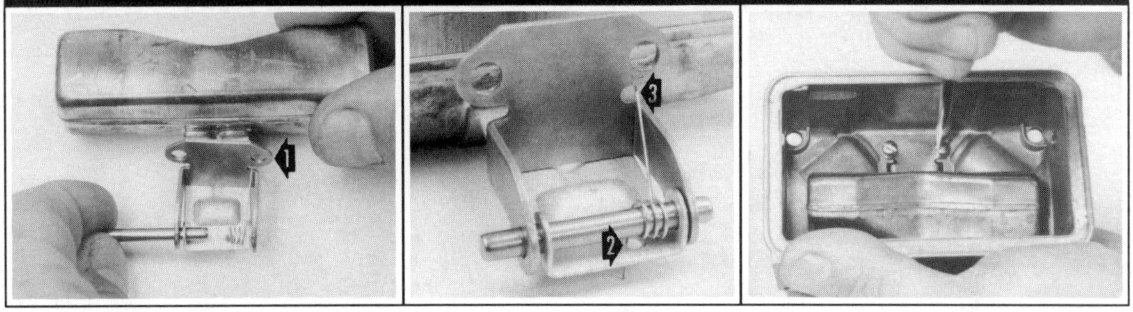

15—If your Holley is equipped with side-pivot bowls, proceed to the next step. Metal and Nitrophyl center-pivot floats use a small counterspring that keeps the float from bouncing and maintains a stable fuel level (white-plastic Duracon floats—and LeMans floats—do not use this spring). Begin assembly by installing the angle bracket (1) onto the float arm. Insert the pivot pin halfway through, as shown on the left. Insert the counterspring into the hole in the float arm (2), and slide the pin through the spring and the other pivot, as shown in the center photo. Finally, make sure the other end of the spring is positioned in the bracket as shown (3), and install the float assembly into the bowl. Install and tighten the two 6-32x1/2-inch screws (some have self-locking threads) to 7- to 10-inch pounds. Jump ahead to **Step 17**, *High-Flow Seats.*

16 Install Side-Pivot Float

16—The floats used in side-pivot bowls have a counterspring to stabilize fuel level. Attach the spring over the float tabs as shown. Now install the float on the bowl mounting pin and slide the counterspring into the recess above the accelerator pump feed orifice (arrow). Locate the small "E" clip and snap it into place.

PERFORMANCE STEP

17 High-Flow Seats

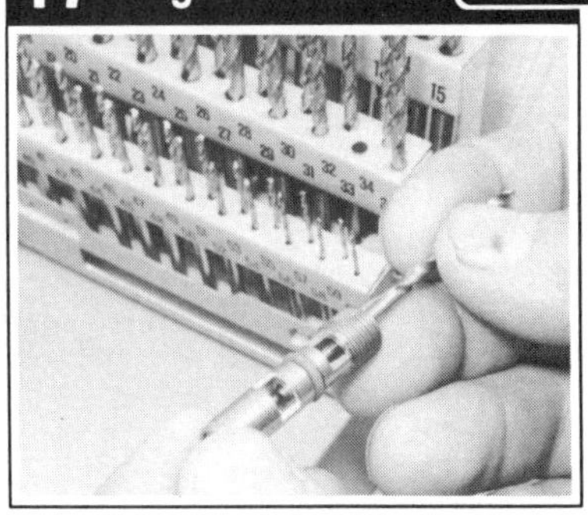

17—If your Holley will be used on a high-performance engine, you should check the orifice size of the needles-and-seats supplied in your rebuilding kit. They should be at least the same size as those removed during disassembly. If your Holley will be used for racing, don't use anything but genuine Holley needles-and-seats. They are machined to give optimum flow for every orifice size. Race applications should use at least .098-inch orifices, preferably .101- or even .110-inch on drag-racing engines, with at least 6psi fuel pressure supplied at all times during engine operation.

IMPORTANT STEP

18 Lubricate Bowl, Needle & Seat

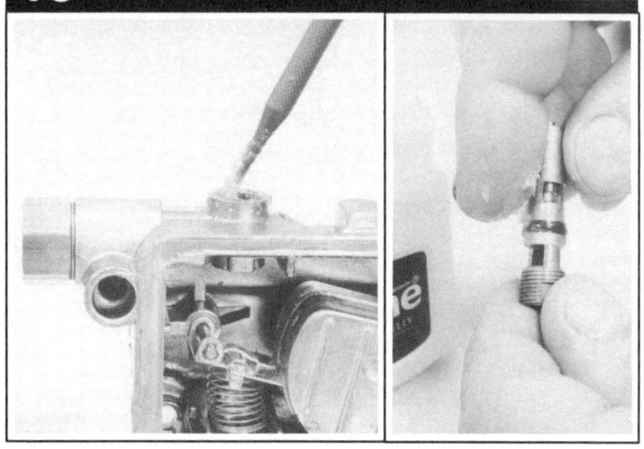

18—If you have LeMans bowls, skip ahead to **Step 22**, *LeMans Float Bowls*, otherwise perform this step. If the needles-and-seats in your rebuilding kit do not have new O-rings installed on them, locate the O-rings and install them now. Lubricate the bore and threads in the float bowl with petroleum jelly; a small probe or stick is helpful, but don't scratch the bore surface. Also apply lubricant to the threads and O-ring on the needle-and-seat. Installing the needle-and-seat without lubricant will damage the O-ring, resulting in unstable float level or flooding.

19 Install Needle & Seat

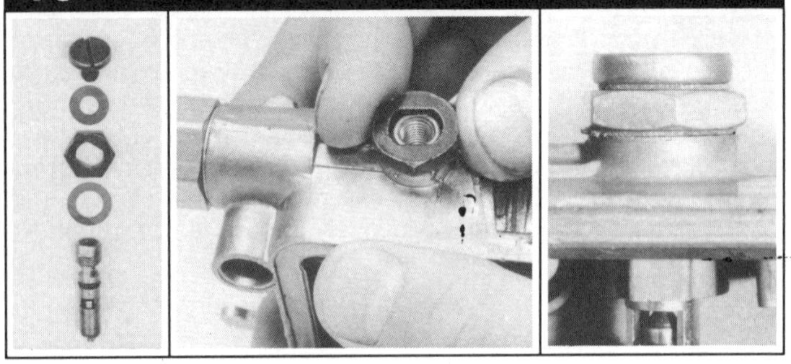

19—If the needles-and-seats in your carburetor kit have a solid head with a screwdriver slot, skip to the next step. Locate a hex nut, locking screw, and two gaskets needed to complete the needle-valve assembly. Begin screwing the needle-and-seat into the float bowl; use the hex nut as a "handle." If it gets hard to turn, STOP! Remove the valve assembly and carefully inspect the threads and bore for damage. Continue to screw the needle valve into the bowl until it's about half way down the adjusting nut (center). Then install the gaskets and lockscrew as shown (right) but don't tighten the assembly yet. Now jump ahead to **Step 21**, *External Float Height Adjustment.*

TORQUING FASTENERS

20 Non-Holley Needles-And-Seats

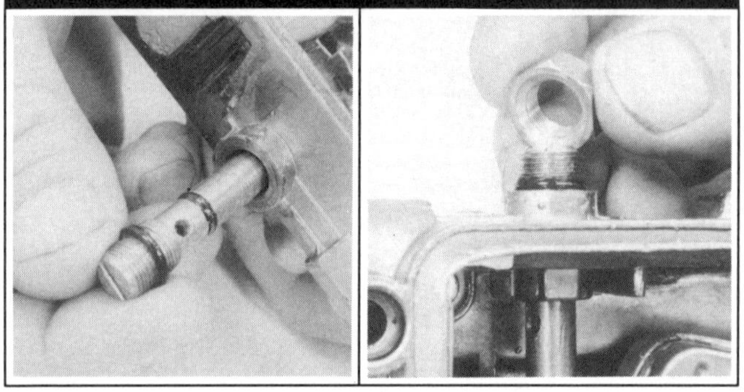

20—If you have needles-and-seats that look like this, they were not manufactured by Holley. (Note: These needle valves should not be used in performance applications.) To install these valves, first lubricate the valve, bowl threads, and O-rings. Then screw the assembly into the float bowl. If it gets hard to turn, STOP! Remove the needle-and-seat and inspect the threads and bore for damage. Screw the valve in until 1/4-inch protrudes from the top of the bowl. Then install the locknut with the inside taper toward the O-ring, but leave it loose. Next, invert the bowl and rotate the valve assembly until the float surface is parallel with the top of the bowl (see photos in next step). Prevent the needle-and-seat from turning and tighten the locknut to 30- to 40-inch pounds. Skip ahead to **Step 25**, *Primary Bowl Vent Valve.*

Vacuum Secondary Basics

The airflow requirements of an engine vary dramatically as engine speed increases. From very-low flow at idle with the throttles closed, to wide-open-throttle air-gulping demands at high speed, a carburetor must supply optimum air/fuel ratios at any engine speed and throttle position. Unfortunately, no carburetor can maintain accurate fuel metering when all four barrels are mechanically opened at low engine speeds. Low airflow velocity through all venturis is insufficient to generate a proper fuel-metering signal, particularly in the secondaries. The result is poor acceleration, backfiring, and engine bog.

An effective way to avoid these problems is to delay secondary opening until sufficient engine speed and airflow can be generated. The vacuum-secondary actuation system provides this function. This mechanism, despite the precision with which it functions, is relatively simple. It is composed of three elements: 1) an airflow pickup tube located in a primary throttle bore, 2) a vacuum diaphragm assembly, and 3) mechanical linkage that connects the diaphragm to the secondary throttle shaft. As airflow increases in the primaries, a greater vacuum signal is generated by the pickup tube. This signal is routed to the vacuum diaphragm that contains a counterspring and an opening-rate restriction or check ball. The diameter of the diaphragm, the tension of the counterspring, and the size of restriction all play a role controlling when and how fast the vacuum diaphragm—and the secondary throttles—are activated.

All Holley carburetors that have vacuum secondaries also have a secondary-closing linkage system. This essential safety device forces the secondary throttles closed when the primaries (connected to the gas pedal) are closed. *Don't eliminate, modify, or prevent this linkage from functioning! If it fails to work properly, it can leave the secondaries wide open when your foot is off the gas.* That is one experience you don't need!

This mechanical-secondary Holley (left) is a great performer on highly-modified race engines, but the vacuum-secondary carburetor (right) is the only practical choice for street engines.

There are several variations in diaphragm and actuating rod design, so be sure of the *list number* on your Holley when ordering a replacement.

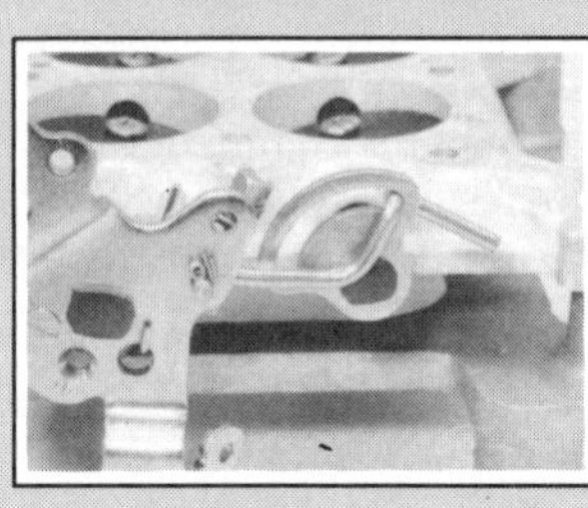

Vacuum-secondary Holleys are equipped with a secondary-closing linkage system. This essential safety device forces the secondary throttles closed when the primaries are closed. *Don't eliminate, modify, or prevent this linkage from functioning!*

TORQUING FASTENERS

21 External Float Height Adjustment

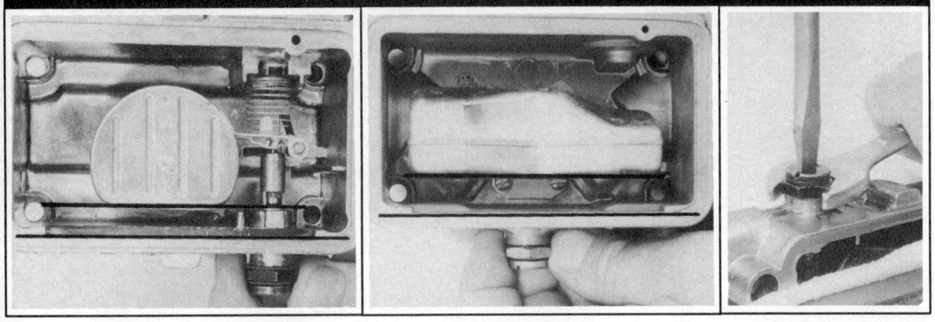

21—Invert the bowl and rotate the hex nut (the lockscrew must be loose) until the float surface is parallel with the inside (top) surface of the bowl. Now, prevent the adjusting nut from turning and tighten the lockscrew to 30- to 40-inch pounds. The final float level setting will be performed on the engine using the sight-plug window. Unless you have LeMans float bowls, skip ahead to **Step 25**, *Primary Bowl Vent Valve.*

PERFORMANCE STEP

22 LeMans Float Bowls

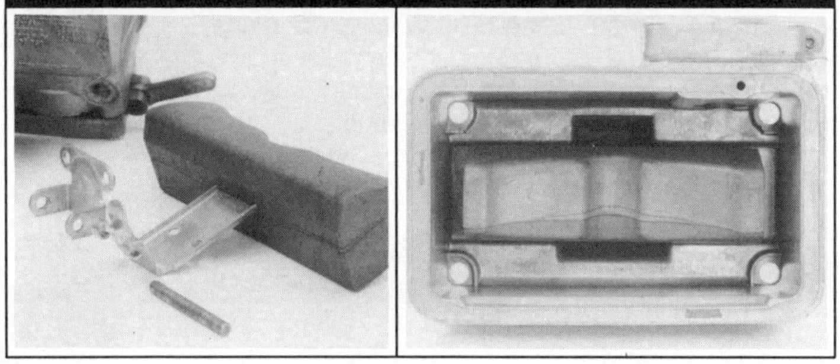

22—The nostalgic LeMans bowls are unique among the Holley modular line. General bowl assembly is as follows: Install the float as described in **Step 15**, *Install Center-Pivot Float.* Install the needle-and-seat and its copper seal ring in the end of the bowl and tighten to 70- to 90-inch pounds. Float height is adjusted by bending the tab on the float arm—you may have to remove the float or the needle-and-seat to bend the tab. Initially set the float to align with the horizontal ridges inside the bowl—further height adjustments may be necessary to obtain optimum performance. Install the access and side plugs. Skip ahead to **Step 26**, *Secondary Bowl Assembly.*

23 Internally-Adjustable Bowl

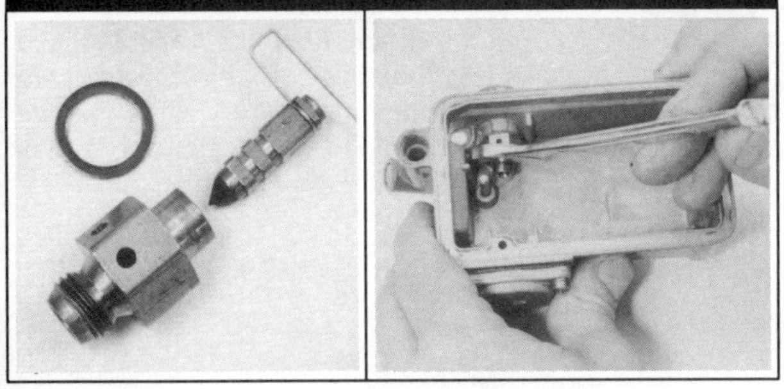

23—Locate the needle-and-seat assembly from your carburetor rebuilding kit. Also find the composite-metal gasket that is required to seal the seat assembly to the bowl (refer to page 126 for help in locating the proper gasket). Screw the seat (with gasket) into the bowl and tighten securely with an open-end wrench. (If you can get your torque wrench inside the bowl, tighten to 60- to 80-inch pounds for 5/16-24 seats and 70- to 90-inch pounds for 7/16-24 seats.)

PRECISION MEASUREMENT

24 Internal Float Height Adjustment

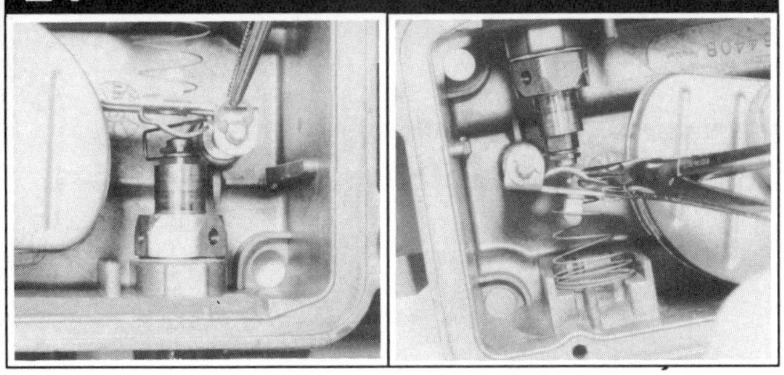

24—If the wire clip is not already installed on the new needle, install it now. Invert the bowl and place the needle into the seat. With the wire clip oriented as shown above, slide the float onto the pivot pin. *Make sure the wire clip goes between the adjusting tab and the float arm.* Now install the "E" clip. Refer to the instructions in your carburetor rebuilding kit for setting the proper float height. To adjust the float height, hold the float against the bottom of the bowl and carefully bend the float adjusting tab. *Don't force the needle into the seat; this can easily damage the Viton needle.*

25 Primary Bowl Vent Valve

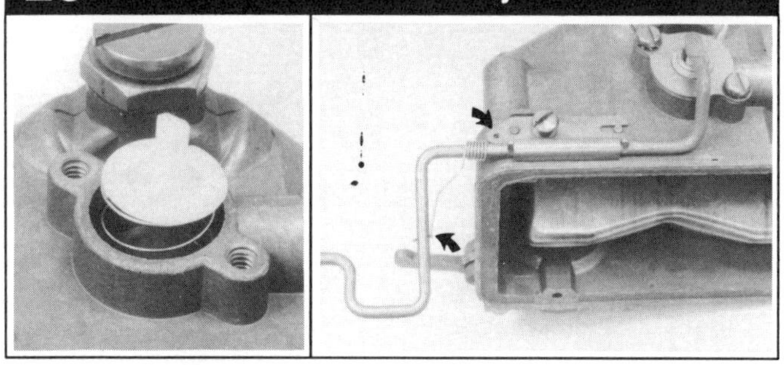

25—If your Holley has a throttle-arm actuated bowl vent, perform this step, otherwise, skip to the next step. Check your Work-A-Long sheet for notes you may have made during disassembly. The steps for installing the vent generally are: 1) install a new rubber gasket onto the plastic valve, or if supplied, use a new valve; 2) install the spring and valve into the bowl; 3) install the cover and two screws (usually 6-32x1/4-inch—torque to 7- to 10-inch pounds); and 4) install the arm, holding bracket, and spring as shown (arrows show spring attaching points), then tighten the mounting screw (usually 8-32x1/4-inch—torque to 12- to 18-inch pounds).

26 Secondary Bowl Assembly

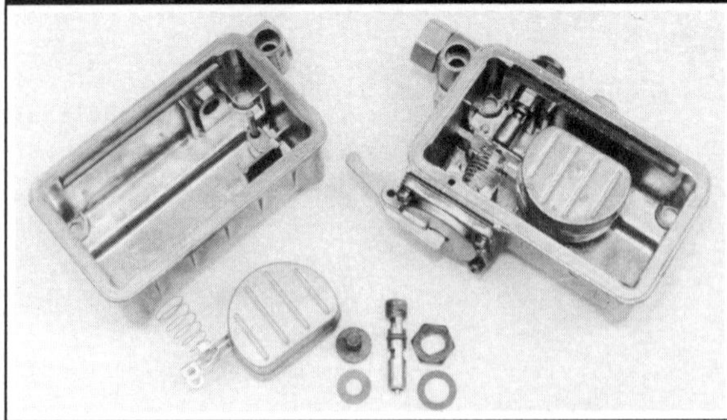

26—Unless you're working on a two-barrel, jump back to **Step 2**, *Begin Float Bowl Assembly* and repeat the process with the secondary bowl. If your Holley has mechanical secondaries, like the 4165, 4500, and several 4150s and 4160s, install the secondary accelerator pump in the same fashion as the primary. If the secondary bowl has internally-adjustable floats, refer again to the specifications in your carburetor kit for setting the proper secondary float height.

27 Install Fuel Baffle

27—Primary and secondary bowl assembly is nearly complete. If your Holley does not have side-pivot float bowls, skip to the next step. Side-pivot bowls are equipped with plastic fuel baffles. The baffles come in right- and left-hand versions, making it impossible to install them in the wrong bowl. However, if your carburetor has a metering plate instead of metering block on the secondary side, you may be wondering why only one baffle has an extended leg. The leg is designed to rest against the metering block and keep the baffle in place. However, a metering plate sticks into the float bowl, making the plastic leg unnecessary. If your Holley is a model 4150 or 4165, both baffles will have an extended leg. Install the baffles now by pressing them in position as shown.

28 Install Sight Plug

28—If your Holley has internally-adjustable floats, skip to the next step. Installing the sight plugs is the final bowl-assembly procedure. Locate the sight plugs and the composite-metal gaskets in your carburetor kit (for help refer to page 126). Install a plug and gasket in each bowl and tighten to 90- to 110-inch pounds.

29 Vacuum-Secondary Diaphragm

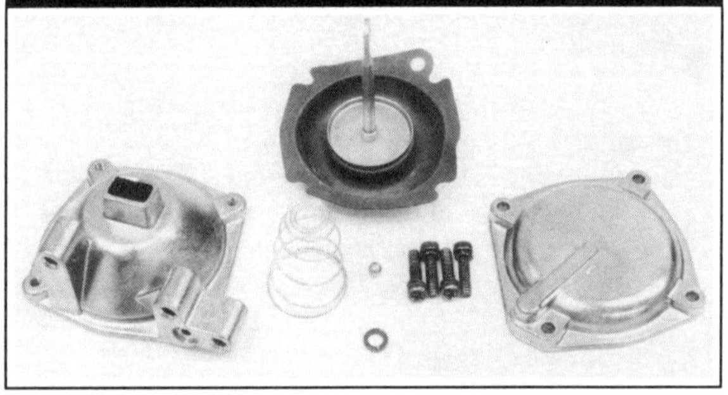

29—If your Holley has mechanical secondaries, skip ahead to **Step 38**, *Assembling Primary Metering Block*. Locate the major components of the vacuum-diaphragm assembly, including the upper and lower halves, the rubber diaphragm, and the color-coded return spring. Refer to your Work-A-Long sheet to determine if your diaphragm assembly uses an opening-rate restriction or ball check. I recommend that you reassemble your housing with the stock restriction; leave the fine tuning for later, when the carburetor is reinstalled on the vehicle and the engine is running properly.

3-BARREL CARBURETOR 3 bbl

30 Rebuilding The 3-Barrel Secondary Diaphragm

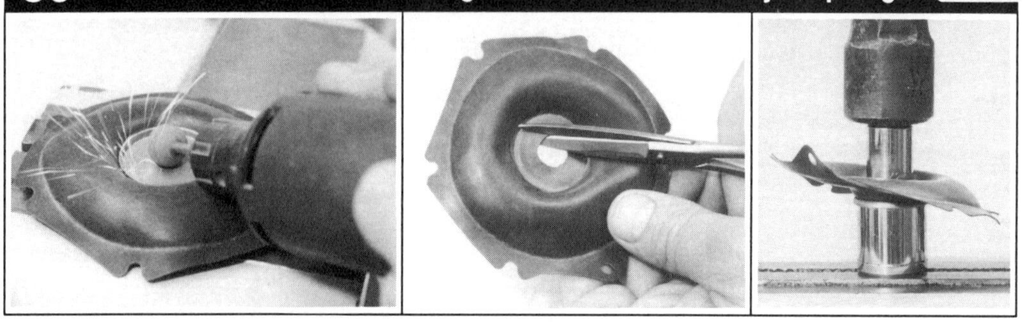

30—The 3-barrel (model 3150/3160) uses a large diameter diaphragm and housing. A replacement diaphragm is not available from Holley, however, **part 135-6** consists of a usable rubber diaphragm, but it must be removed from its metal stem and installed on the 3-barrel stem. First, slightly pry apart the upper and lower washers on the 3-barrel stem and remove and discard the old rubber. Next, grind the rivet end off the new diaphragm stem and remove the upper washer and diaphragm. Carefully enlarge the center hole in the rubber diaphragm with scissors to about 3/8- to 1/2-inch. Stretch the new diaphragm over the upper washer on the old stem. Do a trial assembly of the housing (lubricate the cover screws with petroleum jelly) on the main body to make sure the stem aligns with the arm on the secondary throttle shaft. Then stake the washers back together by placing a 13/16-inch socket under—and a 1/2-inch socket on top of—the washers. Lightly strike the upper socket to force the washers to grip the new diaphragm. Finally, reassemble and test the housing for leaks.

PERFORMANCE STEP

31 Quick Change Cover

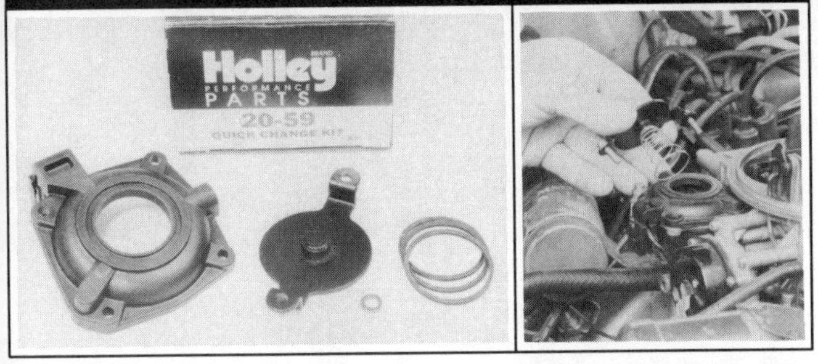

31—Holley has developed a quick-change kit, **part 20-59**, that allows changing secondary springs in a few, hassle-free minutes. The kit includes a special top cover for the diaphragm assembly. It works with Holley's standard assortment of diaphragm springs (**part 20-13**). If you wish to use this accessory, it easily installs in place of the stock cover, just follow the instructions included with the kit.

MASTER MECHANIC TIP PRO TIP

33 Lube Screw Threads

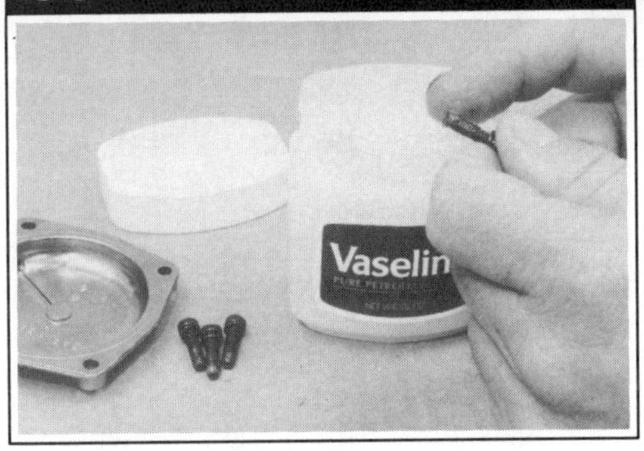

33—Before you proceed, make sure you lubricate the threads of the four 8-32x5/8-inch cover screws with petroleum jelly. If you install these screws dry, the threads can "grab" the rubber diaphragm, causing it to wind around the screw threads. If this happens, the diaphragm will almost certainly leak. A little petroleum jelly lube on the screws will prevent this problem.

CRITICAL INSPECTION

35 Tighten Cover Screws

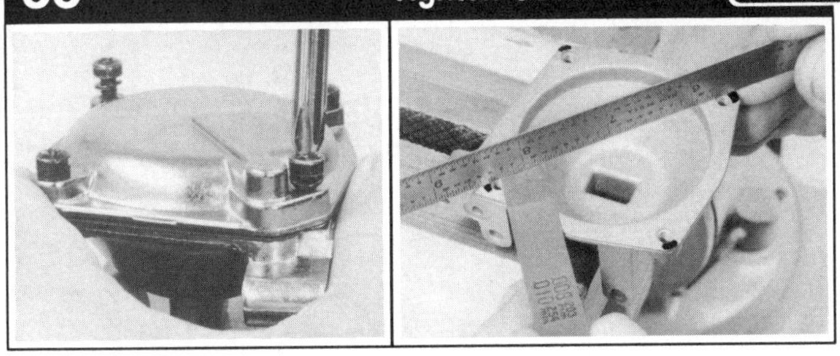

35—As you tighten the cover screws, carefully observe the diaphragm. It should not be dragged into the housing. If this happens, remove the screws and apply a little more lubricant. If the gasket appears to severely squash out as the cover is tightened (a little flattening is normal), remove the screws and cover, and inspect the .008- to .010-inch lower housing tabs to make sure they are intact. Without these tabs the cover would slice the rubber in half. If the tabs are missing or damaged, you'll need to obtain a replacement lower housing.

32 Install Rubber Diaphragm

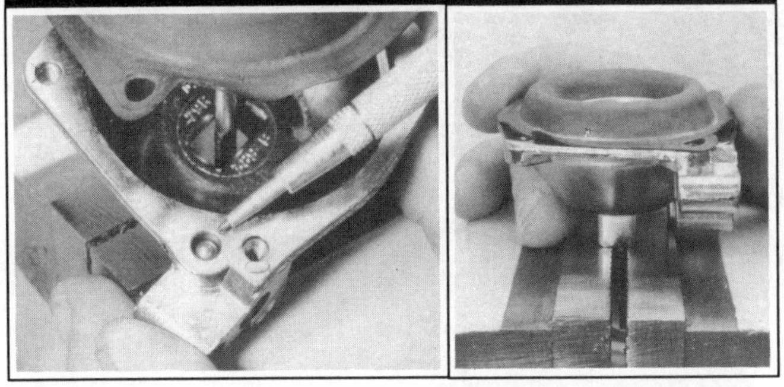

32—While it may seem like a trivial task, assembling the vacuum diaphragm housing often is done incorrectly, resulting in leaks that prevent the secondaries from opening. Here is a technique that is simple and reliable. First, if an opening-rate check ball is used in the lower housing, install it now. Then, dress the vacuum diaphragm so that it is under the least stress, with the edges laying relatively flat (see above). Now, place the diaphragm in the housing, align the holes properly, and grip the stem in the vise so that the diaphragm and housing sit relaxed in/on the vise jaws, as shown on the right.

34 Install Spring And Cover

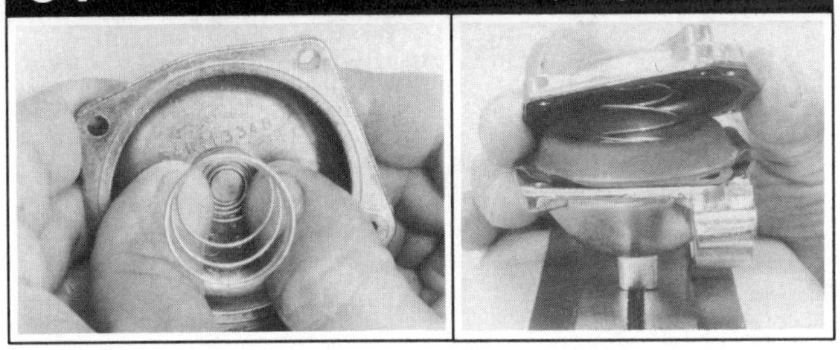

34—Attach the diaphragm spring in the top cover by snapping the small end of the spring onto the short button in the center of the cover. Then, position the spring onto the metal center of the rubber diaphragm, and lower the cover into place. Make sure the screw holes and the vacuum port hole line up. Install the four screws (make sure they are lubricated) and tighten finger tight. Note: Some Holleys have a manual choke cable bracket or ground wire attached to screws on the vacuum diaphragm. If your carburetor has either of these attachments, check your Work-A-Long sheet for their location and install the accessory bracket now.

TORQUING FASTENERS

36 Torque Screws

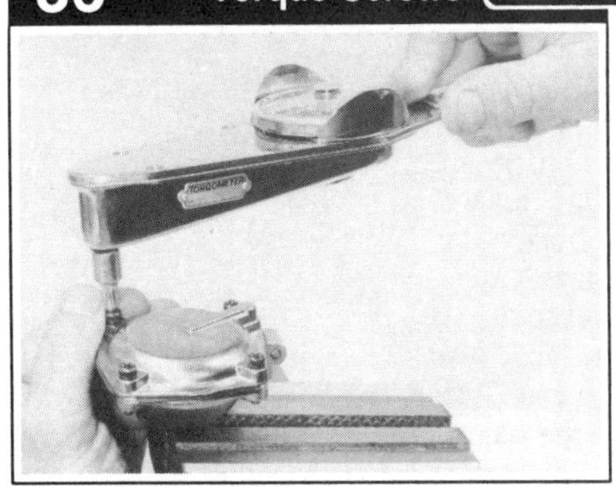

36—When the screws are snug, regrip the housing in the vise on the lower cover as shown here, but don't tighten the vise too tight, you could crack the housing! Tighten the cover screws to 12- to 15-inch-pounds.

IMPORTANT STEP

37 Test Vacuum Housing

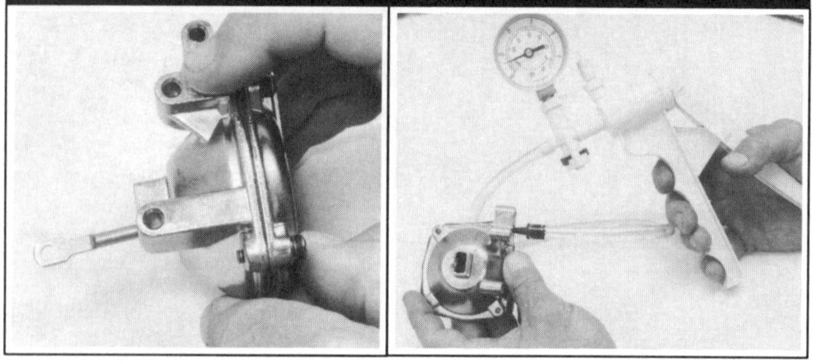

37—The housing should now be tested for leaks. The common method is to compress the diaphragm and seal off the vacuum hole with your finger (left). A more "sophisticated" method is to use a vacuum pump. In either case, the diaphragm stem (and/or the vacuum gauge) shouldn't budge, even after several seconds. If the housing leaks, disassemble the housing and carefully inspect the diaphragm for tears and the housing halves for damage. If none are visible, try reassembling the housing. If it still leaks, you'll probably have to obtain a replacement rubber diaphragm.

38 Assembling Primary Metering Block

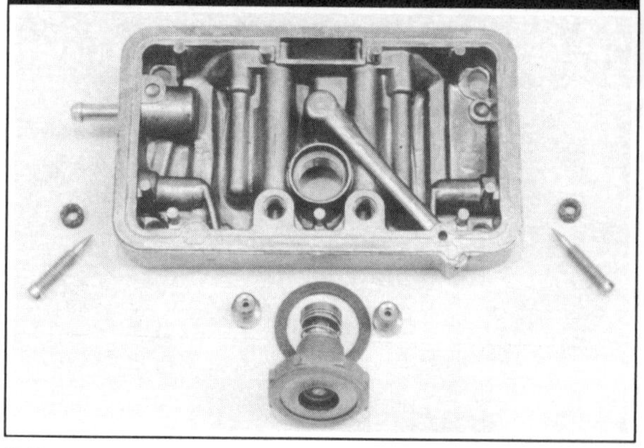

38—Locate the primary metering block. If your Holleys uses dual metering blocks, the primary block has a majority of the following characteristics: 1) a power valve, 2) idle screws, and 3) a vacuum takeoff. (If you followed the disassembly chapter, you marked a small "s" on the secondary metering block.) Also locate the idle needles and their cork gaskets. Refer to your Work-A-Long sheet and locate the primary jets, the replacement primary power valve and gasket.

39 Installing Idle-Needle Screws

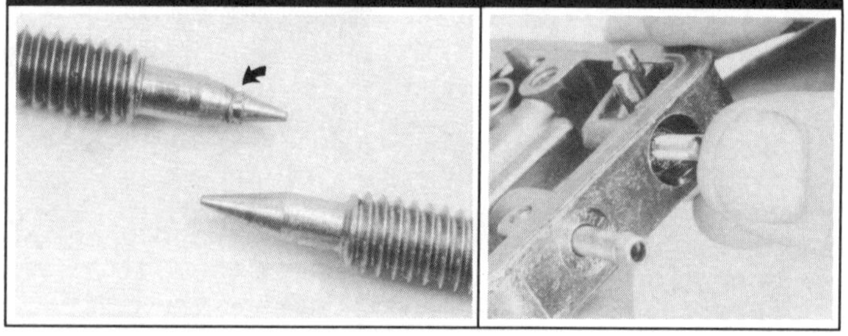

39—First, inspect both idle needles for gouges on the pointed end. If either screw does not have a smooth tapering surface (arrow), replace it, since it will be difficult to obtain a stable idle. Next, screw one idle needle into its cork/composition gasket about 1 to 1-1/2 turns. Then lubricate the gasket and its metering-block counterbore with petroleum jelly. Push the idle needle and gasket into the bore until the gasket is fully seated. Then screw in the needle until it just seats—DON'T TIGHTEN. Repeat this process with the other idle needle. Refer to your Work-A-Long sheet and back out each idle needle the same number of turns you recorded during disassembly. If you didn't record this information, unscrew each needle 1-1/2 turns.

TORQUING FASTENERS

40 Install Main Jets

40—Install the primary jets in the metering block using a screwdriver with a blade at least as wide as the outside diameter of the jets. Then torque them to 30- to 40-inch pounds.

IMPORTANT STEP

41 Power Valve Gasket

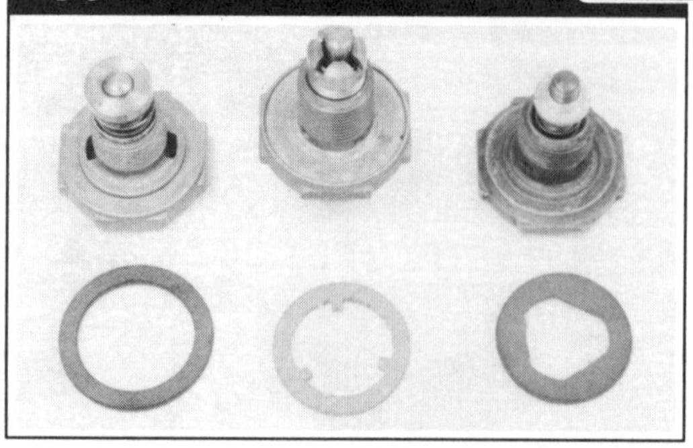

41—Power valves vary in design and so do their gaskets. Valves with no alignment ridges or with an outside ridge (right and center) use gaskets that contain inside alignment tabs, but some of these gaskets are quite restrictive. Power valves that have an inside ridge use a larger diameter gasket with no tabs. To sort this all out and discover other facts about power valves, refer to the sidebar, *Power Valve Basics*, on page 52.

TORQUING FASTENERS

42 Power Valve

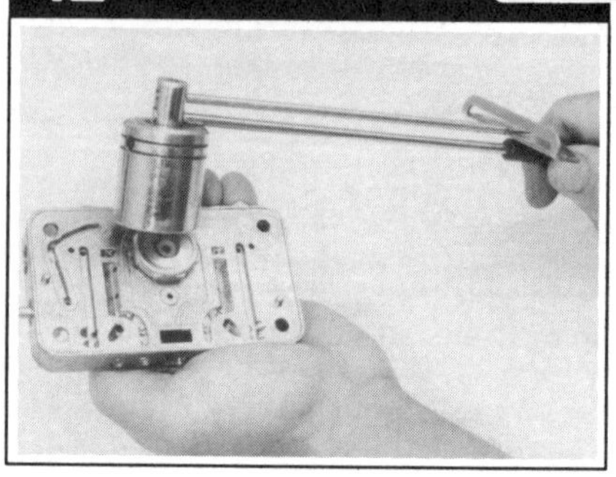

42—Screw the primary power valve into the primary metering block and make sure the correct gasket is centered under the head of the valve. Torque power valves with multiple small feed holes to 40- to 50-inch pounds; power valves with two large feed passages should be torqued to 100 inch-pounds. After the valve is tightened, recheck the gasket to make sure it has not slipped out of place.

43—If your Holley is equipped with an accelerator pump fuel-transfer tube that fits into a counterbore in the primary metering block, perform this step; otherwise skip to the next step. This brass tube and O-ring assembly provides a more positive seal for accelerator pump fuel than the usual flat surface and gasket, but only if it is assembled correctly! Locate new O-rings in your carburetor kit—refer to the *Common Holley Gaskets And O-Rings* illustration on page 126 for help in finding the right O-rings. Inspect the tube, O-rings, and the metering-block counterbore for scratches, damage, or metal burrs. Install the O-rings on the brass tube. Lubricate the tube, O-rings, and counterbore with a little petroleum jelly. Using a side-to-side wiggling motion to press the tube into the metering block. It should slip in smoothly yet firmly. Don't force it! Just a nick on the surface of the O-ring can cause a fuel leak.

43 Install Pump Transfer Tube

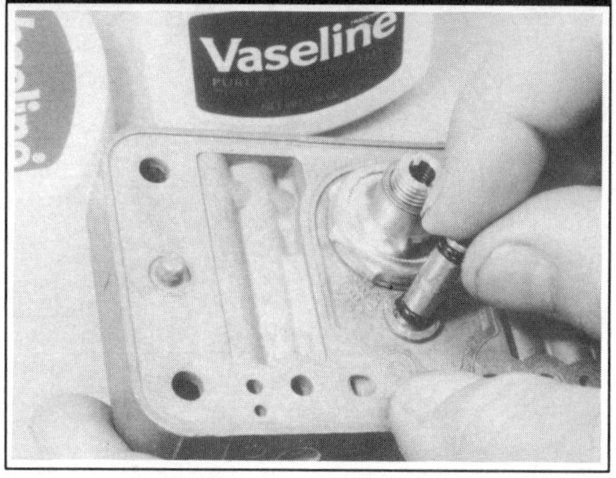

Repairing Loose Throttle Shafts

Carburetors that have been in service for a long period of time, or that have improperly adjusted throttle linkage, may develop excessive throttle-shaft clearance. Because Holley has used several throttle-plate designs, they do not publish a numerical specification for "excessive" clearance. Instead, they suggest that if any of the following problems arise, it usually indicates a worn throttle plate or shafts: 1) variations in idle speed when the throttle is opened and returned to idle position, 2) whistling or other abnormal air noises at idle, and 3) inability to achieve the correct or stable idle air/fuel mixture due to air leaks around the shafts and/or misalignment in the butterflies in the bores.

The procedure for repairing loose throttle shafts depends on the design of the throttle plate. Many late-model Holleys use very loose throttle-shaft bores and depend on Teflon strips—wrapped around the shafts in several places—to seal the shafts and prevent air leaks. Installing new Teflon strips, and replacing the throttle shafts if they are worn, will often correct air leakage in these plates. Earlier designs do not use Teflon strips, instead, they rely on a snug fit between the shaft and the plate to prevent air leaks. These earlier plates, and late-model plates that can't be repaired with new Teflon strips, can be "rebushed" with special kits available from custom carburetor shops, like The Carburetor Shop and Mike Jones Carburetion. Some custom shops also rebuild throttle plates shipped to them from their customers.

Repairing early or late throttle plates requires removing the butterflies. Before the screws are loosened, the butterflies should be scribed with an awl along the throttle shafts to indicate their precise position. If they are not reassembled in the same position, they may leak air and produce an unstable idle. And if they are flipped over or rotated 180-degrees, they won't seal at all!

When the butterflies are marked and the throttle body is mounted securely in a fixture, carefully remove the staked portion of the butterfly screws with a small file or precision grinder. When the staking has been removed, the screws can be carefully unscrewed using a back-and-forth motion. IMPORTANT: The screwdriver must fit the screw heads tightly or the heads can strip, leaving drilling as your only recourse for removal.

Holley recommends always using new butterfly screws to reassemble throttle bodies. They are included in some Holley rebuilding kits and are available separately as **part 5R635** (brass) and **part 5R593** (steel). Begin assembly by repositioning each butterfly in precisely the location from which it was removed. Install all butterfly screws, but do not tighten them. To make sure that each butterfly is precisely centered in the throttle bores, unscrew the primary and secondary idle-stop screws, and allow the throttle shafts to close fully. Apply a slight twisting and side-to-side force to each throttle shaft to center the butterflies in their bores. When the throttle plates are completely seated in their bores, tighten the screws. *Restaking the screws prevents them from loosening, and is an absolutely essential reassembly step.* *Loose butterfly screws will cause the throttle to stick open.* Make sure you support the throttle shaft before staking the screws to prevent bending the shafts. Perform a careful final inspection to ensure that the throttle shafts open and close smoothly and do not stick or bind anywhere within their full range of motion.

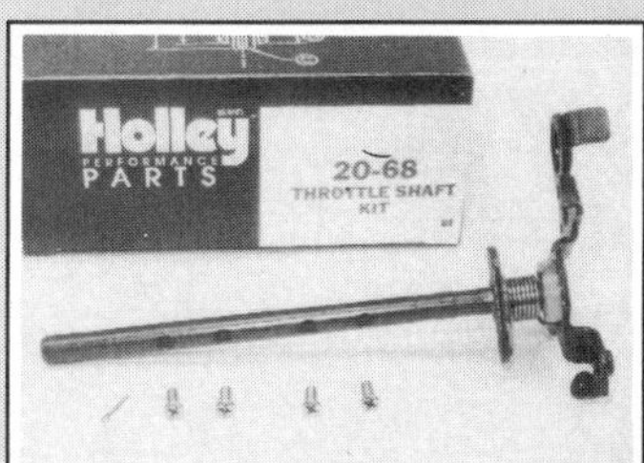

Holley offers several throttle-shaft repair kits that include a new throttle shaft, Teflon strips (if used), new butterfly screws, and instructions.

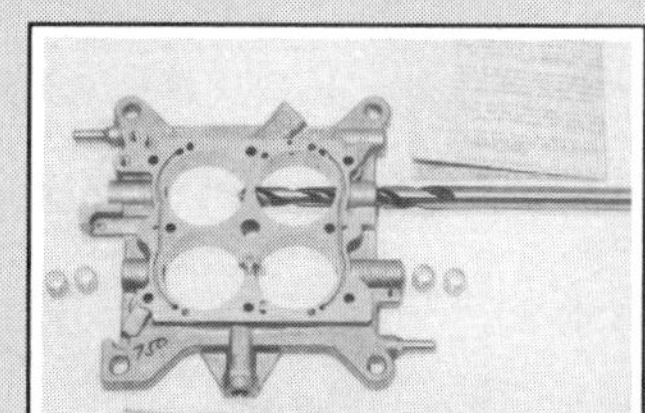

Several custom shops offer rebushing kits; this kit from The Carburetor Shop includes a unique drill with a guide stem.

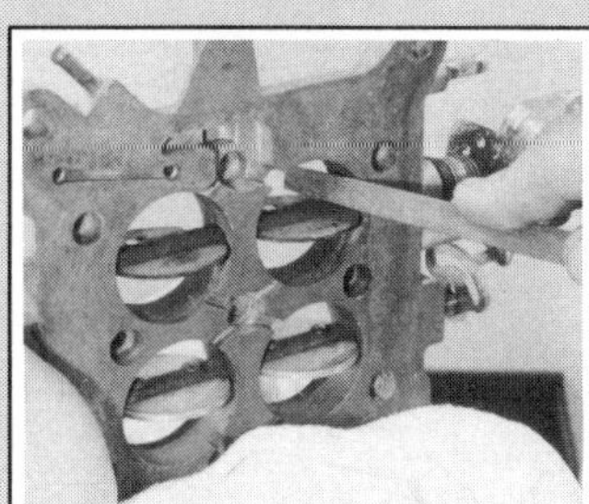

After the position of the butterflies is marked with an awl and the throttle body is mounted securely in a fixture, carefully remove the staked portion of the screws with a small file.

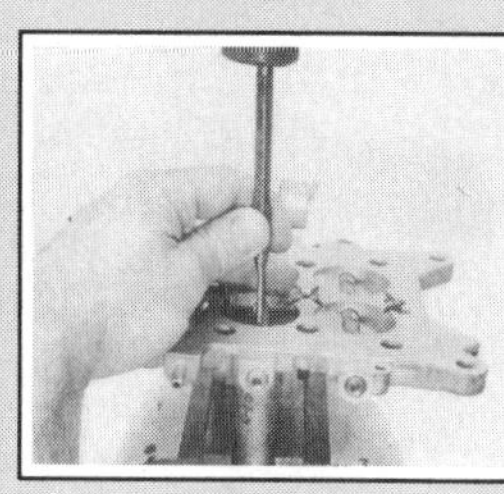

After the butterfly screws are tightened, they must be staked to prevent loosening. Always support the shaft from underneath when staking the screws.

44 Secondary Metering Block

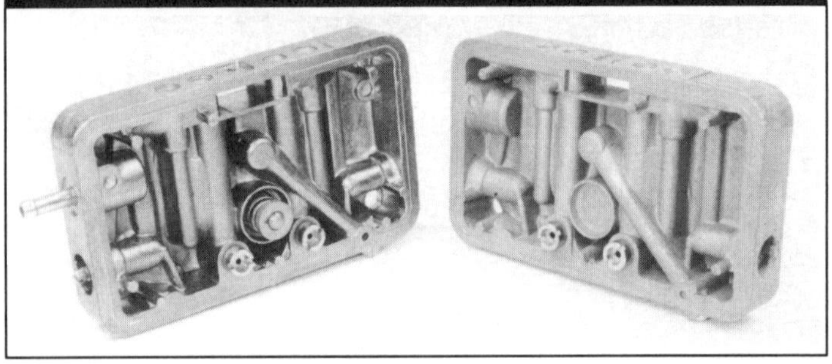

44—Some Holley carburetors use a secondary metering block instead of a metering plate (3150's, 4150's, 4165's, 4500's, etc.). If your carburetor falls into this category, jump back to **Step 38**, *Assembling Primary Metering Block,* and repeat the assembly procedures on the secondary side. Refer to your Work-A-Long sheet for jet size, power-valve number and idle-needle preset (if applicable).

45—Locate the throttle plate assembly. (Note: If engine idle was unstable or the carburetor whistled at idle, refer to the sidebar *Repairing Loose Throttle Shafts* on the previous before you proceed.) Most carburetor kits come with several throttle plate gaskets. Follow these tips for selecting the proper one: 1) first, try matching the gasket up with the old one, 2) the throttle-bore holes in the new gasket should very closely match the bores in the throttle plate and main body, 3) all holes that feed discharge ports and connecting passages should line up, and 4) if some holes/passages don't line up (arrows), it's OK as long as there is no mating hole in the main body, or if there is a mating hole, it is a dead-end passage.

45 Selecting The Correct Throttle Plate Gasket

Accelerator Pump Basics

Carburetors are not perfect, even Holleys. In fact, they have several "flaws" that only are overcome with the addition of compensatory devices or systems. One almost universal problem occurs when the throttles are opened quickly, especially at low engine speed. A short time elapses before the main system can begin delivering fuel through the boosters. This delay can cause a severe bog and backfiring unless additional fuel is "injected" into the airstream to return the air/fuel ratio to the correct balance. This short-term fuel injection system is commonly known as the accelerator-pump circuit.

The operation of the accelerator pump system is easily understood when it is broken down into its three basic elements. The first is the *cam and actuating lever* that converts the rotary motion of the throttle shaft to the up-down motion needed by the pump diaphragm. The second is the *main pump assembly*, consisting of a housing cover, diaphragm, spring, and check valve. The third element in the system is the *discharge nozzle* and anti-pullover check valve located near the top of the venturis. The system functions like this: When the throttle lever rotates open, a cam mounted on the throttle arm actuates the pump lever that, in turn, compresses the pump diaphragm. This forces fuel to flow from the discharge nozzle into the venturis.

The accelerator-pump is located on the bottom of the primary float bowl on vacuum-secondary and "center-squirter" carburetors. "Double-pumper" and Dominator carburetors use dual accelerator pump circuits, one on each of the primary and secondary float bowls. The secondary-side accelerator pumps are driven by the secondary throttle shafts, using separate cams and actuating levers.

Holley offers a range of nozzle sizes and designs, two sizes of pump diaphragms, and several actuating cams to tune the accelerator pump discharge volume and rate to suit any application. For more information on tuning the accelerator pump circuit, see Chapter 11, *Performance And Economy Tuning.*

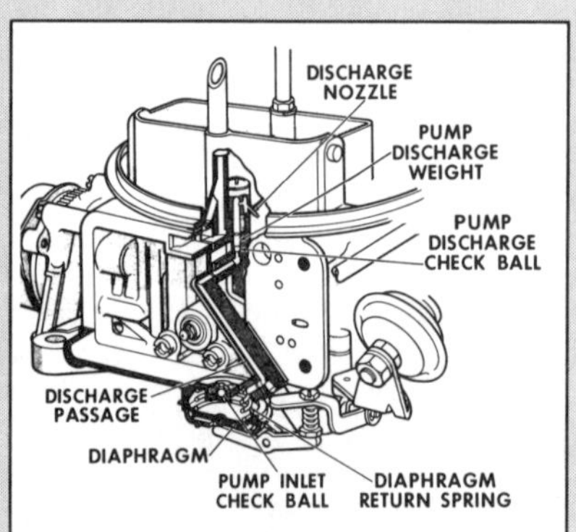

Here is the path of fuel from the float bowl, into the accelerator pump, through the metering block, into the main body, and finally discharged from the nozzle.

The accelerator-pump system functions like this: When the throttle lever rotates open, a cam mounted on the throttle arm (1) actuates the pump lever (3) that, in turn, compresses the pump diaphragm (2). This forces fuel to flow from the discharge nozzle into the venturis.

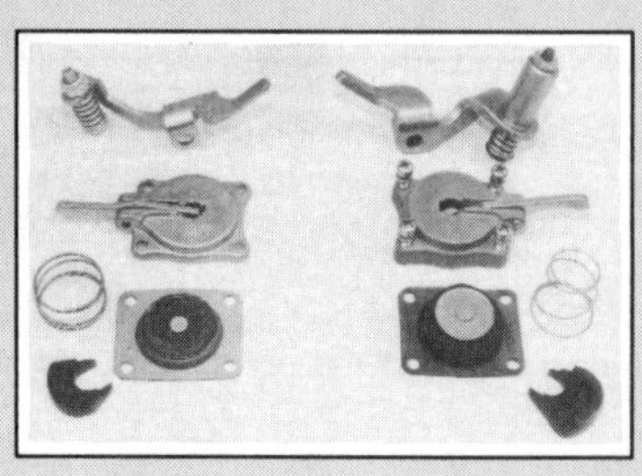

The accelerator pump circuit can be tailored for various engine and vehicle combinations by varying the pump cams, squirters, and pump capacity.

46 Install Throttle Plate

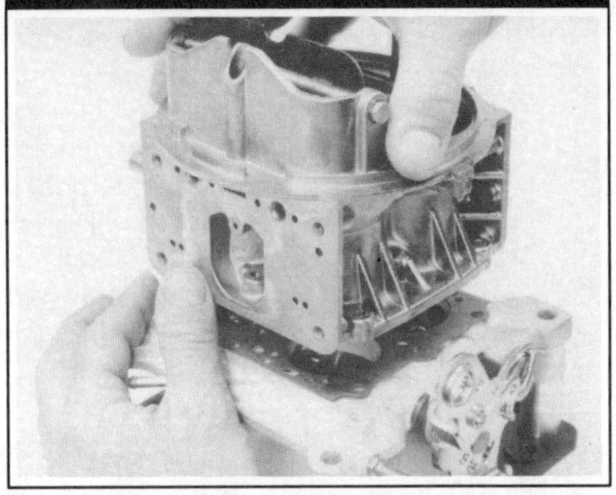

46—If you have a 4165—or any Holley that has mechanical-secondary linkage with a bellcrank idler-arm—proceed to the next step. Otherwise, with the correct gasket on the throttle plate, lower the main body onto position. *Make sure that the primary butterflies are located under the primary venturis.* Now, skip to **Step 48**, *Throttle Plate Screws.*

PERFORMANCE STEP

47 Bellcrank Secondary Linkage Arm

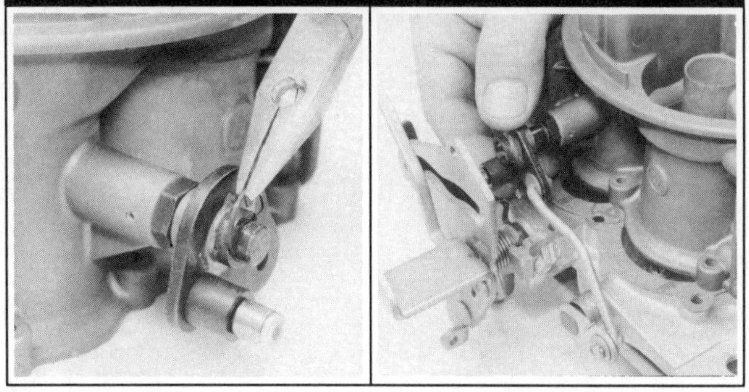

47—If the bellcrank cam-follower arm has been removed from the secondary connecting link, install the arm on the main body now and lock it in place with a large "E" clip (left photo). If the follower arm is still attached to the secondary link (I suggested leaving it attached during disassembly), slide the arm onto the main-body pin as you lower the main body onto the throttle plate (right photo) then install the "E" clip. In either case, make sure the correct gasket is positioned on the throttle plate before you begin. If the connecting link was disconnected, reconnect it with washers and new cotter pins.

TORQUING FASTENERS

48 Throttle-Plate Screws

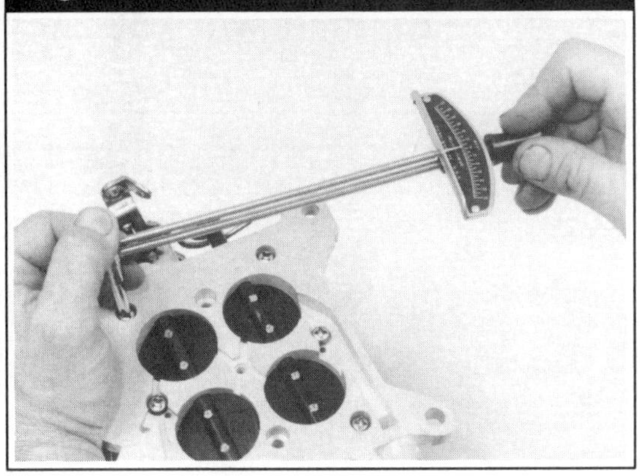

48—Grasp both the main body and throttle plate, and flip them over. Refer to your Work-A-Long sheet for the correct number of 12-24x3/4-inch throttle plate attaching screws and their location. Lubricate the threads and install the screws. Before you tighten them down, open the butterflies and make sure that both the throttle plate and gasket are lined up with the main-body throttle bores. No gasket material should protrude into the bores. Finally, tighten the screws to 30- to 50-inch pounds.

49 Pump Nozzle Check Valve

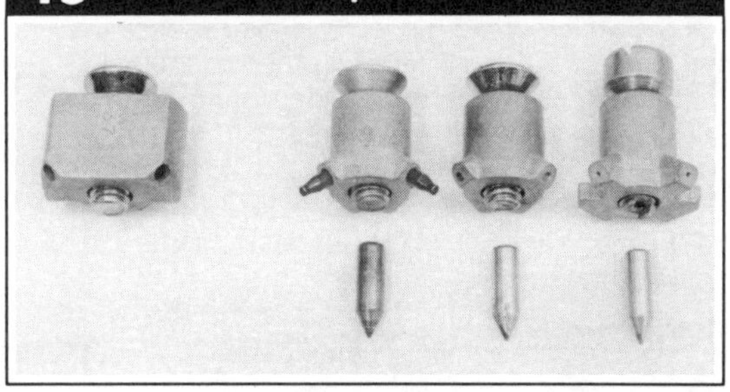

49—Refer to your Work-A-Long sheet to determine if a check valve was located under the primary discharge nozzle. If there was no check needle or ball-and-weight, and your carburetor did not have anti-pullover discharge nozzles—as shown on the left—it was assembled improperly. If your Holley is equipped with any one of the standard nozzles—like those on the right—a check valve must be used to prevent fuel from siphoning from the nozzle. Although some carburetor kits include a small check ball to replace Holley's needle valve, a single ball (or even two stacked on top of one another) may not have sufficient weight to prevent siphoning. If you need to obtain a replacement check needle, stick with a genuine Holley part.

50 Install Check Valve

50—If your Holley uses a check needle, as explained in the previous step, install the needle valve by dropping it—pointed end down—into the nozzle discharge passage in the main body.

PERFORMANCE STEP

51 High Flow Nozzle Screw

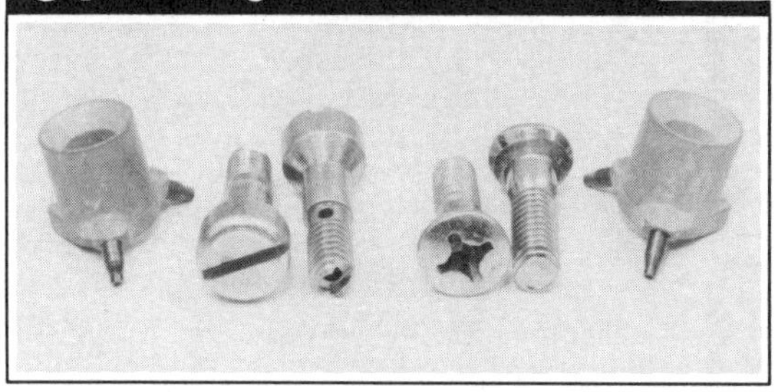

51—Some high-performance Holleys use a high-flow nozzle screw, **part 26-12**, that can be identified by a straight-slot head (instead of Phillips). If your Holley will be used for racing, or is a center-squirter, double-pumper, a Dominator, or other high-performance carburetor, you should use high-flow screws to ensure that the discharge nozzle orifices, not the passages below the nozzle, are the restrictions that control fuel flow.

52 Stick Gasket On Nozzle

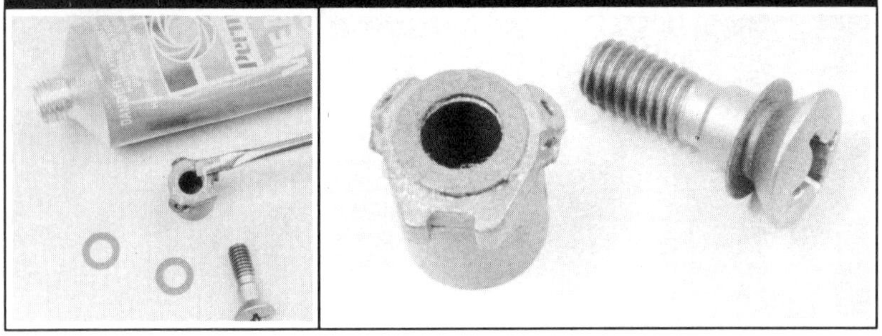

52—Locate the primary discharge nozzle (if your carburetor is a double pumper, refer to your Work-A-Long sheet for the primary nozzle number), the attaching screw, and two new paper gaskets from your carburetor kit. Installing the discharge nozzle and its lower gasket can be tricky on carburetors with airhorns. The job is easier when the gasket is held on the nozzle with a little weather-strip or other high-tack adhesive. Position the other gasket on the screw as shown. If you have a 4165 or 4175 spreadbore carburetor with anti-pullover nozzles, skip ahead to **Step 54**, *Installing The Anti-Pullover Nozzle.*

53 Install Discharge Nozzle

53—With the lower gasket held in place on the nozzle, carefully lower the nozzle onto the main body with needlenose pliers or hemostats. The locating slot should line up with the tab on the main body. When the nozzle is in place, lower the screw and gasket into the nozzle. Then tighten the attaching screw to 48- to 52-inch pounds. Skip ahead to **Step 55**, *Secondary Nozzle.*

Booster Venturi Basics

The main venturis in nearly all Holley carburetors contain smaller *booster venturis* of various shapes and sizes. A booster venturi amplifies the vacuum signal transmitted to the main metering system and provides more accurate air/fuel metering, especially at low engine speeds. Without a booster venturi, the transition from low- to mid-range engine speeds would produce flat spots, hesitation, and generally reduce driveability. In addition, some booster venturis incorporate tabs or wings that redirect airflow and allow the carburetor to optimize fuel distribution for a specific engine and manifold combination.

Holley carburetors use several distinct boosters, and each design offers specific advantages and disadvantages. One of the most important considerations in booster design is its overall effect on carburetor airflow. Generally, boosters that develop the strongest fuel delivery signal also offer the best low-speed performance, throttle response, and fuel economy. Unfortunately, improved signal strength is produced when the booster "interferes" with main-venturi airflow. So, high-signal boosters almost always cause the greatest airflow restriction and produce the lowest high-speed power. As you might expect, more streamlined boosters usually generate less main-system signal and offer less throttle response (especially when used on small displacement engines), but they present the least airflow restriction and can optimize high-speed horsepower.

The annular discharge booster is a relatively recent design. It is used on several "emissions" carburetors, such as the model 4180/4190, and several new high-performance designs, including some Dominator models. The annular-discharge booster offers an exceptionally strong low-end signal combined with excellent fuel distribution and atomization. The main-system signal generated by these boosters is so strong that main jet size must be reduced by several Holley numbers. The main disadvantage is, predictably, considerable airflow restriction, due mainly to their large dimensions.

Holley carburetors utilize several styles of booster venturis. There is often a direct trade-off between airflow capacity and throttle response among various designs. Some performance shops, like The Carburetor Shop, can remove and install boosters for special applications.

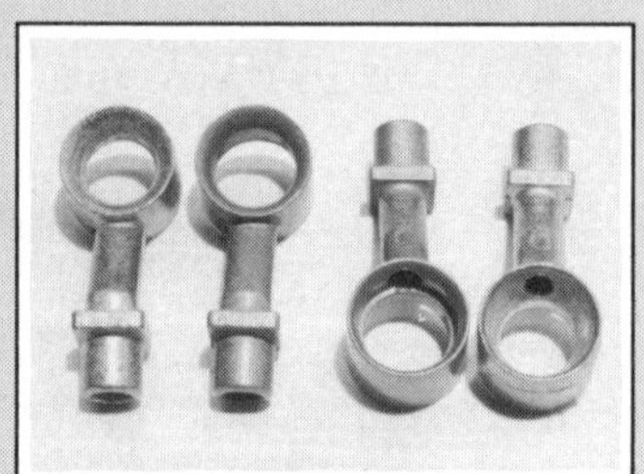

Boosters that develop the strongest fuel delivery signal (outside pair) produce the greatest restriction. More streamline boosters (inside pair) improve airflow and horsepower, but reduce throttle response.

This kit from Holley, *part 34-9*, converts Dominators to annular discharge boosters. Unfortunately, other Holley carburetor models will not accept these replacement boosters.

4165/4175 SPREAD-BORE CARBURETORS

4165

54 Installing The Anti-Pullover Nozzle

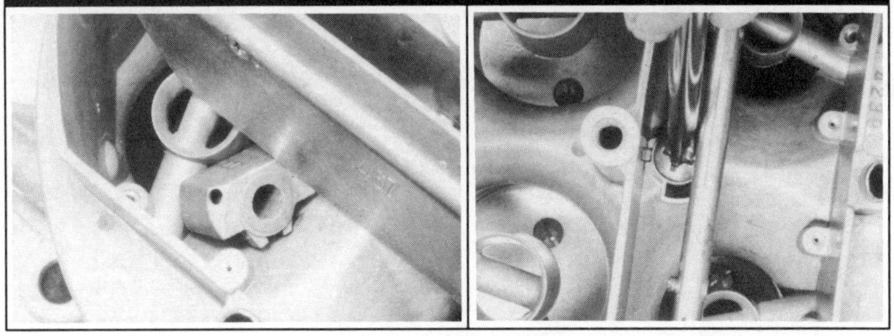

54—Position the discharge nozzle upside down next to the primary booster venturi as far forward as possible with the alignment tabs facing away from the booster. The choke blade should now close over the nozzle. If the nozzle doesn't clear the choke, just wiggle the nozzle back and forth. When it finally clears the choke flap, use needlenose pliers or hemostats to turn it rightside up (make sure the gasket is still in place) and set it on the main body. Align the slot on the booster with the tab on the main body. Position the other circular gasket on the nozzle screw and install the screw in the nozzle assembly. Tighten the screw to 48- to 52-inch pounds.

PERFORMANCE STEP

55 Secondary Nozzle

55—If your Holley has mechanical secondaries, like this 4165, install the secondary discharge nozzle just like the primary, although it is not necessary to glue the gasket to the nozzle. If the nozzle is not an anti-pullover design, make sure to install a needle valve under the nozzle.

56—The 4500 uses a cam/linkage system to open the secondary throttles. Three cam kits are available to vary secondary opening rate—**part 20-5** for 1-to-1; **part 20-6** for progressive; and **part 20-20** for "soft" progressive—installation instructions are included in the kits. Skip ahead to **Step 71**, *Installing Pump Arm.*

4500 DOMINATOR

4500

56 Main Body Complete

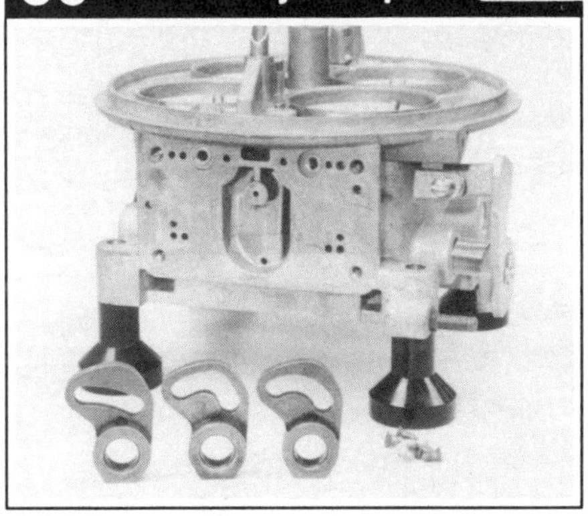

PERFORMANCE STEP

57 Center-Squirter Nozzle

57—Holley center-squirters have 4-way discharge nozzles and special air-cleaner brackets. Make sure that a needle valve has been installed below the nozzle. Position the two circular gaskets as described in previous steps and install the nozzle and high-flow nozzle screw; tighten to 48- to 52-inch pounds. Install the air-cleaner bracket and two 8-32x1/2-inch screws; tighten to 12- to 18-inch pounds. Finally, bend the outside tangs in over the screws as shown. Skip ahead to **Step 71**, *Installing Pump Arm.*

58—The secondary diaphragm assembly is sealed to the main body with a composite-cork gasket. Apply a little high-tack adhesive to hold the gasket in place in the shallow recess on the diaphragm mounting flange. (Note: Some OEM Holley carburetors did not use "E" or "C" clips to attach the diaphragm rod to the secondary throttle shaft arm—see next step. On these carburetors, the diaphragm rod must be installed in the throttle arm before the diaphragm housing is attached to the main body.) Locate the three 8-32x1-inch attaching screws and install the diaphragm housing to the main body; torque the screws to 12- to 18-inch pounds.

58 Install Secondary Diaphragm Assembly

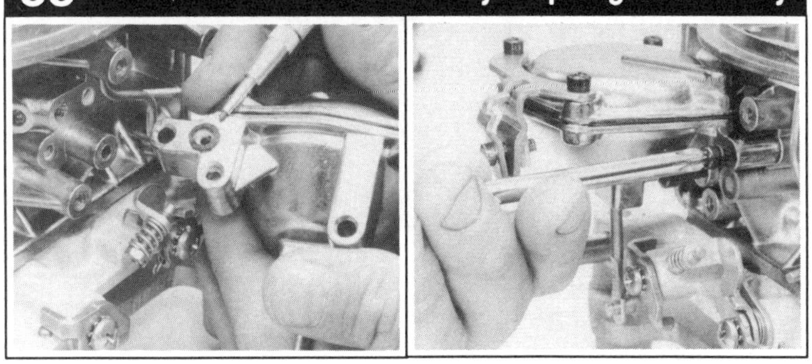

59 Diaphragm Link Clip

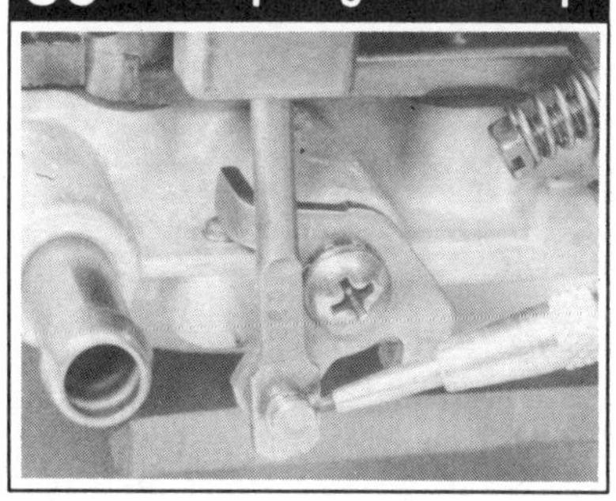

59—Locate the "E" or "C" clip for the diaphragm rod from bin #3 or your new parts in bin #5. Position the diaphragm rod on the actuating link and install the clip.

60 Installing Mechanical Choke

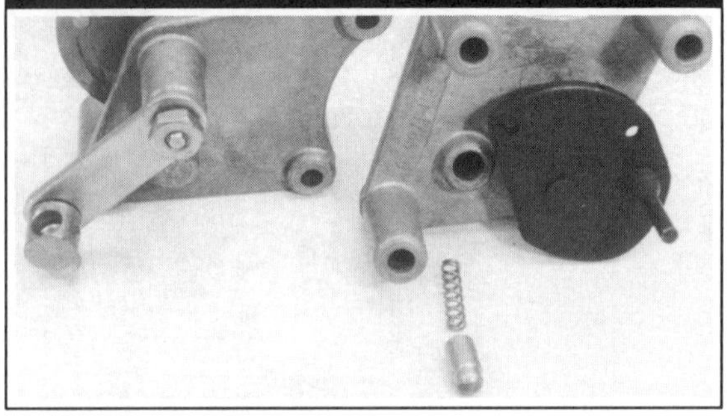

60—If your Holley is equipped with an automatic choke, skip to **Step 64**, *Install Automatic Choke*. Locate the manual choke linkage bracket. If you have immersion cleaned this component, make sure that the cable arm moves freely and is not "gummed up." BUT BE CAREFUL: If you rotate the arm too far, the detent ball and spring may fly out! If the arm does not seem to move freely, disassemble the linkage, clean, lubricate, and reassemble.

61 Install Backup Plate

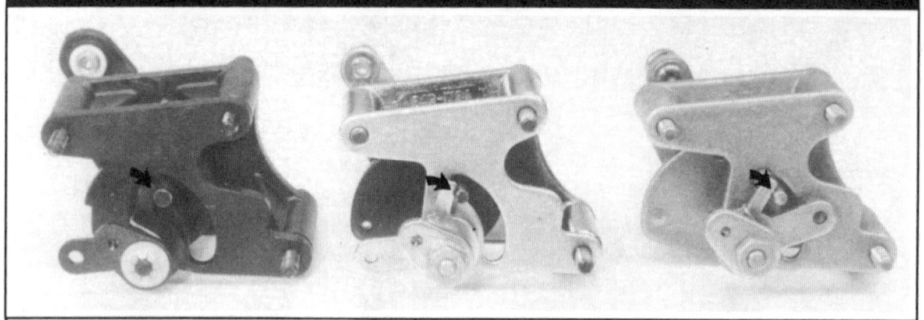

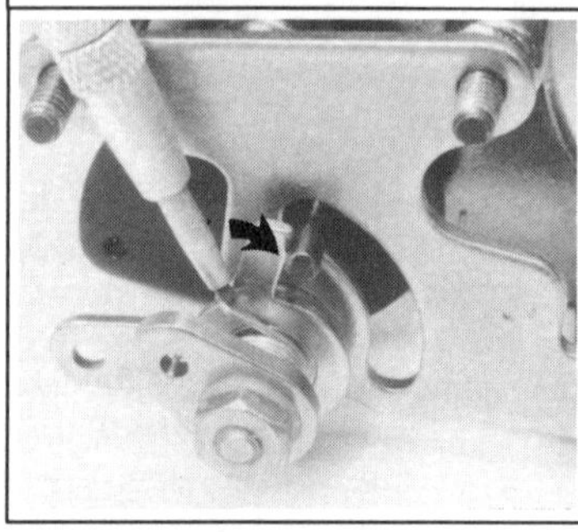

61—Make sure the backup-plate lever also moves freely. If you need to disassemble the lever, refer to the photo (left) for proper choke-tension spring position. Locate the three 8-32x1-1/2-inch screws and install them in the bracket. Then install the backup-plate on the bracket assembly. Although there are several backup-plate designs, *the protruding fast-idle cam pin always rides against the flat surface of the actuating lever* (arrows).

62 Vacuum Passage Seal

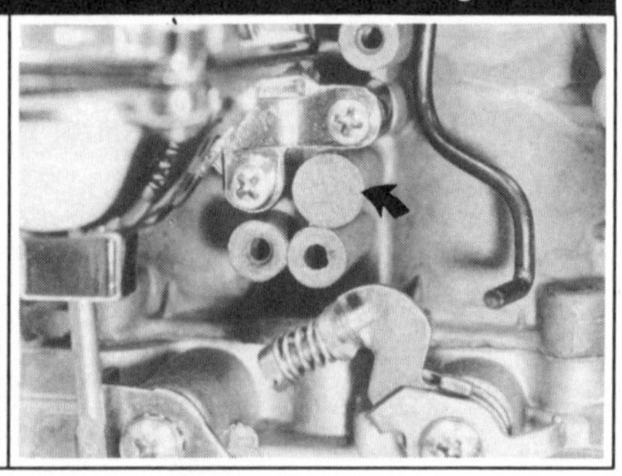

62—Most Holleys that are equipped with manual chokes have one or two vacuum passages (arrows, left photo) in the main body to accommodate the conversion to an automatic choke. These vacuum passages, if they are drilled through, must be sealed off with paper gaskets. Test these passages by applying vacuum or low pressure air. If either are drilled through, locate the circular paper gaskets in your rebuilding kit. Some gaskets have a pressure-sensitive adhesive, others require a little high-tack cement to hold them in place. After the gasket(s) is in place (arrow, right photo), install the linkage bracket and backup plate onto the main body, *but only install the screws about 2 to 3 turns.*

CRITICAL INSPECTION

63 Fast-Idle Cam Alignment

63—Using needlenose pliers, position the choke-plate link in the backup-plate arm (arrow-1). Locate the "R" or "E" clip and install it on the link (arrow-2). Before you tighten the choke linkage bracket turn your Holley upside-down and make sure the fast-idle screw rides on the center of the cam (arrow-3), not behind it! Also check the override bracket tab (arrow-4); it should hold the fast-idle arm as shown. If you tighten the bracket without checking these items, the linkage can be badly damaged. Now tighten the attaching screws to 12- to 18-inch pounds. Open and close the choke a few times to make sure the linkage operates properly.

64 Install Automatic Choke

64—If your carburetor uses a divorced choke, skip ahead to **Step 69**, *Installing An External Choke Pulloff.* (Note: If your Holley uses both an integral automatic choke and an external choke pulloff, jump ahead to **Step 69**, *Installing An External Choke Pulloff,* then come back to this step and install the choke assembly.) Holley used several types of automatic chokes, but many late-model carburetors (and choke conversion kits) use this electric choke mechanism that draws vacuum from the main body. A composite cork gasket (same as the secondary-diaphragm gasket) seals the connecting vacuum passage. Apply a little weather-strip adhesive to the housing recess to hold the gasket in place.

65 Connect Choke-Plate Link

65—Locate the three choke housing attaching screws (usually 8-32x1-1/4-inch). With the vacuum-passage gasket held in place, position the choke housing onto main body, *but only install the screws about 2 to 3 turns.* Using needlenose pliers, position the choke-plate link in the automatic-choke arm. Make sure the plastic fast-idle cam arm is above the link (arrow). Locate the "R" clip and install it on the link.

CRITICAL INSPECTION

66 Fast-Idle Cam Alignment

66—In this step you will tighten the choke housing to the main body. But before you tighten the screws, check the fast-idle cam and adjusting screw alignment. Turn your Holley upside-down and make sure the fast-idle screw rides on the center of the cam (arrow-1), not behind it! Also check the override bracket tab (arrow-2); it should hold the fast-idle arm as shown. If you tighten the bracket without checking these items, the linkage can be badly bent out of position. Now tighten the housing screws. Open and close the choke a few times to make sure the linkage does not stick and is operating properly. Finally, torque the attaching screws to 12- to 18-inch pounds.

Manual/Auto Choke Basics

Starting and operating a cold engine requires a rich fuel mixture, particularly in cold weather. The choke system performs this essential function. Composed of a choke blade and the mechanism to control blade position, it may seem to be the ultimate in simplicity. After all, what could be complicated about *a choke?* Although the basic function is quite simple, it must accurately deliver a wide range of air/fuel ratios before and after starting and during engine warm-up. This requires sophistication and precision, both in the design and adjustment of the choke mechanism.

During cold engine cranking, the choke plate should be fully closed to deliver a very strong vacuum signal to the main, intermediate, and idle circuits, forcing a fuel-rich mixture to be "pulled" from all of these circuits simultaneously. Immediately after cold start up, however, the engine requires a less-rich mixture. To accomplish this, the choke blade must open slightly. This opening—called choke qualification or pull-off—is very critical for proper cold-engine operation on an automatic choke carburetor. Too much qualification causes hesitation and stalling. Too little qualification results in an over-rich mixture, black smoke from the tailpipe, and sparkplug fouling. Then as the engine warms up, the choke must continue to open at just the right rate to match the engine's decreasing demand for fuel. This job is controlled by the precision elements in the automatic choke cover, balancing engine temperature against throttle position and airflow demands.

But that's not all an automatic and manual choke system must accomplish. Without a fast-idle feature, a cold engine would still be unmanageable. A fast-idle cam, controlled by the choke linkage, provides a significant idle-speed increase for a fully closed choke that gradually decreases as the choke opens. (More information on choke adjustment is given in **Steps 67** and **68** in this chapter and **Step 11** in Chapter 10 on page 107.)

With the exception of Dominator and some race-only carburetors, most Holleys are equipped with either a manual or automatic choke. A manual choke lacks the convenience of its automatic counterpart but gives the driver more control over choke operation. Holley kit **part 45-225** converts most automatic choke models to manual operation. You may also need a choke cable, **part 45-228**, and choke-cable bracket, **part 45-229**. Most Holleys can be converted to automatic operation with one of these two kits: **part 45-223,** if the main body has an open vacuum passage, **part 45-224,** when an external hose connection is required for vacuum. The accompanying photos will help you determine which kit you need.

Holley offers two quality automatic choke kits that include all the linkage, hoses, wires, hardware, and instructions to convert most manual-choke Holleys to fully-automatic choke operation.

Choke qualification is one of the most critical adjustments for proper choke operation. A gauge (or drill bit) of the proper diameter is used to measure the choke-blade opening.

This vacuum source (arrow) must be provided on the main body to use the automatic choke kit *part 45-223.* If this boss on the main body does not exist or is not drilled through, use kit *part 45-224.*

PRECISION MEASUREMENT

67 Choke Pulloff Adjustment–Integral Automatic Choke

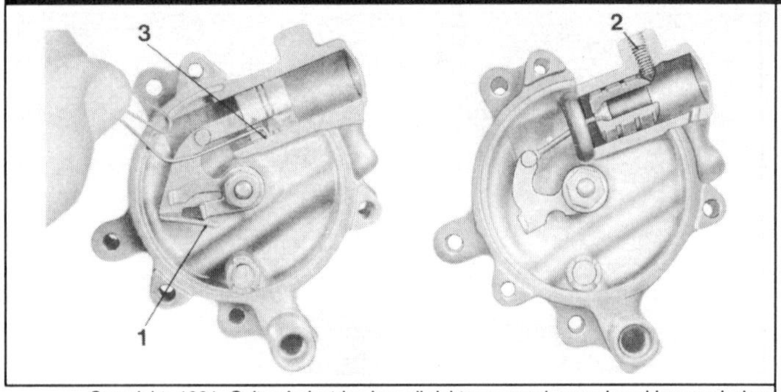

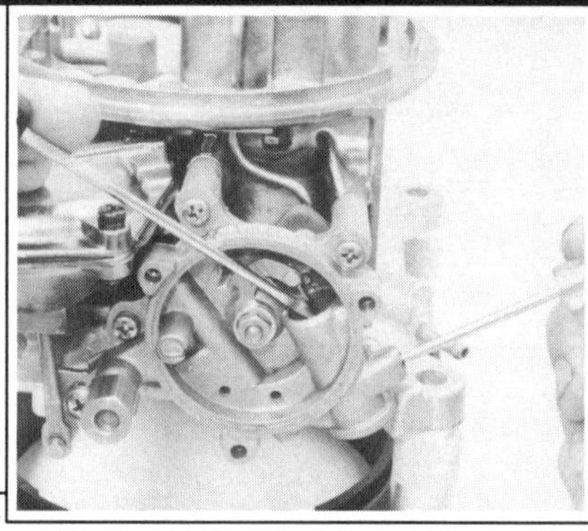

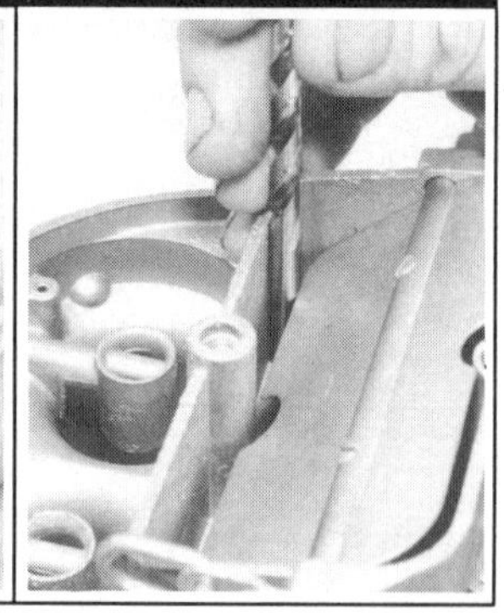

67—The choke pulloff adjustment sets how far the choke will open when the engine starts and vacuum builds up in the intake manifold. The correct setting should be printed in the instructions in your carburetor kit, or you can refer to *Holley Parts And Specifications* manual, **part 36-51**. If you don't have the proper adjustment for your carburetor, either leave the setting as it is, or try adjusting the pulloff clearance to .350-inch. There are two common models of automatic choke housings (left photo) and each has a different adjustment procedure: the early model has an adjustment tab inside the housing (1); the late model has an external adjustment screw (2). If you have the early housing, bend a paper clip as shown and hook it into the end of the slot in the bore (3). If you have the late housing, hold the piston down with a probe (center photo), but be careful not to scratch the bore. With the piston held in place, gently close the choke plate until all free play in the linkage is eliminated. The distance of the choke plate to the inside airhorn wall is the choke pulloff clearance (right photo). Note: Some Holleys require measuring the distance between the choke plate and the tallest part of the airhorn; refer to your carburetor kit for the correct setting and adjustment procedure. Adjust the early housing by bending the tab, while the late housing has an external adjusting screw (inward for less pulloff clearance). Note: Don't back the external screw out too far or you may damage the piston when the screw is retightened.

68 Install Choke Cover

68—Now install the choke cover gasket, cover, retaining spring, and three 8-32x7/16-inch screws. Make sure the spring loop engages the choke-arm tab inside the housing (arrow, on cutaway in left photo). Before the three screws are tightened, rotate the choke cover back to the position indicated in your carburetor kit or on your Work-A-Long sheet (the choke should close as you rotate the cover to the correct position). If you did not record this setting, position the choke cover at the center adjustment (right photo). If your choke cover uses a ground wire, connect it to the negative (minus) cover terminal now. Finally, tighten the choke cover screws to 12 inch-pounds. Skip ahead to **Step 71**, *Installing Pump Arm.*

CRITICAL INSPECTION

69 Installing An External Choke Pulloff

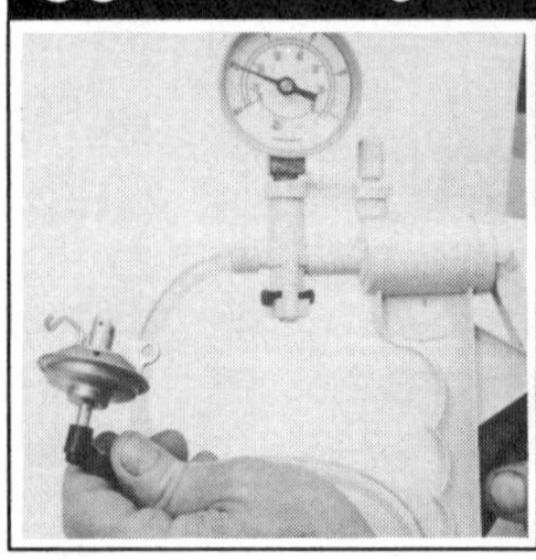

69—Locate the choke pulloff diaphragm, linkage, and linkage clips. Check the diaphragm with a vacuum source, like the *Mity-Vac*, or compress the diaphragm and hold your finger over the vacuum port; there should be no detectable leakage. Note: Some pulloffs have an internal bleed and cannot be tested using these methods. Install the diaphragm using two 10-32x9/16-inch screws; torque them to 20- to 30-inch pounds. Install the choke control lever with a large "E" clip.

PRECISION MEASUREMENT

70 Adjusting An External Choke Pulloff

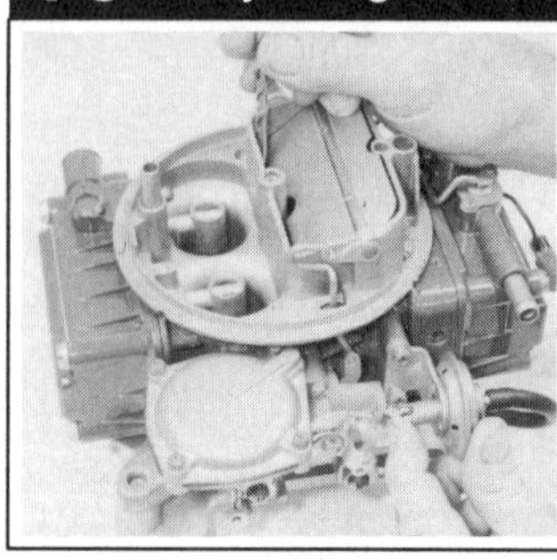

70—The pulloff adjustment for an external diaphragm is similar to that for an integral choke. First, review **Step 67**. Then hold the vacuum diaphragm retracted either by applying vacuum or by pushing in the diaphragm with your finger or a small screwdriver. Remove slack from linkage and measure choke plate-to-airhorn clearance as explained in **Step 67**; adjust by bending the loop in the link rod (right photo).

71 Installing Pump Arm

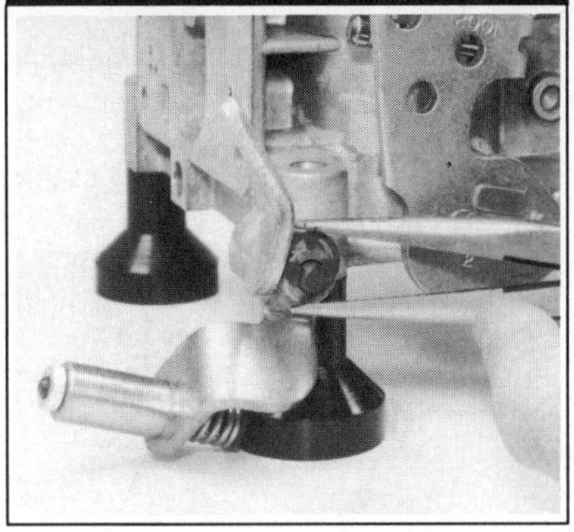

71—If you removed the accelerator pump arms from the throttle plate (or main body on a Dominator as shown here), reinstall them and their "E" or "C" clips now. If your carburetor uses dual accelerator pumps and you marked the secondary pump arm with a small "s," remount the secondary arm on the pivot stud closest to the secondary throttle shaft. Reinstalling the pump arms in the same position they were originally located can make pump linkage adjustments easier. A heavier pump arm (like this one used on Dominators and other high-performance Holleys) should be used with the taller 50cc pump, when the pump is actuated by one of the large 50cc plastic cams.

72 Installing Accelerator Pump Cam

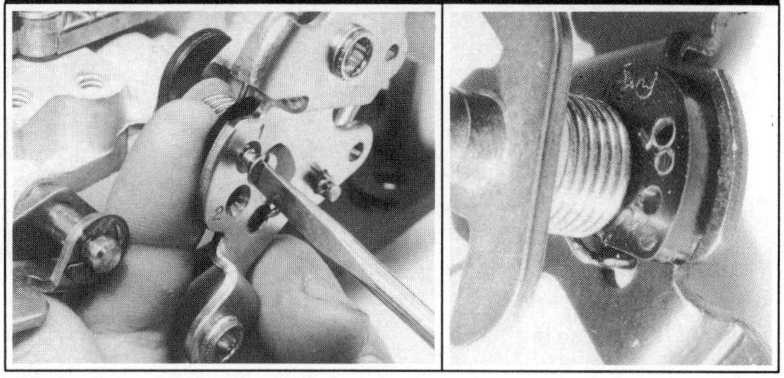

72—Locate the accelerator pump cam(s) from your parts bin. Refer to your Work-A-Long sheet and confirm the shape (30cc or 50cc type), color, ID number, and the indexing screw position of the primary and secondary (for double-pumpers) cams. Also locate the self-tapping, taper-head screws used to hold the cams in position on the throttle arms. Install the cams on the throttle shafts with the flat side of the cams facing the throttle arms. Insert the taper-head screws into the #1 or #2 holes on the throttle arms and screw them into the corresponding holes in the plastic cams (the holes in the cams are also numbered—right photo). Install both primary and secondary cams in this step.

Spacer Power Secrets

Carburetor spacers, in one form or another, have been around for decades. In some cases, they can offer an inexpensive method to obtain modest horsepower gains. Spacers often provide an increase in top-end power while reducing some bottom or midrange torque. Whether this is advantageous depends on engine configuration, vehicle weight, transmission type, and rear-end gear ratio. Determining what adds power and what doesn't on any particular vehicle combination can only be resolved through experimentation. Either an engine dynamometer, dragstrip, or a sensitive in-vehicle accelerometer is needed to measure horsepower and evaluate the effects of a carburetor spacer plate (see sidebar, *The Vericom VC-200A* on page 109 for more information on in-car tuning).

Spacer plates are produced in several configurations and a range of thicknesses. *As a very general rule-of-thumb*, dual-plane manifolds work best with open spacers between 5/8- and 1-1/2-inches thick. Single-plane manifolds on engines with radical cams usually respond well with two-inch or taller spacers. In all installations, hood clearance must take precedence; under no circumstances should the carburetor be moved close to the hood; a short, restrictive air cleaner installed; or the air cleaner removed to facilitate the installation of a spacer plate. Moving the carburetor airhorn too close to the hood can have an adverse effect on airflow and fuel-metering.

Spacers are also available that incorporate an insulating material. These spacers reduce the flow of intake manifold heat into the carburetor throttle plate and main body. The benefits are often improved hot-weather starting and increased performance. Spacer/insulators are generally made of gasket material or a hard plastic, although they are sometimes produced from a stack of gaskets or alternating metal plates and gaskets. If the spacer is primarily a compressible material (like those made from gasket material), make sure it has hard inserts at each bolt hole. These inserts are <u>essential</u> to prevent bending and cracking the throttle plate on your Holley. Never use a gasket-type spacer that does not have hard inserts.

Spacers are available in a variety or thickness and configurations. What works best can only be determined by experimentation and accurate performance measurements.

This Holley insulator gasket incorporates a built-in heat shield and reduces the transfer of heat to the carburetor. It is available for both square bore and spreadbore configurations.

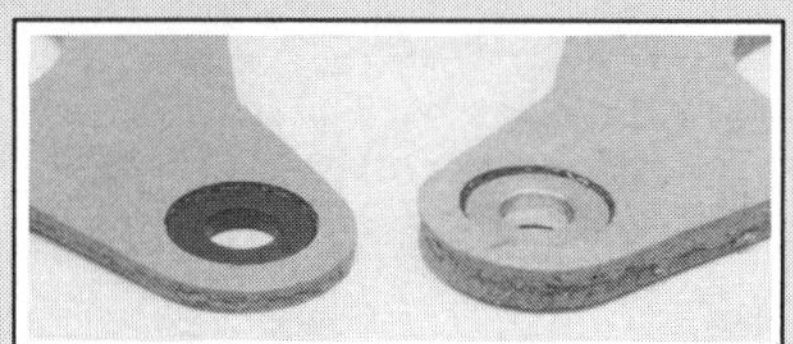

Make sure any compressible spacer has hard inserts at each bolt hole; they prevent throttle plate damage.

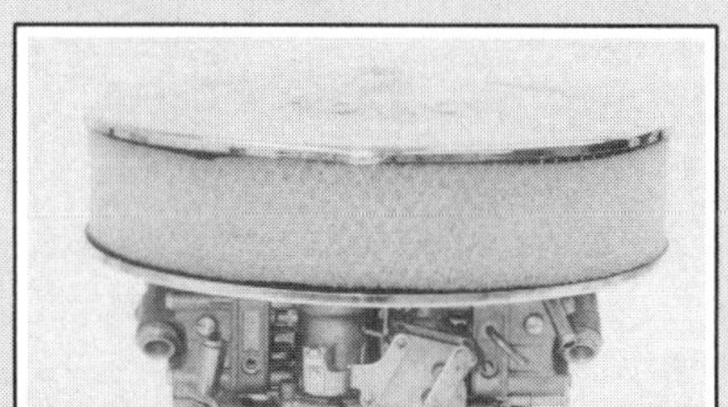

Reducing air-filter size to accommodate a spacer is a mistake. A free-flow filter, like this Holley unit, is the best way to improve performance.

73 Install Balance Tube

73—If your Holley does not use a small, brass fuel-balance tube (arrow) located below the larger aluminum fuel transfer tube, skip to the next step. This tube transfers excess fuel and vapors from the secondary bowl to help maintain the correct air-fuel ratio on emission-calibrated carburetors. Begin installing the tube by inserting it into the main body. Locate the two O-rings from your carburetor kit and the two small brass washers from your parts bin. Apply a little petroleum jelly on the tube and slide an O-ring over each end. The tube must protrude the correct amount into each float bowl (usually a maximum 2-1/4-inch from the primary side of the main body and 1-inch from the secondary; refer to your carburetor kit for the precise dimensions) and move the tube in the main body to adjust. Reposition the O-rings snugly against the main body to hold the tube in the correct position. Finally, slide a brass washer on each end of the tube; apply a little more petroleum jelly to hold them in place.

74 Primary Metering Block Gasket

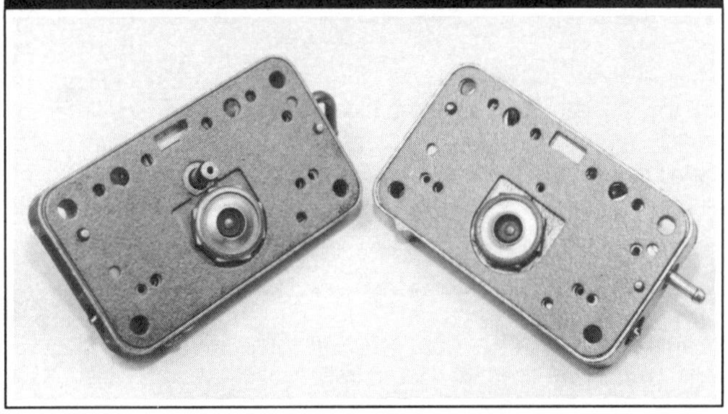

74—Holley carburetors use several different metering block gaskets (between the metering block and the main body). If your Holley uses a brass accelerator-pump fuel transfer tube, the block gasket has a cutout to clear the tube; only use this gasket on metering blocks equipped with brass transfer tubes. *Gaskets without the cutout must be used on blocks that have flat fuel-transfer surfaces.* There are several other subtle differences in metering block gaskets; if necessary, match up the old gasket with the ones supplied in your carburetor kit to find the correct match. Align the gasket over the cast-in dowels on the metering block and press it down firmly (some gaskets may not fit tightly on the dowels).

75 Primary Float-Bowl Gasket

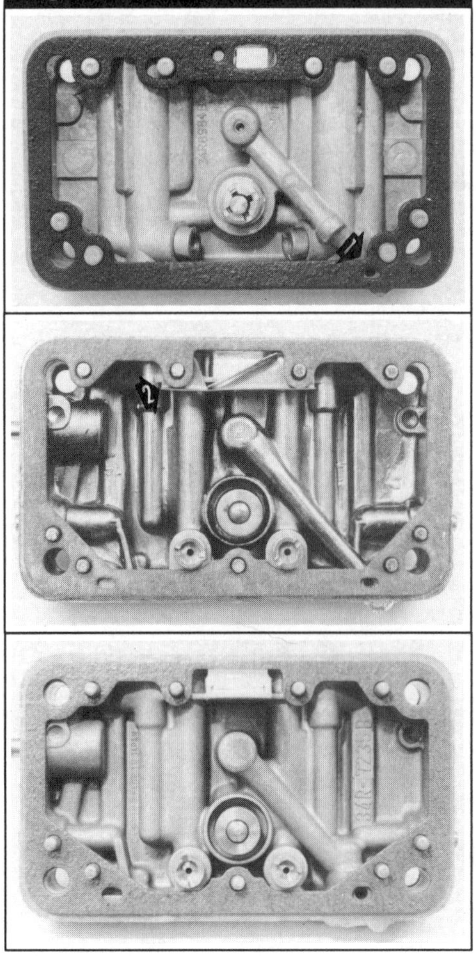

75—The other side of the metering block also uses a gasket that seals the float bowl. There are several different float-bowl gaskets. One important variation is in the punching of the accelerator pump delivery slots. Some gaskets have one slot (arrow-1); these gaskets can only be installed one way. Other gaskets have two slots, allowing the rebuilder to install the gasket swapped right-to-left. In either case, make sure the fuel slot in the gasket lines up with the hole in the metering block. If your Holley uses a small metal or plastic fuel baffle like the one pictured here (arrow-2), install it on the metering block dowels before you press the bowl gasket into position.

IMPORTANT STEP !

76 Fuel Transfer Counterbore

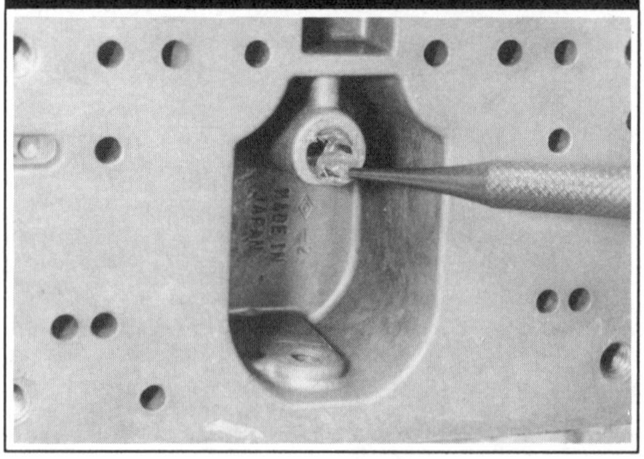

76—If your Holley uses a brass fuel-transfer tube with O-rings for accelerator pump fuel, read on; otherwise jump to the next step. The brass tube (should be installed in the metering block) slides into a mating counterbore in the main body. Take a close look at this bore and make sure there are no defects that may damage the O-ring or prevent it from sealing. Apply a little petroleum jelly to the counterbore and to the brass transfer tube and O-ring. When you install the metering block in the next two steps, don't press it straight against the main body. Instead, wiggle and twist it slightly to allow the O-ring to slip into the counterbore. Don't apply too much force; if you do, the edge of the counterbore can easily "shave" the O-ring and cause leaks.

77 Prepare Bowl Screw Gaskets

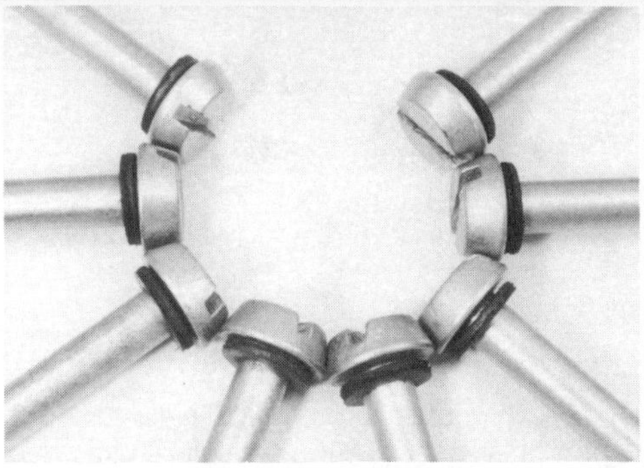

78 Install Primary Float Bowl

77—Locate all eight bowl screws and sealing gaskets. Install the gaskets on the screws as shown here. Never stick the gaskets on the bowl first, since the screw threads can cut chips out of the gaskets when you slide them through. Small gasket fragments can clog jets or other critical orifices.

78—Position the metering block (with gasket) against the main body. Move the bowl into position onto the metering block (bowl gasket should be installed on the metering block). Don't align the accelerator pump linkage yet. Hold the bowl in place and insert all four fasteners. If your Holley uses different length bowl screws, select the screw that will provide thread engagement—measured by how far the head sticks out—of between 1/4- and 3/8-inch. For now, screw the bowl fasteners in only about 2 turns; just enough to hold the bowl in place. Release the bowl and align the accelerator pump cover arm and cam-follower linkage (arrow). When the linkage is in place, tighten the bowl screws in a criss-cross pattern to 25- to 30-inch pounds.

Carburetor Linkage Basics

What often separates a professional carburetor installation from that of an amateur? A quick glance at the throttle linkage is usually all it takes. A carefully designed and installed linkage system is not only attractive, it ensures safe carburetor operation.

The first step is to establish optimum geometry. The link rods should push/pull in a straight line; perpendicular to actuated shafts. Misalignment of throttle cables or linkage rods with the throttle-shaft arm (on the carburetor) can apply excessive side loads and cause premature throttle shaft wear. Also, take care to ensure that the linkage does not force the throttle plates to rotate more than vertical (wide-open). Excessive travel in cable systems is easily corrected by adjusting the cable; solid linkage may require shortening, lengthening, or fabricating new connecting rods to obtain proper travel.

Another common installation problem is insufficient travel to allow the throttle to fully open and/or close. The best solution to this dilemma is to design the linkage from the start so that the attachment points on the carburetor are the same distance from the throttle-shaft centerline as the original carburetor installation.

There should be enough throttle-return spring tension to draw the throttles closed reliably while not producing excessive pedal pressure. The spring should have enough coils to prevent overstretch at wide-open throttle. Experiment with different springs and anchoring locations until you locate the setup that offers the lightest pedal feel and still *reliably* closes the throttle. It may be necessary to use a double (inner and outer) spring, where the small inner spring provides light tension to keep the throttle closed, and the heavier outer spring stretches at wide open throttle to rapidly "snap" the throttle closed. Dual springs also add a safety margin, since each spring has enough tension to close the throttle if either one should fail.

Pay particular attention to the spring anchor points. Purchase or fabricate a sturdy mount; never hook the spring to a fuel-line, clamp, or another inappropriate item. Open and close the throttle several times to make sure that the return spring doesn't rub on linkage, brackets, or the carburetor body. And make certain that it can't go over-center at full throttle; if it does, it will help keep the throttles wide-open, not closed!

When everything is properly set up, have a helper operate the accelerator pedal (with engine off, of course) while you verify that the butterflies just reach a fully-open position when the throttle contacts the floorboard. Also verify that, with the choke held open, the throttle lever returns to the fully closed (idle) position. Make sure all automatic-transmission and cruise-control linkage operate properly.

The time spent in precisely designing and fabricating throttle linkage will be repaid many times over by a professional-looking installation that operates effortlessly and dependably. So do it right the first time; you'll save time!

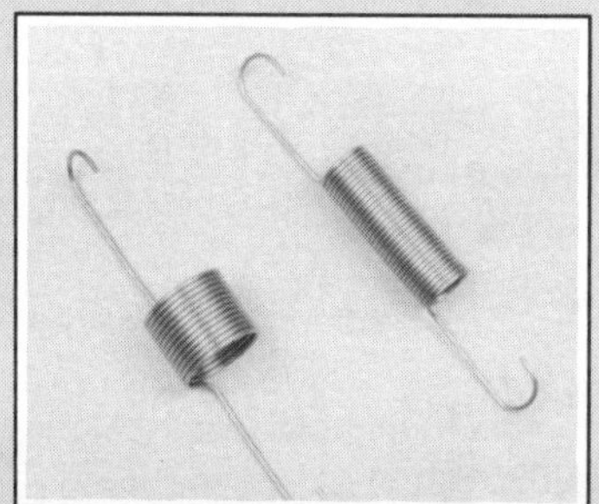

Holley produces this handy throttle-return spring and bracket kit that works on most vehicles. Dual springs add a margin of safety.

With a helper operating the accelerator pedal, make sure that the carburetor fully opens and closes. Adjust the linkage so that the butterflies reach wide open simultaneously with the gas pedal. Misadjustment will overstress the throttle shaft causing rapid wear and possible linkage failure.

79 Install Secondary Metering Plate

79—If your Holley has a secondary metering plate (non-replaceable jets) perform this step, otherwise jump to the next step. A steel support plate and thin gasket must be used with the secondary metering plate when the secondary side of the main body has a central cavity. The steel plate adds the support needed to seal the metering plate in the span over the cavity. Main bodies that have no cavity in the secondary surface do not require—but still can use—a steel support plate and gasket (see left photo). Now, let the metering plate dowel pins slip through the holes in the bowl gasket and register into the main body. Install the six 8-32x1/2-inch clutch-head screws and torque to 12- to 18-inch pounds.

80 Install Secondary Fuel Transfer Tube

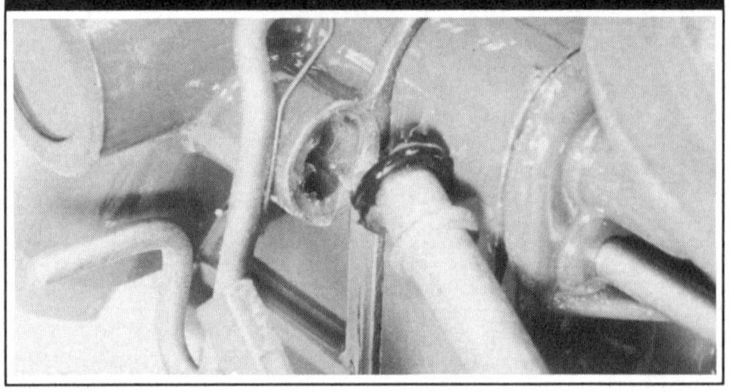

80—If your Holley does not use an aluminum fuel transfer tube to feed the secondary float bowl, skip to the next step. Locate the fuel transfer tube and O-rings. Install an O-ring on one end of the tube. Apply a little petroleum jelly to the edge of the primary bowl counterbore and the O-ring. Press the tube and O-ring into the primary bowl counterbore. You may have to gently push and wiggle the tube to help the O-ring and tube slide in all the way. Now place another lubricated O-ring on the exposed end of the tube.

TORQUING FASTENERS

81 Install Secondary Bowl

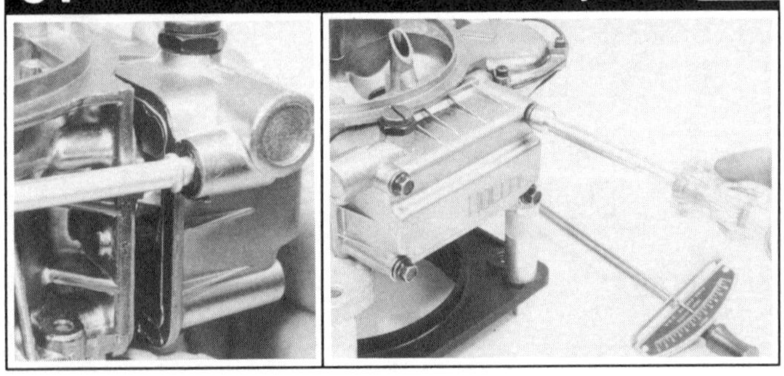

81—With the same technique you used to install the primary bowl, position the secondary bowl against the main body. If your Holley uses a fuel transfer tube, lubricate the bowl counterbore and carefully align the tube with the bowl. Gently press and wiggle the bowl until the fuel tube and O-ring slide completely into the bore. Install all four bowl screws (again, make sure thread engagement will be between 1/4- and 3/8-inch) and screw them in only about two turns. Now align the secondary pump linkage (if so equipped). Finally, screw the bowl fasteners in all the way and torque them in a criss-cross pattern to 25- to 30-inch pounds.

SAFETY STEP

82 Check Fuel Tube

82—If your Holley uses a fuel transfer tube, perform this important safety step. If the tube and O-rings are installed correctly, you will be able to twist the tube with your fingers after the bowls have been installed and torqued in place. The tube should be snug but not excessively hard to turn. If it won't turn, the O-rings are probably jammed out of place and the tube will almost certainly leak fuel. Remove the secondary bowl, check the O-rings, and reinstall the tube and float bowl.

83 Hot-Idle Compensator

83—If your Holley is equipped with a hot-air idle compensator valve in the main body, locate the valve, cover, two 6-32x5/16-inch screws, and the 13/32-inch O.D. composite gasket. Apply a little high-tack adhesive to the composite gasket and place it in the main body recess. Then slide the valve into position. Install the cover and two cover screws (there is no gasket for the cover). Torque the screws to 7- to 10-inch pounds.

PRECISION MEASUREMENT

84 Adjust Accelerator Pump Arm

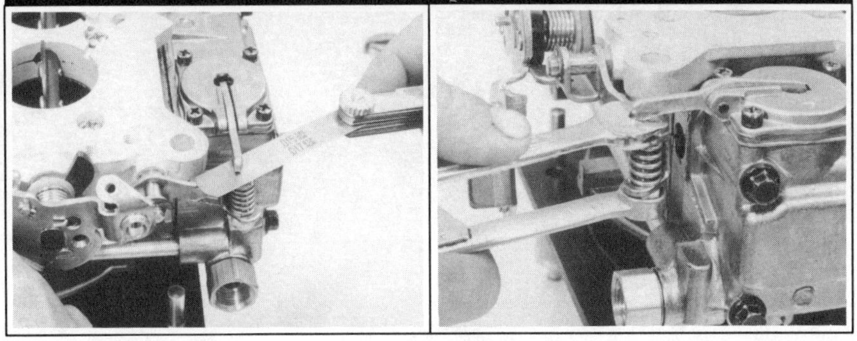

84—The accelerator pump override spring must be adjusted to provide a minimum of .015-inch clearance at wide-open throttle. Here's how to accomplish this: First, open the choke and unscrew the idle-speed adjuster until the primary butterflies are fully closed (make sure the fast-idle cam does not hold the throttle open). Turn the carburetor upside down and fully open the throttle. Insert a .015-inch feeler gauge between the pump-cam arm adjusting nut and pump lever. It should easily slide in without compressing the override spring. If there is not enough clearance, tighten the override screw (or nut) until you can begin to see movement in the pump lever when the cam-follower arm is wiggled up and down; if there is too much clearance, loosen the adjusting nut. Now, remove the feeler gauge, close the throttle (make sure it's *completely* closed), and check the pump-cam arm for any slack. If any free play exists, the pump-cam arm may be bent or the wrong plastic cam may be installed. Recheck these items and repeat the adjustment procedure until both .015-inch clearance is obtained at full throttle and there is no free play in the pump-cam arm when the throttle is closed (free play will cause hesitation when the throttle is opened).

85 Adjust Idle Speed Screws

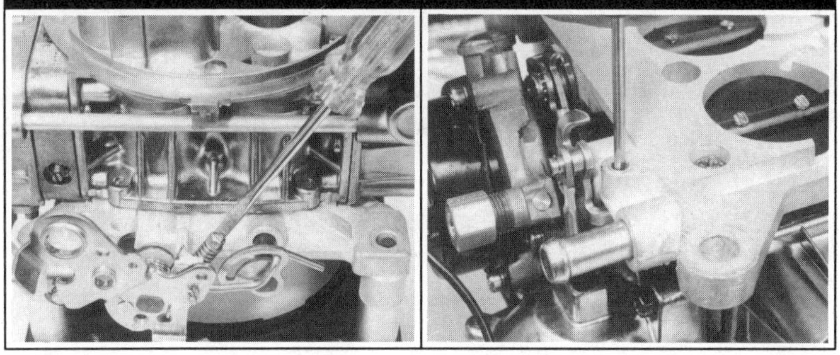

85—To establish a preliminary curb-idle setting, first make sure the choke is open, the primary idle speed screw is backed out, and the throttle is completely closed. Now screw in the idle-speed adjuster until the throttle arm just begins to move (primary butterflies just unseated), then turn the screw in one additional turn. The secondary butterflies on many Holleys also have an "idle" stop adjustment. This adjustment improves idle and prevents the secondary butterflies from sticking closed. The secondaries are adjusted by first backing out the adjusting screw (right photo) until the butterflies are fully closed. Then turn in the adjuster until the butterflies just begin to open. Finally, turn the adjusting screw in an additional 1/4 turn. This screw is self-locking and can be difficult to turn; use only a quality, tempered steel screwdriver. On the other hand, if the screw turns easily, stake it in place after it's adjusted; or remove the screw, apply red Loctite to the threads, and redo the adjustment.

PRECISION MEASUREMENT

86 Bowl Vent Adjustment

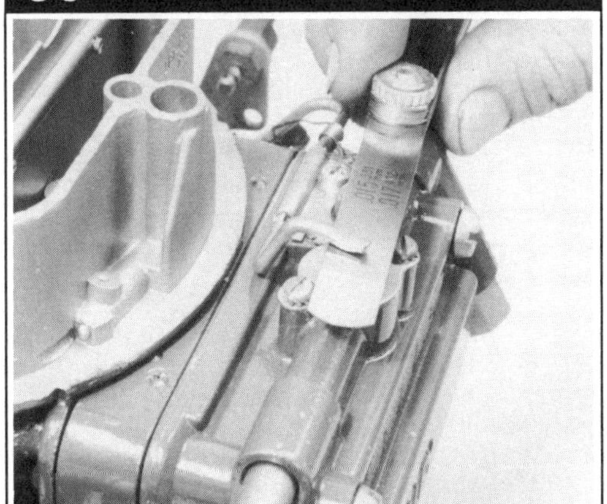

86—If your Holley is equipped with a primary bowl vent, the vent-valve clearance should now be checked. First, make sure the choke is open and the primary throttle arm is resting on the curb-idle stop screw (adjusted in the previous step). There should be at least .015-inch clearance between the vent valve stem and the operating rod (or between the rubber valve and bowl surface on earlier atmospheric vents). Adjust the clearance by bending the actuating rod.

87 Assembly Complete!

87—Congratulations! With the addition of any vacuum hoses—like an external choke pulloff line—other electrical or linkage components (see your Work-A-Long sheet for notes), and the air cleaner gasket, the assembly of your Holley is complete! In the next chapter we'll reinstall the carburetor and make the necessary on-car adjustments. But before you proceed, recheck all of the external fasteners, since they tend to lose tension as gaskets slowly compress. And don't forget to retorque the fasteners (at least the float bowls) after the carburetor has been in operation for a few days. And if you're interested in getting the most performance or economy from your Holley, take a look at Chapter 11.

HOLLEY Rebuilding & Modifying

10—Carburetor Installation

This chapter provides detailed procedures that will guide you through the step-by-step installation of virtually any Holley modular carburetor. In order to properly apply the information in this chapter, you should have a basic familiarity with hand tools (refer to the sidebar, *Underhood Tools And Work Tips* on page 37 for more information). In addition, you should be able to identify the individual component parts of your Holley, as discussed in Chapter 3.

The Step-By-Step Procedures Used In This Chapter

Not all of the step-by-step procedures in this installation chapter apply to every vehicle or model carburetor. For example, if your vehicle does not have an automatic transmission, steps related to installing automatic transmission linkage should be skipped. In many cases, instructions are provided in the step-by-step captions that direct you to "jump ahead to..." when upcoming steps do not apply to a specific carburetor model. However, all steps are clearly titled and in the proper sequence, so if you come upon a step that does not relate to your project, just skip ahead to the next step.

Several steps in the step-by-step chapters in this book are labeled with icons—symbols—that represent important information or special help, such as **IMPORTANT STEP, MASTER MECHANIC TIP**, or **SAFETY STEP**. Carefully read these steps unless you are specifically directed to skip over them. These steps contain important information that can often make the difference between a successful rebuild and disappointment.

Finding Additional Information

If you would like additional information on the following subjects before you begin installation, refer to the indicated chapters.

How This Book Is Organized .... **Chapter 1**

What Holley Models Are Covered In This Book .... **Chapter 1**

Shop/Workspace Considerations .. **Chapter 2**

Required And Optional Tools .... **Chapter 2**

Identifying Holley Carburetors And Components .... **Chapter 3**

Understanding Holley Part Numbers .... **Chapter 3**

Diagnostics Before You Begin ..... **Chapter 4**

Buying The Right Carburetor Kit .... **Chapter 4**

Carburetor Removal .... **Chapter 5**

Disassembling A Holley Modular Carburetor .... **Chapter 6**

Carburetor Cleaning And Parts Inspection .... **Chapter 7**

Custom Finishing For Show .... **Chapter 8**

Reassembling A Holley Modular Carburetor .... **Chapter 9**

Every stage of carburetor rebuilding requires patience and care to achieve professional results. In this chapter it all comes together; now that your Holley is assembled and bench adjustments have been made, it's time for installation and basic on-car tuning.

MASTER MECHANIC TIP — PRO TIP

1 Installing A New Fuel Filter

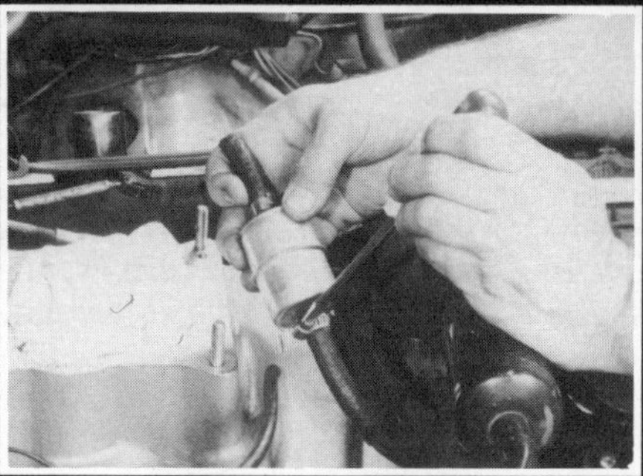

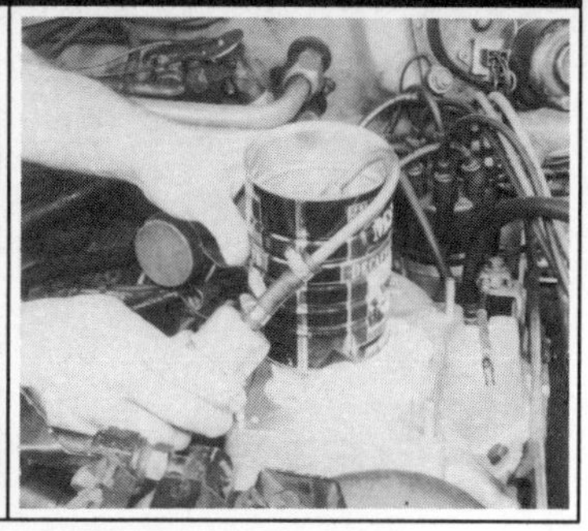

1—Install a new brand-name fuel filter before reinstalling your rebuilt Holley. Flush the fuel lines after the filter is installed to purge the system of small particles that may have been dislodged from the new filter and hose connections. Insert the fuel line into a suitable container, *and with the primary side of the ignition disabled*, have an assistant crank the engine until several ounces of fuel have been discharged. *Caution! Always use extreme care whenever handling gasoline. Wear goggles and have a fire extinguisher handy.*

2 Manifold Surface Preparation

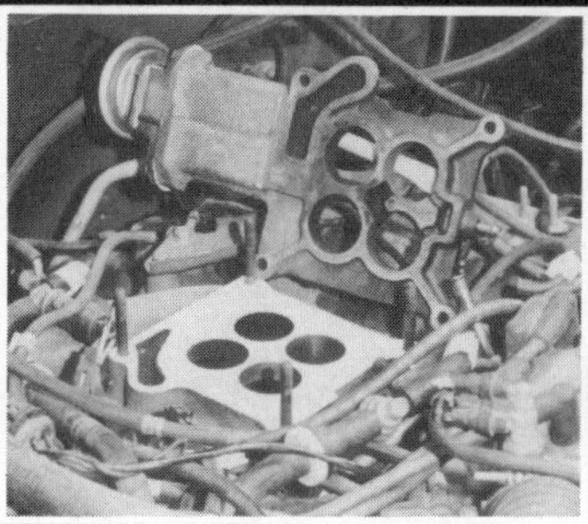

2—Make sure the carburetor mounting surface is flat, clean, and that all old gasket material is removed. Check the studs and/or manifold threads for damage. Ensure that the studs or bolts are the correct length for your carburetor and gasket combination. If the engine uses an EGR plate under the carburetor (left photo), thoroughly clean all mounting surfaces and use new gaskets.

PERFORMANCE STEP

3 Special Base Gasket

3—Check that the carburetor mounting gasket properly fits the throttle bores on the carburetor and the manifold opening. A frequent mistake is to use a gasket with throttle holes smaller than the carburetor throttle-plate bores. When the throttle is forcibly opened, it can jam the butterflies into the gasket, sometimes bending the butterflies or even holding the throttle wide open! Don't let this happen to you. Be sure the gasket is the correct size and positioned properly on the manifold. This Holley insulating gasket for square-bore carburetors, **part 108-51**, helps prevent fuel percolation and can improve performance. Similar gaskets are available from Holley for spreadbore and other models.

TORQUING FASTENERS

4 Install Mounting Bolts

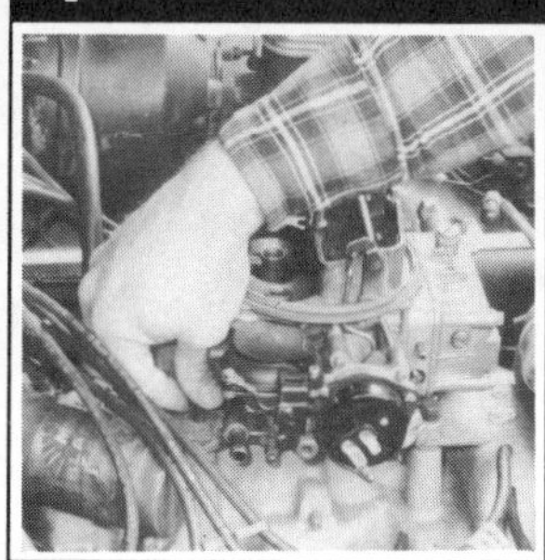

4—Install the carburetor on the manifold. Fully open the throttle to check for free movement. Then install the mounting nuts or bolts and torque them in several steps, using a crisscross pattern, to a final tightness of 60- to 80-inch pounds (for 5/16-inch fasteners). *Do not overtighten the fasteners!* Holley throttle plates are easily warped or cracked from overtightening or unevenly tightening the mounting bolts.

IMPORTANT STEP

5 Check Throttle

5—Once again, fully open the primary and secondary throttle arms to make sure they do not stick or jam against the manifold. If there is a problem, the manifold may be the wrong style for your carburetor. The throttle shafts also can bind if the mounting bolts are overtightened or torqued incorrectly.

6 Connect Fuel Lines

6—Connect and tighten the fuel line fittings. Use stainless worm gear clamps for rubber lines rather than the coil-spring type used by the factory. Use a suitable line wrench for flare-nut fittings. Always use a second wrench on the carburetor inlet fitting to prevent it from turning and possibly twisting the fuel line or damaging the fuel bowl threads (right). *Never use Teflon tape (or silicone rubber adhesive) on any fuel inlet fittings. Small pieces of the tape will find their way into the needle-and-seats, causing carburetor flooding.*

7 Vacuum And Electrical Connections

7—Closely inspect all vacuum and electrical connections. Replace any hoses that are cracked, hardened, or damaged. Repair or replace any electrical wiring or connectors that are loose or worn. Refer to your Work-A-Long sheet and reinstall all vacuum and electrical connections to your Holley.

8 Connecting Throttle Linkage And Return Springs

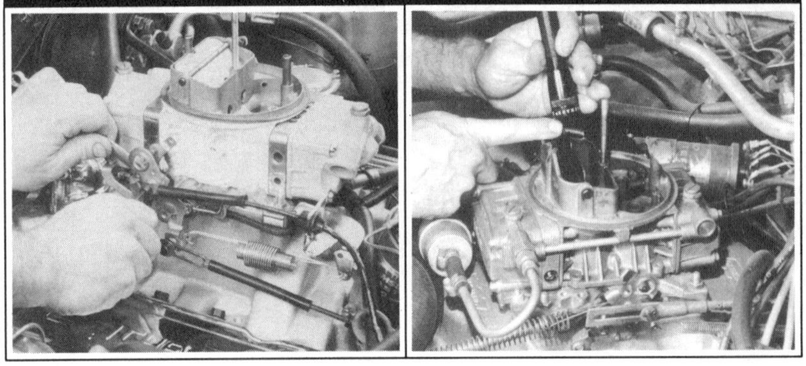

8—Connect the throttle linkage and return springs. Have a helper operate the accelerator pedal and *make sure that the return springs do not swing "overcenter," potentially holding the throttle open rather than closing it!* Ensure that the return springs are strong enough to "snap" the throttle closed, but not so strong that excessive pedal pressure is required. Have your helper depress the throttle to the floorboard. Then verify that the throttle blades are completely opening by looking into the carburetor with a small flashlight. If necessary, adjust the linkage or cable. After adjusting—with the choke held open—make sure the throttle arm returns to a fully closed position (resting against the throttle-stop screw). For more information on throttle linkage see the sidebar *Carburetor Linkage Basics* on page 101.

9 Automatic Transmission Linkage

9—Next, install and adjust transmission kickdown linkage and cruise control linkage, if used. Refer to any notes made on your Work-A-Long sheet and to your vehicle service manual for information on the proper installation and adjustment procedures. Finally, make sure that the throttle still operates smoothly and fully returns to the closed position.

SAFETY STEP

10 Checking For Fuel Leaks

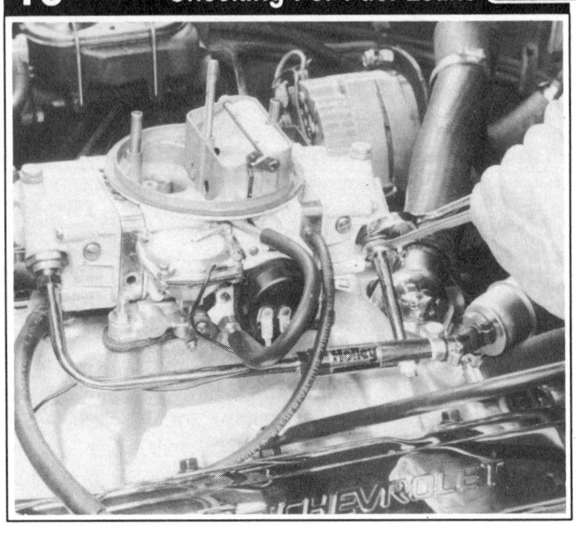

10—Start the engine and immediately check for fuel leaks. Closely inspect bowl gaskets, needle-and-seat adjuster screws, and all hose and fuel-line fittings. If any leaks are found, stop the engine and retighten and/or correct the problem. Clean up any spilled fuel and restart the engine; verify that all fuel leaks have been corrected.

PRECISION ADJUSTMENT

11 Choke Pull-Off Adjustment

11—Now check the choke pull-off (qualifying) adjustment (see **Steps 67** and **70**, on page **98** for the adjustment procedure). The choke should be open enough to prevent rough idling and over-richness, yet it should be closed enough to provide adequate throttle response and prevent stalling. The fast-idle cam follower should be adjusted to manufacturer's specifications or about 1500rpm (if necessary, refer to your service manual for the adjustment procedure). If the choke opens too slowly as the engine warms up, adjust the choke cover by rotating it slightly clockwise.

12 Setting Idle Speed

12—After a sufficient warm-up period, when the fast idle cam is fully released, adjust the main idle-stop screw to the factory recommended idle speed. If you have a modified engine, adjust idle speed to the lowest rpm that provides a stable idle.

PRECISION ADJUSTMENT

13 External Float Height

13—If your Holley has externally-adjustable float bowls, they should be adjusted now. With the engine idling, remove the sight plug from the primary float bowl. Slightly loosen the lock screw (or lock nut for non-Holley needles-and-seats) and rotate (counter-clockwise to raise fuel level) the needle-and-seat adjustor so that the fuel level just reaches the bottom of the sight-plug hole. *Important: Turn the adjustment screw in small increments and wait several seconds to allow the fuel level to stabilize.* Increase the engine speed then let it return to idle several times; the fuel level should remain fixed. Install the sight plug and repeat the procedure for the secondary float bowl.

PRECISION ADJUSTMENT

14 Setting Idle-Mixture Screws

14—Connect a vacuum gauge or tachometer to the engine. First, adjust one primary idle-mixture screw to provide the highest vacuum or rpm reading. Then repeat the process with the other primary screw. Go back and forth several times to achieve the smoothest idle and highest readings. If your carburetor has secondary mixture screws, repeat this procedure on the secondaries. Finally, readjust the idle-stop screw to reestablish the desired idle speed. Note: Satisfactory idle fuel-mixture should be obtained when the mixture screws are between 1/2 to 2 turns out. If this is not possible, refer to the sidebar *Racing Cams Idling Problems* on page 119, and Chapter 4 for additional information.

15 Perform A Road Test

15—You are now ready to road test your Holley. For road testing tips, or if you are having problems with your Holley or engine operation, see Chapter 4. After testing, retorque the carburetor mounting bolts, float-bowl fasteners, and make a final check for fuel or vacuum leaks. When you are satisfied with the basic operation of your Holley, refer to Chapter 11, *Performance And Economy Tuning*, for help in optimizing your induction system.

11—Performance/Economy Tuning

This chapter contains information that will help you optimize your Holley for fuel economy or engine performance. It does not contain troubleshooting information; if you need help getting your carburetor or engine to run properly, see Chapter 4. In fact, you shouldn't apply the tuning techniques presented in this chapter until all engine and carburetion malfunctions are corrected. If necessary, you should perform a complete electrical and mechanical tune-up. Be sure to check sparkplugs, compression, ignition wires, vacuum hoses, and filters. It is impossible to optimize fuel metering if your engine has a vacuum leak or intake restrictions, worn rings, weak valve springs, or other maladies. You must have a stable mechanical "platform" in order to reliably detect the results of small changes to jets, power valves, or secondary opening rates. Trying to fine tune a sick engine or carburetor is, frankly, a waste of your time.

WHAT TO DO FIRST

Almost all Holley carburetors perform remarkably well in stock condition on a wide range of engines and vehicles. This is particularly true of the universal vacuum-secondary and two-barrel models, providing carburetor airflow capacity is carefully selected to match engine airflow requirements. This is not to say that improvements can't be found from custom tuning. However, randomly installing jets and tuning by seat-of-the-pants can easily move you away from, instead of closer to, optimum fuel metering.

Begin your tuning efforts by taking advantage of the extensive research and testing that Holley does on each carburetor it sells: Start with an unmodified carburetor adjusted to original Holley specifications. In other words, install stock jets and power valves; try to verify that the air bleeds have not been drilled (if they have, you may never get your Holley to work properly!); and adjust the choke, idle speed and mixture, and fuel level in the bowls. If you have problems with any of these adjustments, refer to Chapter 4 for troubleshooting information and to Chapter 9 and 10 for step-by-step adjustment procedures. With the engine in optimum mechanical condition and the carburetor as close as possible to its original condition, you are almost ready to begin methodically closing in on optimum fuel metering for your vehicle and driving patterns. You should do just two more things: 1) decide on your tuning goals and what segment of engine performance you're willing to compromise to reach those goals, and 2) obtain some method of precisely measuring the results of modifications. The next two sec-

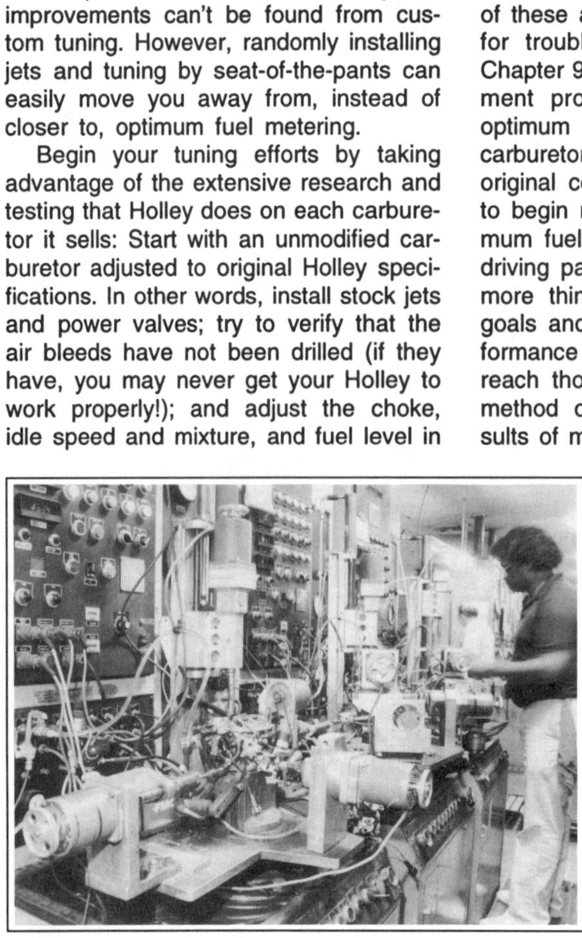

Although Holley carburetors are "mass-produced," few consumers realize that each carburetor has been carefully tested on a wet-flow bench before it was placed in its shipping container. The initial settings for idle, intermediate, and main-system fuel flow are carefully selected for the intended application, and this is the best place to start your tuning efforts: with stock jetting, power valves, and adjustments.

The Vericom VC-200A

The Vericom VC-200A is a performance measuring device that incorporates a highly accurate internal accelerometer and a high-speed computer. Data from the accelerometer is processed by software stored in the unit to calculate vehicle speed, distance, elapsed time, and G-forces. The VC-200A is compact, lightweight, and only requires a 12-volt connection (a cigarette lighter adaptor is included); some Vericom models have rechargeable batteries and require no external connection. Installing and removing the VC-200A is extremely easy (it can be transferred from one car to another in less than a minute); all that's required is positioning four suction cups on the interior of the windshield.

The secret behind the operation of the Vericom lies in the theory of "inertial navigation." If acceleration (G-force) and elapsed time are precisely known, then speed and distance traveled can be calculated. After each "run," the Vericom is capable of displaying: 1) the time it took to travel a specific distance, 2) the time it took to achieve a certain speed, 3) the G-forces (positive or negative) produced in real time, 4) peak G's generated during a test, 5) peak horsepower per 1000lbs of vehicle weight, and 5) the time and distance to decelerate from a given speed (a measure of friction and aerodynamic effects).

Many carburetor modifications are more easily evaluated by noting their effects over a narrow range of acceleration rather than, for example, through full 1/4-mile runs. Because of this reason and other practical road-testing requirements, the wide range of possible test distances and speeds that the VC-200A is capable of performing make it particularly well suited to carburetor tuning. For example, by using the front-panel touch pad, the user can instruct the device to measure 0 to 60-foot acceleration times that can be helpful in tuning the accelerator pump circuit. On the other hand, a measurement of acceleration in only one gear (both peak and instantaneous) after the vehicle is moving can reveal horsepower variations from changes in main jetting, while eliminating the effects of shifting and wheel spin.

The VC-200A is an effective alternative to some of the more traditional methods of engine tuning, such as engine dynamometers, fifth wheels, and race tracks. But most importantly, its reasonable price makes it available to just about anyone to *quickly and accurately* evaluate adjustments and modifications to carburetors and other engine and chassis components. (Important! Never use of the VC-200A in violation of any federal, state, or local laws. Never observe the display while the vehicle is in motion or use it in any other unsafe manner.)

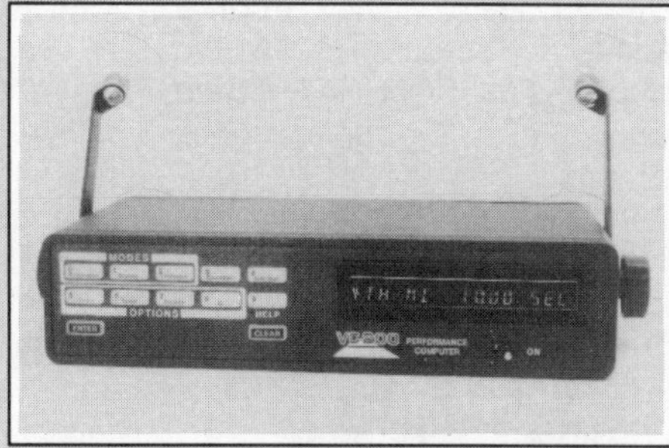

The Vericom VC-200A is an accurate and affordable test instrument that can help the serious tuner evaluate adjustments and modifications to Holley carburetors.

The Vericom has a touch-panel that allows the user to select from a wide range of possible test distances and speeds. Among many tests, the Vericom can display 0 to 60-feet and 1/4-mile times, peak G's, and much more!

In order to obtain results that are meaningful, keep an accurate record of each test run and maintain consistent engine temperature, road conditions, etc.

tions offer some advice on each of these important considerations.

FINE TUNING GOALS

There are many specific procedures used to fine tune a carburetor and induction system, but the intended result of these procedures can be classified in only three general categories: 1) increasing power, 2) improving fuel economy, and 3) reducing emissions. While it's true that some modifications will improve one category at the detriment of others—such as the inevitable reduction in economy from the addition of a racing cam or mechanical-secondary carburetor—in many cases, fine tuning of idle, transition, main, power-valve, and accelerator-pump circuits can provide coincident improvements in power, economy, and emissions. For example, many universal Holley 4-barrel carburetors are calibrated slightly rich in order to safely operate on a wide range of engines. With these carburetor models, improvements in both power and fuel economy are not unusual when the transition, main, and accelerator-pump circuits are slightly leaned. Unfortunately, obtaining *maximum power and maximum fuel economy* is usually not possible, although optimizing fuel economy and minimizing emissions usually co-exist quite well.

Engine and carburetion systems are complex and interrelated, making exact modification choices difficult to determine. While the previous example of tuning a universal-fit Holley is more or less a matter of leaning the fuel metering systems until no further benefits are obtained or until driveability is compromised, such clear-cut, fine-tuning procedures are uncommon. Consider this real-life situation: I came across an emissions-calibrated Holley carburetor on an engine just equipped with headers, a performance cam, and an open-plenum intake manifold. Initial tests revealed a lazy throttle, surging, and an off-idle stumble; all the symptoms of running too lean. Your first reaction might be to install richer primary jetting, despite the fact that the vehicle owner had already increased primary and secondary main-jet size. That would have been a mistake. In this situation, the performance equipment produced a drop in manifold vacuum, reducing signal strength on the idle and transition circuits. This caused the engine to run lean at idle and part-throttle cruise, although full-throttle power—due to the installation of larger main jets—was acceptable. The problem was resolved by enlarging the idle-feed restrictions. The increase in fuel flow through the idle and transition circuits improved throttle response and eliminated surging and the off-idle stumble.

This confusing tuning problem points out the complicated interrelationships between induction systems components and engine mechanical characteristics. Real-world problems, like this modified engine, can make clear-cut decisions about fine-tuning goals and practice very challenging.

Despite the hundreds of induction variables you may encounter, there is one that stands out as *the* most critical factor affecting the overall success or failure of your tuning efforts. It is the compatibility of the specific Holley model with the engine in question. You'll never get excellent fuel economy from an 850cfm double-pumper, and you'll never get excellent race-track performance from a 4180 emissions carburetor. Review the chapter *Selecting The Right Holley Carb* in S-A Design's *Holley Carburetors* book by Dave Emanuel before wasting your time and money tuning a carburetor that is poorly matched to your vehicle, engine, or induction requirements.

MEASURING AND EVALUATING TUNING RESULTS

Carburetor fine tuning almost never produces startling gains in power or economy but rather optimizes the performance of a carburetor that is already functioning within acceptable standards. This means that the required changes and adjustments will be small, and their result will make *small changes* in performance and/or economy. It is impossible to evaluate the effectiveness of your tuning efforts without some way of accurately measuring these changes. Fortunately, developments in electronics and computer technology have made it practical for many enthusiasts to purchase sophisticated measuring tools that can accurately measure small performance improvements on street-driven and even race vehicles. Devices like the *Vericom* and the Banks *Dynafact* offer an accurate means of measuring and comparing the results of carburetor tuning. These devices use an internal accelerometer that measures how fast the vehicle is accelerating or decelerating. A computer and specialized software stored in permanent memory "crunches" this information (the *Dynafact* also directly measures vehicle speed) to calculate overall vehicle performance throughout the rpm range. They are so accurate that it is possible to evaluate even the most subtle tuning changes, such as optimum opening rates for vacuum secondaries, or even changes in vehicle aerodynamics by measuring coast-down speeds.

As good as accelerometer-driven devices are, most can't measure the amount of fuel consumed per mile and, therefore, are limited in their ability to evaluate modifications that can affect fuel economy. However, Holley manufactures a reasonably-priced instrument designed specifically for this purpose. Called a *Driving Computer®*, this small, dash-mounted device receives data from two sensors: 1) a pulse generator installed in the fuel line

The Banks *Dynafact*

The Banks *Dynafact®* is an innovative device that measures and displays rear-wheel horsepower using two displays: an analog dyno "meter" and a digital readout. The *Dynafact* combines state-of-the-art computer technology and a uniquely accurate accelerometer to create an instrument that shows—in real time—gross and net horsepower, G-force, speed, and more.

Installation of the *Dynafact* first requires mounting the accelerometer unit on firm, level surface. Under the front seat, the engine compartment, and trunk are all acceptable locations. Next, the control electronics—incorporating the LCD display—and the analog meter are mounted on the dash with wide Velcro tape. The Velcro offers both a secure mounting and easy removal when the *Dynafact* is not in use. Since the electronics must also analyze vehicle speed, electrical connection to a speed sensor is required. Many late-model vehicles use a "pulse-motor" speed sensor in place of a speedometer cable, and the *Dynafact* will work with most of these factory sensors. However, if the vehicle is not equipped with an electronic sensor, one must be installed in the speedometer cable. Finally, the accelerometer, the *Dynafact*, and the speed sensor are interconnected with the supplied wiring harnesses.

Since the accelerometer mounting position and the properties of the speed sensor are unique for every car/driver combination, the *Dynafact* must be calibrated before use. After this calibration procedure (explained in the installation instructions), the *Dynafact* will generate horsepower and G-readings for this *individual* vehicle. Those readouts are stable at

The Dynafact uses two displays. An analog meter indicates true horsepower just like an engine dyno, and a digital readout allows accurate data recovery.

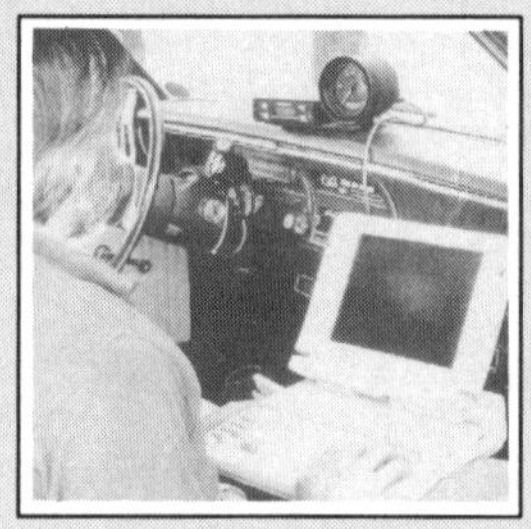

The Dynafact can be easily connected to a laptop computer to produce a powerful performance analysis tool.

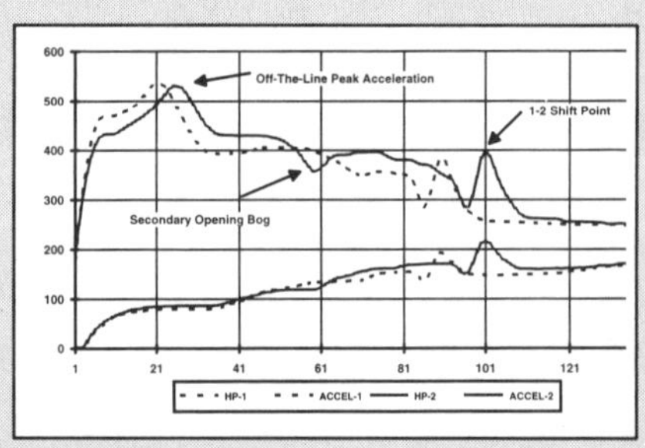

Data from a Dynafact can be analyzed after a "run" by various computer programs, including Lotus 123, Excel, or others. This is a plot of two test runs. Note the bog as the secondaries opened on the second run.

all temperatures and accurate to 1/1000 of a G!

When a test run with the *Dynafact* is complete, the memory-recall function can display the following information at the touch of a button: 1) peak positive and negative (braking) horsepower, 2) peak G-force and the speed at which it occurred, 3) peak Banks Power Factor (a unique measurement of overall vehicle performance), and 4) peak miles-per-hour during the run. Although this information is very useful in evaluating overall performance, small changes that occur from carburetor tuning require an even higher degree of analysis. For these applications, the *Dynafact* is equipped with a computer interface.

The RS-232 serial interface, cable, and software kit for the *Dynafact* adds considerable power to the system. When connected to a portable laptop computer, a direct data stream passes from the *Dynafact* to the computer, logging horsepower, G-force, mile-per-hour, and Banks Power Factor twenty times per second. The resulting data can then be used to produce a graph or "log-book" of exactly the horsepower produced, in 1hp steps, throughout the run. Comparing this data with previous tests, reveals in precise detail the effects of subtle carburetor modifications.

The *Dynafact* is a high-precision instrument that makes it possible for a driver to evaluate the effect of a variety of performance modifications in much the same way as a conventional dynamometer, but at a fraction of the cost. In addition, the *Dynafact* yields information about actual vehicle acceleration that is not possible to obtain from any testing fixture.

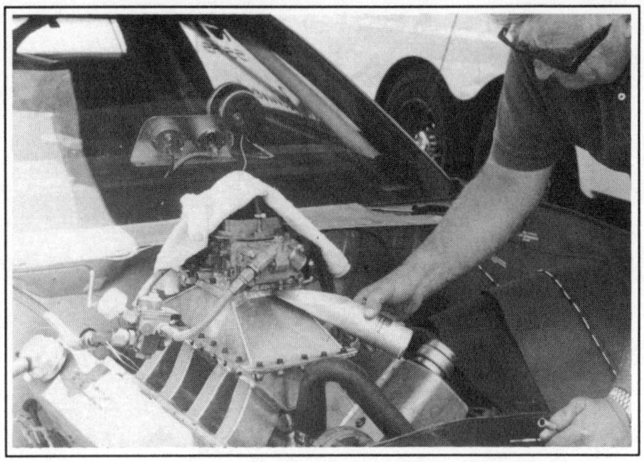

Racing engines often require high idle speeds but may still develop extremely low manifold vacuum. This low vacuum level can allow the power valve to open and begin feeding an over-rich (acceleration-rich) mixture to the engine too soon. This is usually corrected by richening the idle and transition circuits and installing a lower number power valve.

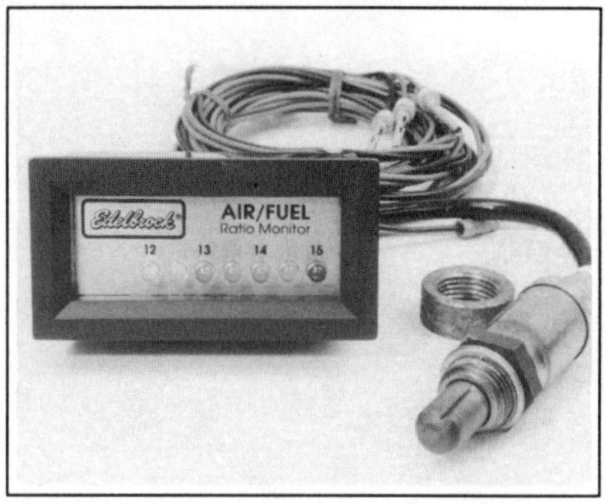

The rich-lean monitor is a very useful tuning aid. A series of LEDs display the air/fuel ratio being delivered to the engine. This monitor is available from Edelbrock and uses an O_2 sensor in the exhaust system. The unit is relatively inexpensive and easy to install.

that works like a miniature gas meter, sending electrical signals to the computer when the impeller is rotated by passing fuel, and 2) a magnetic pickup located near the driveshaft that lets the computer know how fast the vehicle is moving. With a steady stream of data from these two sensors—and a precise internal clock—the *Driving Computer* is able to display gas mileage (instantaneous and average), speed, driving distance on remaining fuel, and when the unit is not in use, the display indicates the time of day.

Although the *Driving Computer* works on many vehicles, there are some exceptions. Most fuel-injected cars have a fuel return line that directs excess fuel back to the fuel tank. Because there is no (simple) way to determine how much of the fuel that enters the injector body—and is measured by the pulse generator—is actually injected into the engine, the *Driving Computer* cannot be used on most fuel-injected engines. However, the instrument works on any vehicle equipped with a Holley carburetor and most other carbureted vehicles.

Another drawback is the limited flow capacity of the fuel-metering pulse generator. For most street-driven vehicles, especially those with smallblock engines, the pulse generator does not present a significant restriction in fuel delivery system. On high-performance or large-displacement engines, however, the pulse generator can introduce enough restriction to cause high-speed fuel starvation. The best advice may be to use the *Driving Computer* to measure and optimize fuel economy, and when that task is accomplished, remove the pulse generator from the fuel line to allow peak performance.

Another useful tuning instrument is the rich-lean monitor. Using a series of LEDs (light-emitting diodes) it provides an instantaneous display of the air/fuel ratio being delivered to the engine. Rich-lean monitors are currently sold by Edelbrock, MSD, K&N, and others. Monitor designs vary, but each consists of an O_2 sensor and a display to indicate the current air/fuel ratio. They are inexpensive and easy to install, requiring only dash mounting of the display and the welding of a mounting nut for the sensor in the exhaust system. Rich-lean monitors offer the easiest and least expensive way to optimize jetting and power valves. The virtually instantaneous response to changes in air/fuel ratios allows this tool to display the entire fuel-delivery curve, from idle, through off-idle, cruise, and full-throttle acceleration. Even variations in accelerator pump delivery can be detected.

If you just can't afford any of these high-tech devices, don't give up. Although not as accurate, there are still effective methods of evaluating performance, and many require little more than a stopwatch. By accurately timing acceleration over a measured distance, you can obtain information that can help you evaluate many carburetion modifications. In order to obtain reliable data, make sure you use the same start and stop locations, try to duplicate your driving technique, and repeat the test several times before making any further modifications. And to really check your results, return to the original carburetion setting after several modification tests; if you are able to repeat the original times, your measuring technique is probably valid.

You can also measure fuel economy without high-tech instruments by dividing the distance traveled by the amount of fuel consumed. To optimize your accuracy, follow these guidelines: 1) make sure to fill the fuel tank to the same level before and after each test, 2) perform each of your tests over the same course and in the same direction, 3) drive at the same speed and use the same amount of throttle to accelerate, and 4) drive over as long a distance as possible, preferably a sufficient distance to consume at least 1/4 of the fuel in your tank.

THE ACCELERATOR PUMP

When accelerating from an idle, engine performance is greatly influenced by the volume, timing, and duration of injected accelerator pump fuel. The requirements for each of these variables can change considerably from one engine or chassis combination to another. For example: Engines with racing cams and open-plenum intake manifolds require more discharge volume than stock engines, heavier cars require longer pump duration than light vehicles, and drag-race vehicles that stage with partially open throttles need accelerator pump fuel de-

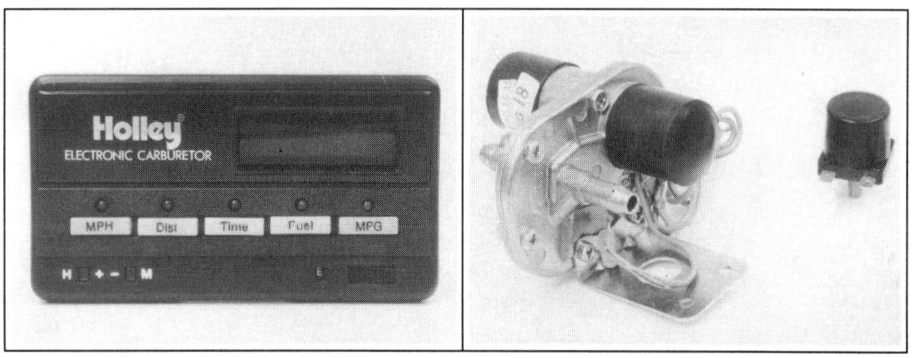

Holley's *Driving Computer* is a small dash-mounted device that uses two sensors (right) to monitor fuel flowing to the carburetor and vehicle speed. With a steady stream of data from these sensors—and an accurate internal clock—the *Driving Computer* is able to display gas mileage (instantaneous and average), vehicle speed, driving distance on remaining fuel, and even the time of day.

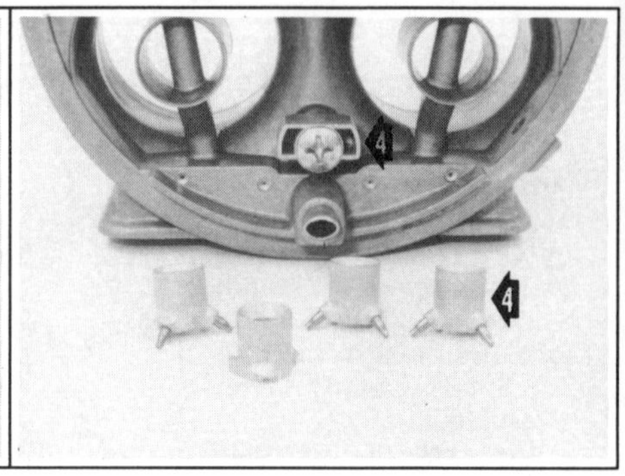

Holley accelerator pumps consist of four main elements: 1) The plastic cam located on the throttle shaft that determines when, and affects how much, fuel volume is discharged into the engine... 2) The pump, located on the bottom of the float bowl, consisting of diaphragm, cover, and spring assembly. The size of the pump determines the maximum volume of discharge fuel... 3) The pump lever that actuates the pump assembly; and 4) The discharge nozzle, mounted in the airhorn area between the throttle bores, contains orifices that determine how long the fuel "shot" will last.

livered later than normal.

Accelerator pumps on Holley modular carburetors are made of four main elements: 1) the plastic actuation cam mounted on the throttle shaft that determines when, and affects how much, fuel volume is discharged into the engine; 2) the pump, located on the bottom of the float bowl, consisting of diaphragm and cover assemblies. The pump determines the maximum volume of discharge fuel; 3) the pump lever that rides on the cam and actuates the pump assembly; and 4) the discharge nozzle, located in the airhorn area between the throttle bores, contains orifices that determine how long the fuel "shot" will last.

Holley offers a wide variety of actuating cams, discharge nozzles, and two pump assemblies that permit a considerable range of adjustment in fuel volume, duration, and timing.

The pump itself is available in two sizes: A 30cc (delivering 30 cubic centimeters of fuel when the pump is actuated through ten full strokes) is standard equipment on most vacuum-secondary carburetors, and a larger 50cc version is often used on double-pumpers and Dominator carburetors. The larger pump can be installed on any modular Holley carburetor by using the Holley installation kit, **part 20-11**.

The larger pump volume and duration can be a big help in reducing throttle-transition problems (bog), especially on all-out racing engines. On the other hand, installing a 50cc pump can create a bog when the added pump volume is not required. Since the 50cc pump package injects a substantially larger volume of fuel than the standard 30cc pump each time the throttle is opened, this modification should be avoided in any fuel-economy application.

The plastic actuating cam determines *the amount of fuel delivered by each stroke of the pump and at what point fuel begins to flow in the rotation of the throttle shaft.* Varying these pump characteristics is accomplished by changing actuating cams and by using the different ("1," "2" or sometimes "3") mounting holes located in the plastic cam and throttle-linkage arm. The Holley kit, **part 20-12**, includes several pump cams that vary delivery volume of the standard 30cc pump from about 17cc to over 35cc. These 30cc cams can also be used on 50cc pumps to reduce the amount of fuel discharged. But never use 50cc plastic cams with 30cc pumps; they will bottom out the pump diaphragm and bend or damage the actuating linkage.

Holley manufactures three different styles of pump discharge nozzles: straight (no tube extensions), an extended tube type, and an anti-pullover type. Spread-bore carburetors must use anti-pullover squirters, since they do not use a discharge check valve normally located under the nozzle that prevents fuel from "dribbling" into the airstream (for more information on this check valve, refer to **Step 46**, on page 51 in Chapter 6 or **Step 49** on page 93 in Chapter 9). All square-bore models incorporate the check valve and may use any style squirter.

Nozzles are available with discharge diameters from .025- to .052-inch. The discharge diameter is stamped on the squirter. *Note: Whenever the size squirter exceeds .040-inch, a high volume nozzle screw,* ***part 26-12****, should be used in place of the standard screw to ensure that the only limiting restriction in the pump system is the discharge nozzle.* Along with the size of the pump and the shape of the actuating cam, *the size of the discharge nozzle is important in controlling the length of time accelerator-pump fuel is delivered when the throttle is opened.* When nozzle flow is limited—by using a smaller nozzle diameter—total injected fuel volume remains unchanged, but fuel is squirted for a longer period of time. Larger nozzles increase the flow rate but shorten pump-discharge duration.

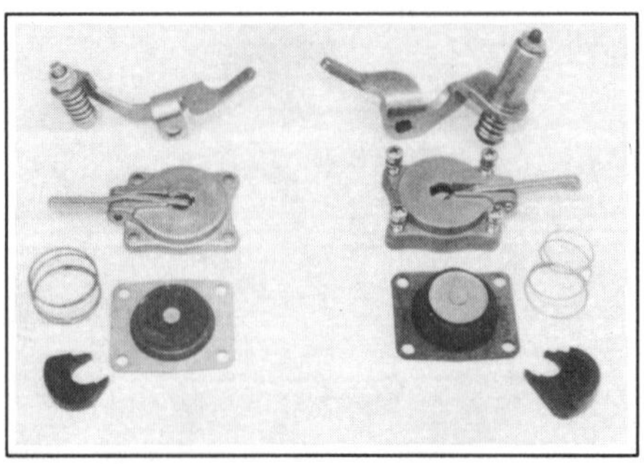

Holley pumps are available in two sizes: a 30cc standard pump (left), and a larger 50cc version often used on double-pumpers and Dominator carburetors. The larger pump can be installed on any modular Holley carburetor by using the Holley installation kit, *part 20-11*.

Accelerator Pump Practical Tuning Tips

Since all of the tunable elements in the accelerator pump circuit have interrelated functions, you should carefully think out a performance problem before you

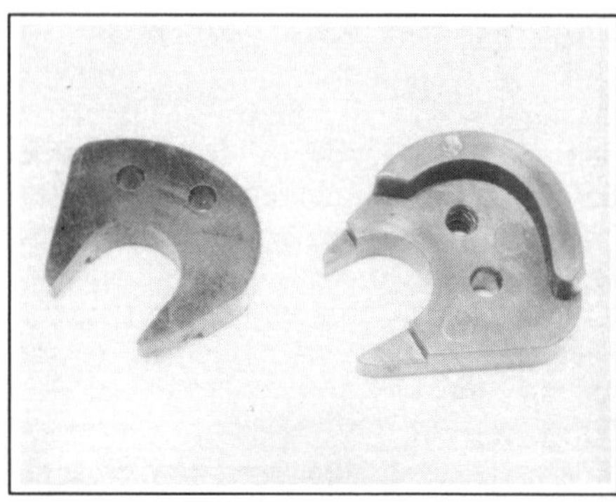

Holley manufactures two distinct types of pump cams: 30cc (left) and 50cc types. While 30cc cams can be used with 50cc pumps to reduce the amount of fuel discharged, 50cc plastic cams can never be used with 30cc pumps; as they will bottom out the pump diaphragm and bend or damage the actuating linkage.

attempt a modification.

The remedy often applied to a vehicle that begins accelerating strongly but then bogs and finally picks up again is to drill-out or install a larger "squirter" (discharge nozzle). If the vehicle is heavy, it may not need the greater discharge rate that a larger nozzle diameter produces. The engine bog may be caused by the need for a longer discharge of the same or, perhaps, slightly more volume. The best cure in this situation is often to use a smaller discharge nozzle! The reduced flow rate will lengthen the time pump fuel flows and allow the main system to activate before pump flow ceases. In some instances, you may need to use a higher displacement pump cam to extend the discharge duration even further. However, a light car with a high-performance engine often builds engine speed very quickly. In this case, large fuel volume delivered quickly may be the best cure for a hesitation occurring immediately off the starting line. Here, a larger diameter squirter—possibly combined with a larger-displacement cam or even a 50cc pump—would optimize standing-start acceleration. *Anytime you change pump cams, make sure the pump lever-to-cam clearance is correctly adjusted* (for step-by-step instructions on this adjustment, see **Step 84** on page 103, in Chapter 9).

Excessive fuel delivery can also degrade throttle response and produce a bog. This condition frequently causes sluggish off-the-line performance, lazy throttle response, and even a puff of black smoke from the exhaust every time the throttle is rapidly depressed. The cure is to install a squirter several sizes smaller, or, if you are using a 50cc pump, you may need to install a 30cc pump cam to reduce fuel delivery volume.

When tuning for optimum economy, use the smallest squirter and cam that prevents hesitation. Excessive pump output is a pure waste of fuel, and a poorly adjusted accelerator pump can make a substantial difference in fuel economy, especially in stop-and-go traffic where the throttle is constantly moving. The relationship of carburetor flow capacity to engine displacement has a major effect on the amount of accelerator-pump fuel required to prevent hesitation. A small carburetor installed on a large displacement engine usually requires minimum pump discharge volume. However, a large carburetor on a small engine will need considerably more pump fuel before the main system is activated, and that will decrease fuel economy. Once again, the importance of matching carburetor type and size to engine and vehicle needs can't be overemphasized.

VACUUM SECONDARY TUNING

Before tuning the vacuum-secondary system, it is essential to verify that all parts of the mechanism are functioning properly. Start by checking the secondary throttle connector-link and the secondary throttle-stop screw (if present). If you suspect that the diaphragm is malfunctioning, remove it, and check it as described in **Step 37**, on page 90 in Chapter 9. Finally, verify that the throttle shaft rotates freely and is not bent or carbon fouled. Once you are sure that the mechanical elements of the system are in good shape, you can proceed with fine-tuning.

Choosing An Intake Manifold

Selecting the single "best" intake manifold for any particular engine/chassis combination can be an awesome task, especially if you are unable to test each manifold on the dyno and carefully evaluate its effect on horsepower, torque, engine smoothness, fuel distribution, throttle response, and other variables. Usually, performance enthusiasts make their choices based on advertising, word-of-mouth, or just on what manifold "looks" the most powerful. But manifold selection does not have to be a hit-and-miss process.

The basic design of an intake manifold greatly influences whether it will be better suited for street or racing, high or low engine speeds, or provide an economy or performance boost. There are only two basic categories into which *most* intake manifolds can be classified: *dual-plane and single-plane* designs. Sometimes called a "180-degree" design, the dual-plane manifold has a divided plenum that separates the intake runners into two groups. This design increases signal strength at the booster venturis, improving low-speed torque and throttle response. The increased signal strength also allows the carburetor to meter fuel more accurately, and that can improve fuel economy. Heavier street machines, particularly those equipped with automatic transmissions and low-stall-speed torque converters, will usually perform better with a dual-plane manifold. But the dual-plane design produces more airflow restriction than other high-output designs, and this, combined with that fact that any one cylinder is fed by only one-half of the carburetor, can reduce peak horsepower.

The *single-plane* manifold, sometimes referred to as a "360-degree" design, has an open plenum that allows all the cylinders to draw air/fuel mix from all the carburetor venturis. Restriction is minimized in a single-plane manifold but so is airflow speed, and that reduces signal strength at the booster venturis. The reduced signal makes it more difficult for the carburetor to meter fuel precisely and maintain an optimum air/fuel ratio at lower engine speeds. This often reduces torque, throttle response, and fuel economy. However, at higher engine speeds, the lower restriction—and because any single cylinder can receive air/fuel mix from the entire carburetor—often allows the single-plane manifold to produce higher peak horsepower. Lighter vehicles with modified engines and stick shift transmissions—or those with automatics and high-stall convertors—often perform best with a manifold based on a single-plane design.

A high performance dual-plane manifold provides good throttle response and top end power. Contrary to popular "street lore," a performance dual-plane manifold is the best choice for all but the most radical street machines.

However, for racing, open-plenum manifolds are hard to beat. Notice the differences in runner shape between the Edelbrock and Holley manifolds (both for a Chrysler). Variations in the size, shape, and length of the intake runners have an important effect on the power potential.

Holley Electronic Fuel Bowl Kits

Holley's *Quarter-Mile-Dial*® and *Mile-Dial*® electronic bowl kits allow you to adjust the main metering over a range of approximately 10 to 12 Holley jet numbers from the driver's seat, even while the engine is running! These innovative kits include special float bowls and metering blocks that incorporate an electronic mixture-control solenoid and additional fuel pathways.

The *Mile-Dial* system adds a measured amount of fuel directly into the main wells (this allows adjustment of air/fuel ratio at both light and heavy throttle for street applications), while the *Quarter-Mile-Dial* system supplies fuel through the power valve circuit (affects air/fuel ratio only at heavy throttle, principally for racing applications). By simply turning a knob on a dash-mounted control unit, the amount of added fuel can be easily varied. The kit includes all the components required to complete the installation on one or two bowl assemblies, including the wiring harnesses, gaskets, various jets, and miscellaneous hardware.

Single-bowl kits are available for 600cfm, **part 34-105**, and for 750cfm, **part 34-104**, vacuum-secondary carburetors. Dual-bowl kits are also available for 600-750cfm, **part 34-112**, and for 800-850cfm, **part 34-113**, mechanical-secondary carburetors. Complete two- and four-barrel carburetors factory fitted with electronic bowls are also available from Holley in a variety of sizes. Installing any of the kits involves little more than replacing the stock float bowl(s) and metering block(s) with the electronic unit(s), mounting the control box(s), and making the appropriate wiring connections. Holley recommends selecting the initial primary and secondary jets sizes from a chart provided with the kit. These starting jet sizes approximate stock jetting.

The value of instant jet changing is obvious in drag and road racing, where quick rejetting for different track and weather conditions can be a big plus. The *Quarter-Mile-Dial* system is also helpful in a professional dyno shop, as it allows optimum fuel-mixture settings for each dyno run, saving time and reducing engine wear. But what value does quick rejetting of the *Mile-Dial* offer for a street machine? If a vehicle is frequently driven from sea level to high altitudes, jetting may be too rich at altitude or too lean at sea level. Also, normal mixture settings may be too lean for extreme cold conditions. In these situations and to facilitate fine-tuning, Holley's mixture control can adjust the air/fuel ratio for improved driveability and economy.

The *Quarter-Mile-Dial* and *Mile-Dial* systems also can be used with Holley's *Mileage Computer*, **part 34-200**, discussed in the text. Using feedback from the *Mileage Computer*, the air/fuel ratio can be adjusted to obtain the optimum gas mileage while cruising at a steady speed on the highway. These two Holley accessories can even add fun to a long highway drive. The Holley *Quarter-Mile-Dial* and *Mile-Dial* electronic float bowl kits are quality components that provide a reliable and repeatable way of easily adjusting air/fuel ratios.

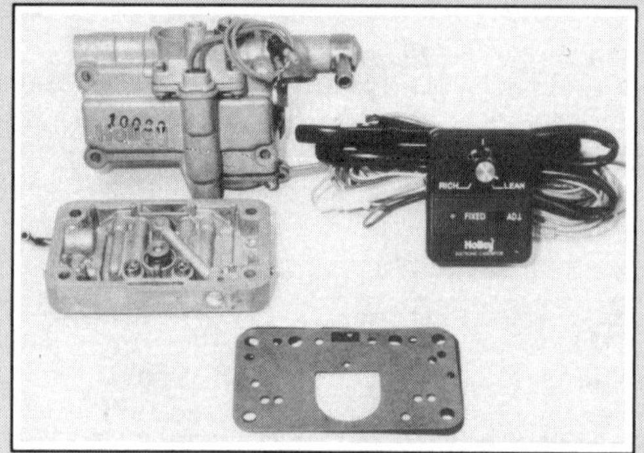

The *Quarter Mile Dial* and *Mile-Dial* conversion kits consist of a special metering block(s), float bowl(s), signal generator (control unit), and the necessary wiring harnesses. Installation is straightforward and can be accomplished in a short time.

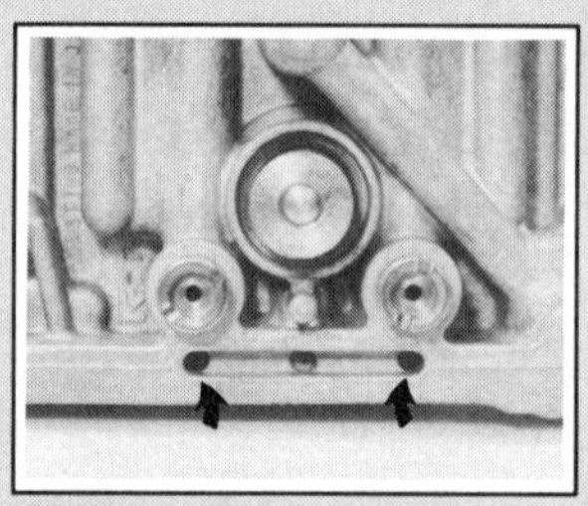

Holley offers two separate kits that deliver fuel through different added pathways: the *Mile-Dial* system adds fuel directly into the main wells and allows air/fuel adjustments at both light and heavy throttle for street applications, the *Quarter-Mile-Dial* system supplies fuel through the power valve circuit and affects air/fuel ratio only at heavy throttle, best for racing applications.

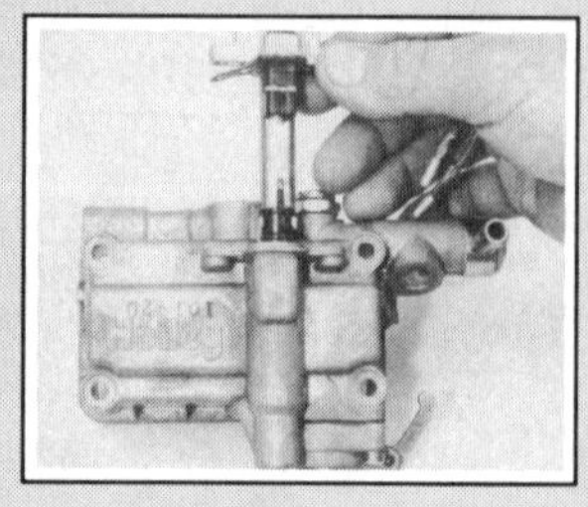

The heart of the system is this solenoid. The signal generator "pulses" the solenoid, controlling the amount of added fuel.

Changing The Opening Rate

Holley recommends changing vacuum-diaphragm springs (not changing or plugging vacuum passages or restrictions) as the only method of tuning vacuum secondaries. Holley offers a kit, **part 20-13**, that contains an assortment of diaphragm springs. Several of these springs are also included in Holley *Trick* carburetor rebuilding kits. A *Quick Change*® Kit, **part 20-59**, makes changing spring a snap (see **Step 31** in the reassembly chapter on page 89 for more information about the *Quick Change* kit). But changing diaphragm springs is not the only way that enthusiasts have altered secondary opening rates; removing or changing diaphragm restrictions and/or plugging secondary "killer bleeds" have also been used, often to the consternation of Holley engineers.

Removing the check ball or drilling the metered restriction in the secondary diaphragm housing will, in many cases, cause the secondaries to open too quickly, adversely affecting acceleration. These ineffective modifications are popular because they allow the driver to "feel" the secondaries open, but what is usually being felt is an engine bog before the secondary main system can deliver the proper air/fuel ratio. The only exceptions can be found on some vehicles with very high power-to-weight ratios using small vacuum-secondary carburetors. In these unusual circumstances, the vehicle may benefit from removing the check ball or enlarging the metering restriction in the secondary diaphragm housing. But as a rule, despite the vehicle and engine combination, if you can distinctly feel the secondaries open, they're probably opening too early.

Some vacuum-secondary carburetors incorporate a "killer bleed." This is a small-diameter passage drilled into the secondary venturi that "bleeds off" part of the vacuum signal generated by the primary vacuum pickup. Its purpose is to delay the full opening of the secondary throttles until airflow through the secondary venturis reaches a predetermined level. Enlarging this passage will delay the full-opening of the secondaries, while decreasing or plugging this passage will cause the secondaries to open earlier. Although secondary killer-bleed modifications are not recommended by Holley engineers, this modification usually has a less-pronounced effect than modifying the diaphragm restriction or removing the

The signal that actuates the secondary diaphragm is generated by a small hole just below the minimum diameter point in the primary venturi. Some vacuum-secondary carburetors incorporate a "killer bleed," a connected passage drilled into the secondary venturi that delays the full opening of the secondary throttles until airflow through the secondaries reaches a predetermined level.

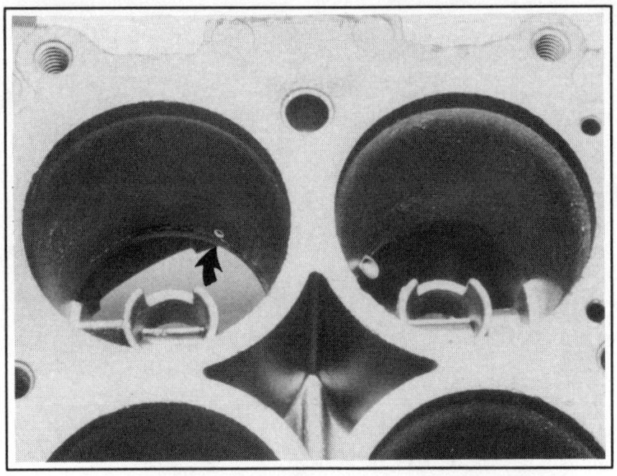

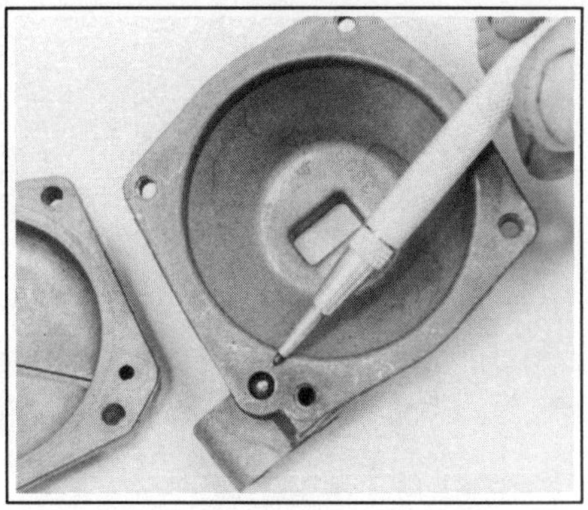

Removing the check ball or drilling out a metering restriction in the secondary diaphragm housing will, in almost every case, allow the driver to "feel" the secondaries open. But what is often being felt is an engine bog. As a rule, despite the vehicle and engine combination, if you can distinctly feel the secondaries open, they're probably opening too early.

check ball.

Secondary Fine Tuning Tips

The trick to fine tuning the secondary opening rate is to locate a diaphragm spring that opens the secondaries as quickly as possible without producing a bog in acceleration. *However, with the optimum spring installed, you may not be able to feel the secondaries open.* If you can clearly tell when the secondaries kick in, it often means that they have opened too soon. A noticeable lurch halfway through first gear is really engine bog, and is caused by insufficient airflow through the secondary venturis. If this is how your Holley seems to work, change to the next heavier diaphragm spring until the bog is no longer felt. If the existing spring does not cause an engine bog, change to the next lighter spring until you begin to feel the secondaries open, then move back to the next stiffer spring (refer to the appendix, page 126, for a list of spring stiffness and their color codes).

If the Holley you have selected for your engine is too large, you may find that a spring light enough to allow the secondaries to fully open also creates a noticeable bog. Without replacing the carburetor with a smaller one, you will probably find better overall performance and driveability by using a spring that eliminates bogging instead of one that allows full secondary opening.

Finally, make sure to perform all testing with the air cleaner installed. The slight airflow restriction from the filter element affects the secondary opening rate.

TUNING THE MAIN METERING SYSTEM

Air Bleeds

The next important air/fuel metering circuits that can be tuned to improve performance and economy are the primary and secondary main metering systems, including the power-valve enrichment circuits. But before we begin to explore the modification possibilities, I must draw your attention to an important part of these circuits; the idle and main system air bleeds. These are small restrictions located on the top of the carburetor in the airhorn area. Although some enthusiasts casually modify these orifices, these changes have two major effects: 1) it raises or lowers the airflow and vacuum required to cause the main and idle circuits to begin delivering fuel, and 2) it alters the overall fuel delivery curve. The impact that these changes may have on performance and economy is not easy to evaluate, since only portions of the fuel delivery curve may have been altered. In fact, Holley engineers spend many hours of research with sophisticated wet-flow benches to determine the sizes of these air bleeds and other minute "corrector" restrictions in the metering blocks. Without similar equipment or an engine dyno, it is unlikely that modifications to the air bleeds will result in any improvements. It is much more likely, however, that the fuel delivery curve will be altered in some way that may render the carburetor "untunable" by any normal methods. I have only three words of advice if you're considering some "trick" change to the air bleeds on your Holley: DON'T DO IT!

The four primary idle and main system air bleeds are located on the top of the carburetor inside the airhorn (arrows). This Holley has additional intermediate circuits that use two additional air bleeds in front of the airhorn (small arrows). Although some enthusiasts modify these orifices, the results can be unpredictable. I have only three words of advice if you're considering some "trick" change to the air bleeds: DON'T DO IT!

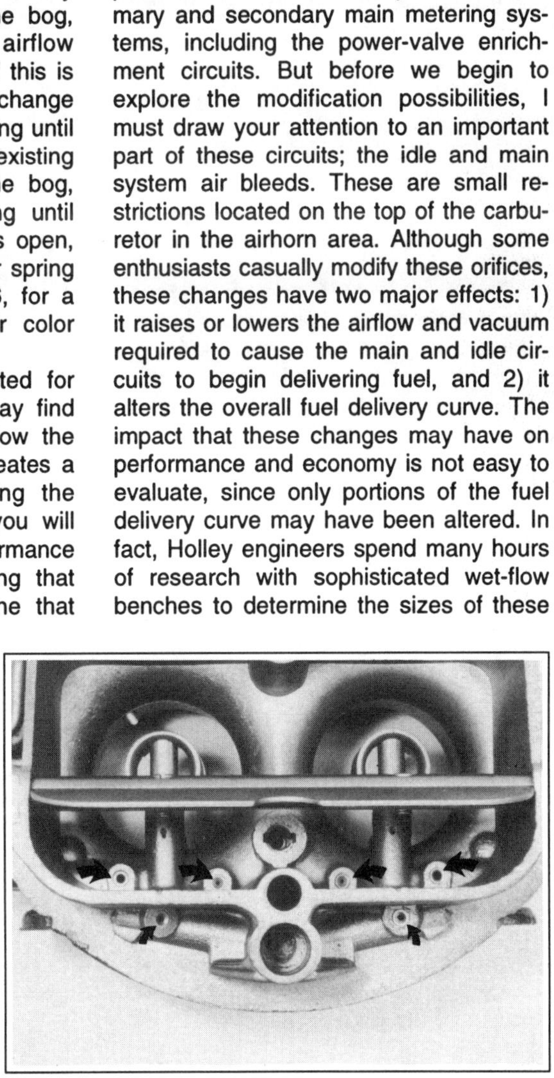

Power Valves And Jetting

You cannot effectively fine tune the main-metering system without understanding the relationship between the power-valve circuit orifices (called Power Valve Channel Restrictions, or PVCRs) and the main-jet orifices. These independent flow channels work together to establish the air/fuel ratio during part- to full-throttle operation.

All gasoline engines require a lean mixture for good fuel economy during cruise and a richer mixture for good performance during heavy acceleration. The power valve is a vacuum-actuated valve that performs this enrichment function. It allows additional fuel to enter the main-metering system when the vacuum level in the intake manifold drops to a preset value. The vacuum level at which the valve opens is fixed by internal spring tension, and its opening point is stamped somewhere on the valve assembly.

When the power valve opens, additional fuel is allowed to flow through the PVCRs into the main wells. Fuel coming through the main jets also enters the main wells and mixes with PVCR fuel before it is delivered to the engine. *So, the PVCRs are actually a second set of*

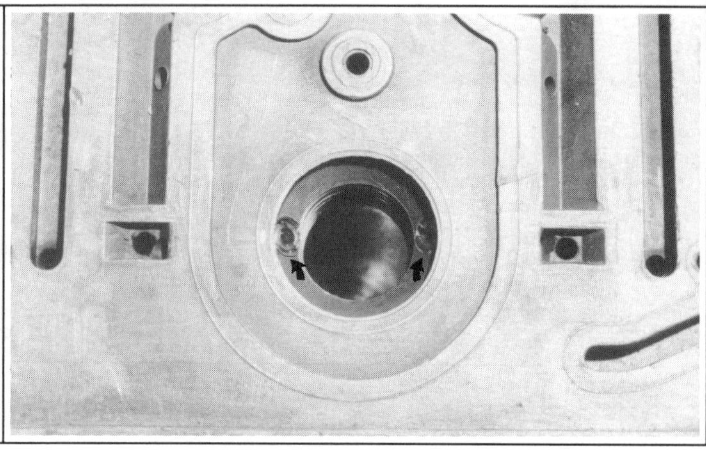

You cannot effectively fine tune the main-metering system without understanding the relationship between the power-valve circuit orifices (called Power Valve Channel Restrictions, or PVCRs) and the main-jet orifices. These independent flow channels (arrows) work together to establish the air/fuel ratio during part- to full-throttle operation. When the power valve opens, additional fuel is allowed to flow through the PVCRs into the main wells. Fuel coming through the main jets also enters the main wells and mixes with PVCR fuel before it is delivered to the engine. *So, the PVCRs are actually a second set of main jets that determine heavy-load air/fuel ratios.* Unfortunately, PVCRs are a fixed restriction and are not designed to be easily changed or modified. The easiest way to tune this circuit is to modify your metering block(s) to accept replaceable PVCRs jets (right photo) with a kit available from *The Carburetor Shop*.

main jets that determine heavy-load air/fuel ratios. Unfortunately, PVCRs are a fixed restriction and are not designed to be easily changed or modified. The easiest way to tune this circuit—and there are often respectable power and economy benefits from doing so—is to modify your metering block(s) to accept replaceable PVCRs jets. *The Carburetor Shop* offers a kit for this purpose or you can send your metering block(s) to them to have the modification done for you.

PVCRs Or Just Main Jets?

When most enthusiasts consider power tuning the main system on a Holley carburetor, they think about changing—usually richening—main jets until optimum power is measured at wide-open throttle operation. However, the same primary jets that produced optimum peak power will often allow too much fuel to flow at part throttle, even when the PVCRs are closed at high manifold vacuum. The result is good full-throttle power and poor part-throttle economy.

Economy tuning typically involves the same procedure, except smaller and smaller main jets are installed until engine experiences part-throttle cruise misfire, ping, or surging. Then the main jets are richened just enough to eliminate these problems, often producing an air/fuel ratio of about 15:1 or 16:1. But the air/fuel ratio at full throttle may be too lean, causing backfiring, severe surging, etc. In this case, richer main jets have to be installed to return adequate performance to the engine. The overall result is poor full-throttle power and less than optimum part-throttle economy.

In a nutshell, without modifying the PVCRs you can only use the main jets to optimize either wide-open throttle power or part-throttle cruise economy, not both, and sometimes neither. This is an unavoidable drawback to fine tuning only main jets. When the PVCRs are included in tuning process, *the main jets can be tuned to deliver the optimum air/fuel ratio during light load, while the PVCRs are adjusted to add sufficient fuel to provide an optimum air/fuel ratio during heavy-load operation.*

Hands-On Tuning Tips

Finding the exact primary and secondary main-jet and PVCR sizes for your engine depends on three main factors: 1) your fine-tuning goals, 2) the effects of temperature and altitude, and 3) the efficiency of the engine and the exhaust system. Because of these, and several other minor considerations, primary and secondary main jetting for a street-driven vehicle has to be a compromise. There is no single jet size or one perfect fuel supply curve that will produce optimum horsepower or economy in all situations. Despite this fact, a little fine tuning can match your Holley's fuel delivery curve closer to your engine's requirements. Here are some practical tips and techniques that can help you with your specific tuning efforts.

General Tuning Tips—Use a Holley power valve plug to close off the PVCRs and isolate main-jet fuel flow. Then tune the main jets to provide the optimum air-fuel ratio at cruise (vacuum levels above about 10in-Hg). When the optimum main jet size is determined, reinstall the power valve and tune the PVCRs for optimum power at wide-open throttle.

Maximum Power—To calibrate your Holley for maximum power on gasoline, main jet and PVCR sizes must be chosen that produce an air/fuel ratio of about 12:1 at wide-open throttle. As the mixture is enriched beyond this point, power gradually falls off. When jetting is extremely rich, acceleration can be quite sluggish. The power loss from rich jetting is gradual—making it easy to miss the optimum power point—fortunately, an

Tuning a Holley carburetor does not just mean changing the main jets. But when the PVCRs are included in the tuning process, the main jets can be tuned to deliver the optimum economy at light load, while the PVCRs are tuned to produce good performance at heavy throttle. Here a power valve plug is being used to isolate main-jet fuel flow and optimize part-throttle performance before adjusting the PVCRs (see text for details on this procedure).

over-rich mixture rarely leads to engine damage. It does, however, reduce power, ruin fuel economy, and may even carbon-foul sparkplugs. If you are modifying PVCRs, adjust the main jets to produce an air/fuel ratio of 14.7:1 with the PVCRs plugged off. Then adjust the PVCR sizes to produce a full-throttle mixture of about 12:1.

Tuning the PVCRs along with the main jets to produce maximum power is more important for road-race or high-performance street vehicles. Drag race engines do not require superior part-throttle operation or economy; mixtures lean enough to prevent the sparkplugs from loading up in the pits and staging lanes is all that's required. Stock PVCR restrictions are often adequate for that purpose.

Maximum Economy—Optimum economy on gasoline is achieved at fuel mixtures near 15.75:1. Mixtures leaner than this usually cause misfire, hesitation, and severe surging. Using very lean mixtures at other than light-load cruise can cause high combustion temperatures and engine damage. Lean mixtures can burn valves and cause detonation (engine knock) that erodes pistons and may cause stress cracks in crankshafts, rods, pistons, cylinder heads and blocks. If you jet your Holley for super-lean mixtures, make sure that the PVCRs sufficiently enrich the mixture when manifold vacuum drops under heavy loads. If you're modifying the PVCRs along with the main jets, plug the power valve and reduce main jets until the air-fuel mixture approaches 15.75:1 at light throttle. Perform easy acceleration and cruise tests at different engine speeds. If there is surging, excessive pinging, or hesitation, jet richer until throttle response is acceptable. Then reinstall the power valve and jet the PVCRs to provide full-throttle air/fuel ratios of about 14.7:1 for optimum economy or near 12:1 for the best economy/power compromise.

Altitude Modifications—Changes in altitude have a small effect on the air/fuel mixture delivered by the carburetor. Most Holley carburetors are calibrated for operation at sea level. Since less oxygen is available at higher elevations, fuel flow from the main system must be reduced to maintain the optimum air/fuel ratio. Holley engineers recommend decreasing the jet size one Holley number for every 2000-foot increase in elevation. If you modify main jetting for higher altitudes, don't forget to richen the jets when the vehicle returns to lower altitudes.

Temperature Variations—Changes in ambient and/or inducted-air temperatures also have a small effect on the fuel mixture. Holley carburetors are calibrated at 70 degrees(F). For every 35 degrees drop in ambient operating temperature, one Holley number increase in main jet size will restore optimum jetting. Similarly, main jets should be decreased one Holley number for every 35 degrees(F) increase in ambient operating temperature. Again, the relatively small adjustment for temperature variations is only a consideration for race cars or street vehicles that are operated for extended periods of time in extremely cold or hot environments.

Engine Efficiency Considerations—Variations in engine design and overall intake, combustion, and exhaust efficiency

Power Valve Alternatives

Gasoline engines require a richer fuel mixture under heavy loads, and it is the power-valve circuit that allows Holley carburetors to make this adjustment in the air/fuel ratio. Despite the fact that the power-valve circuit is only slightly more complex than an on-off switch—"on" to supply additional fuel, limited by the size of the power-valve-channel restrictions (PVCRs), "off" to limit fuel flow at light throttle—its function is often misunderstood. As a consequence, many enthusiasts have eliminated or replaced the power valve. However, there are only two options to the standard on-off power valve: 1) The *power valve plug* that replaces the regular power valve and blocks off the enrichment circuit, and 2) A *two-stage power valve* that provides a partial "half-on" enrichment as vacuum decreases, in addition to the standard full enrichment at low engine vacuum. Let's take a close look at each of these alternatives.

Power Valve Plugs—A Holley power-valve plug, **part 26-36**, shuts off fuel flow through the enrichment circuit. Plugging the *primary* power valve circuit is rarely necessary, except for tuning purposes (see text), and often creates unsolvable mixture problems. Without enrichment fuel, it is necessary to increase main jetting from five to ten Holley numbers to regain adequate full-throttle performance. The increased main-jet flow at light- and part-throttle produces an extremely rich mixture, plug fouling, and terrible gas mileage. Road race and street driven vehicles (that need excellent part-throttle response) cannot be jetted correctly with the power valve removed. An engine set up for drag racing does not need outstanding part-throttle response, but fouled plugs can still be a serious problem when the power valve is eliminated.

Model 4150 Holley carburetors sometimes have a power valve installed in the secondary metering block. Installation of a power-valve plug in the *secondary* metering block, along with appropriate secondary main-jet enrichment, is an acceptable modification for racing applications and will not significantly affect part-throttle operation. This modification can simplify secondary jetting on a race car; it is of questionable value on a street driven vehicle.

Two-Stage Power Valves—Replacing the standard power valve with a two-stage valve can, in some cases, increase throttle response and fuel economy. The first stage (partial enrichment) opens at relatively high vacuum levels (9- to 12.5-in[Hg]). This partial enrichment sometimes allows the use of smaller main jets that would otherwise be too lean for a single-stage valve at very-light throttle. At lower vacuum levels (2.5- to 6.5-in[Hg]) the valve opens and provides full enrichment, exactly like a single-stage valve.

Improvements with two-stage power valves are most noticeable with heavy vehicles that have low power-to-weight ratios. On these vehicles, such as trucks, heavy cars, or RVs—especially when operated in stop-and-go traffic or on hilly terrain—the two-stage power valve can improve throttle response and economy.

A two-stage power valve may also *decrease* the fuel economy when used in light vehicles with a high power-to-weight ratios. In these situations, the partial enrichment at high vacuum just isn't needed, since the engine is so lightly loaded. In addition, two-stage power valves are restrictive and not capable of supplying fuel to PVCRs larger than .060-inch (each), and then only at moderate engine speeds. *Because of this limitation, two-stage valves should never be used on performance or competition high-rpm engines.*

Use A Power Valve?—The power valve circuit allows Holley carburetors to adjust the air/fuel ratio to suit engine requirements. Disabling this feature almost always results in poor fuel economy and power. Furthermore, two-stage power valves—the only other alternative—have flow limitations and must only be used on low-power, low-speed applications. The bottom line: Using a genuine Holley single-stage power valve is an essential step in obtaining the best performance and economy in any *high-performance* application.

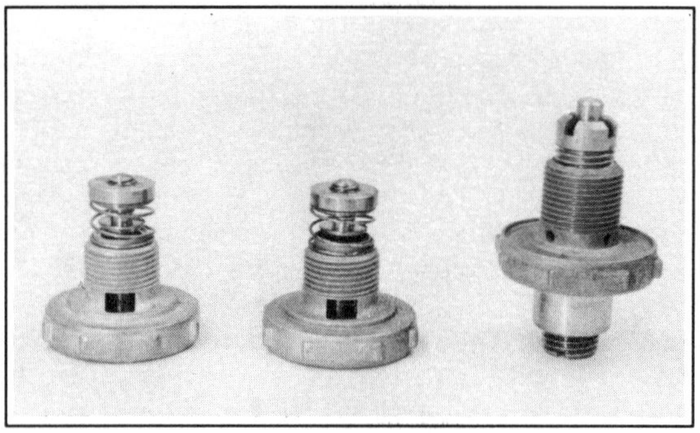

Holley standard-flow power valves (left) accommodate PVCRs up to .090-inch and are available at opening points from 2.5- to 10.5-in(Hg). Holley also manufactures two high-flow power valves (center), for carburetors with larger PVCRs, that open at 6.5- and 10.5-in(Hg). Two-stage power valves (right) are also available that can, in some cases, improve gas mileage on stock street engines.

also affect mixture and jetting requirements. Open-plenum intake manifolds, high compression ratios, ported cylinder heads, radical cams, and headers generally reduce signal strength at the carburetor and often require richer main-system metering to deliver optimum air/fuel ratios. Because there are so many interrelated variables when the induction and exhaust system of an engine is modified, the only effective way to locate optimum economy and/or performance jetting is to use an air/fuel ratio monitor to indicate mixtures and a sophisticated G-meter to measure absolute performance, like the Vericom or Dynafact.

Selecting Power Valves

As I've mentioned, power valves act as a switch controlled by manifold vacuum that turn on the flow of additional fuel into the main metering system. The point at which the power valve opens is determined by the spring tension built into the valve assembly, and this value—in inches of mercury (in[Hg])—is stamped on the valve assembly. For example, a power valve that opens at 6.5-in(Hg) will have the number "65" stamped on the valve. Holley standard-flow power valves, that accommodate PVCRs up to .090-inch, are available from 2.5- to 10.5-in(Hg). Holley also manufactures two high-flow power valves, **part 125-165** and **125-1005**, that open at 6.5- and 10.5-in(Hg), respectively. High-flow valves are recommended for carburetors with PVCRs larger than .090-inch. Two-stage power valves are also available that can, in some cases, improve gas mileage on stock street engines (see the sidebar *Power Valve Alternatives*, on the previous page for more information on using two-stage power valves).

Selecting the correct power valve can only be accomplished with a quality, non-damped vacuum gauge. Observe manifold vacuum readings during various steady cruising speeds. Choose a power valve that opens at about 2 in(Hg) below the lowest steady cruise reading. A number 65, 75, or 85 power valve often works well on stock or slightly modified engines. The higher-number power valves begin adding enrichment fuel sooner, often at the detriment of fuel economy. On the other hand, low-number power valves may open too late, causing hesitation and lean surging at medium throttle.

Power valve selection for race engines equipped with high-overlap cams is a different process. These vehicles often idle at speeds high enough to draw fuel from the main metering system and develop manifold vacuum levels of 10-in(Hg) or less. This low vacuum at idle can open the power valve, resulting in a rich mixture, carbon fouled plugs, and poor throttle response. Observe manifold vacuum at idle (in gear for automatics) and select a power valve that opens about 2-in(Hg) below this minimum level.

In some rare instances, exactly the opposite phenomenon can take place: The engine may develop enough manifold vacuum near full throttle to close a low-number power valve at high engine speeds. This usually occurs with a large-displacement engine equipped with a two-barrel or small four-barrel carburetor. When this happens, power is substantially reduced, and possible engine damage can occur from lean-mixture induced detonation. The power valve rating should be above the highest vacuum observed at heavy-throttle, high-rpm operation.

IDLE FEED RESTRICTIONS

While the main jets and PVCRs are the air/fuel ratio controlling restrictions in the main system, the idle-feed restrictions (IFRs) control the idle and transition circuits. Holley carefully matches the size of the IFRs in each carburetor for its intended application. In most circumstances, this circuit will not need "fine tuning," however, if your Holley is used in an application for which it was not intended, here are the indications that the IFRs may need modification: 1) little or no change in idle quality when the idle-mixture screws are adjusted from full in to full out, and 2) an off-idle stumble that occurs when the throttle is slowly opened from an idle position; accelerator pump tuning has little effect on this stumble. A good example is a race-calibrated double-pumper—designed to work with radical cam profiles and free-flowing exhaust systems—installed on a street engine. Holley fitted the double-pumper with large IFRs to accommodate a weak idle signal commonly found on modified engines. When this carburetor is installed on a street engine with a strong signal at idle, the engine will idle rich even though the mixture screws are almost completely screwed in. At the other end of the scale of misapplication, as I mentioned earlier, is an emissions carburetor installed on a modified engine. In this case, the IFRs are too small to deliver sufficient idle fuel, even when the mixture screws are completely unscrewed. This application will commonly suffer from off-idle hesitation and stumbling.

After reading this you may be convinced that your Holley needs IFR adjustments. But wait! These same symptoms can be caused by many other maladies, including: 1) engine mechanical defects, including weak valve springs, low compression, etc.; 2) electrical systems malfunctions, like a weak coil or electronic control box; 3) the primary throttle blades are not in the correct idle position (see the sidebar below); 4) vacuum leaks; 5) excessively worn throttle shafts (see the sidebar *Repairing Loose Throttle Shafts*, on page 91); 5) incorrect

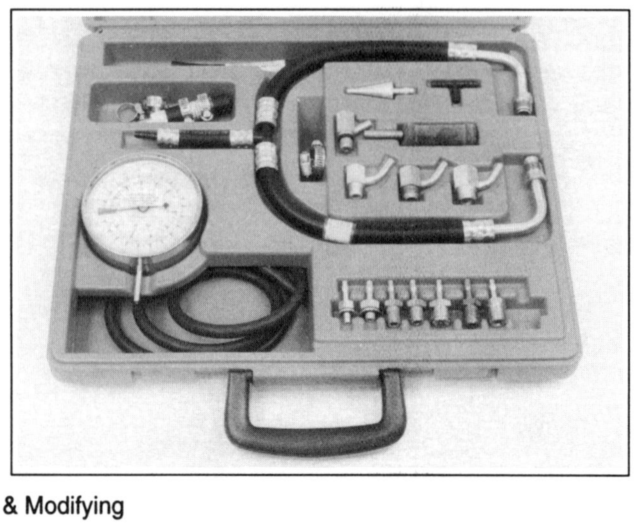

Selecting the correct power valve, and several other engine tuning and troubleshooting procedures, can only be accomplished with a quality, non-damped vacuum gauge. This *Snap-on* vacuum gauge kit includes several adaptors, various lengths of tubing, and an easy-to-read, high-quality gauge.

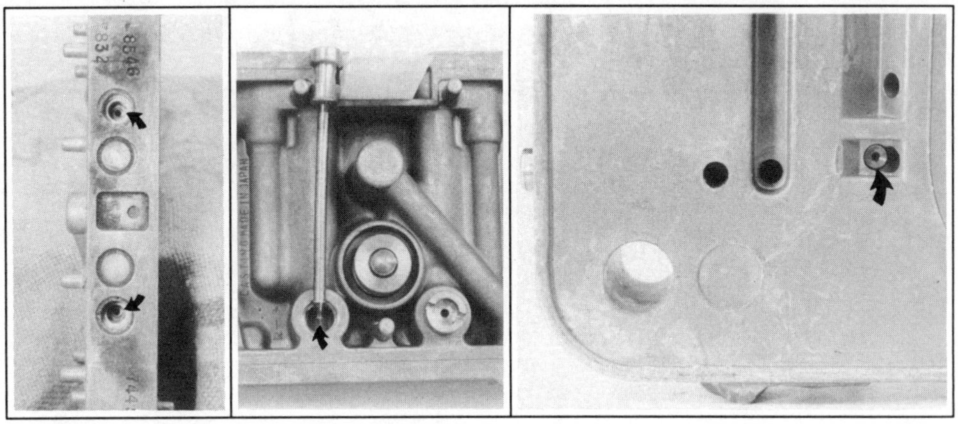

The IFRs are very small orifices, often located in the face of the metering block (1), in the top of the idle wells (2), or at the end of tubes in the main wells (3). A small increase or decrease in size will cause a substantial change in idle-fuel flow. Change the IFRs in increments of one or two thousandths of an inch, and check the results after each modification. As you approach optimum IFR size, the mixture screws will produce a good idle when adjusted between 1/2- and 1-1/2-turns, and off-idle hesitation and stumble will disappear.

or damaged metering block gaskets; and 6) idle wells or IFRs plugged with debris (see the sidebar *Metering Block Cleaning*, on page 63).

When you are certain that no other malfunctions are causing the idle problems you are experiencing, only then consider IFR modifications. The IFRs are very small, and a small increase or decrease in size will cause a substantial change in idle-fuel flow. Change the IFRs in increments of one or two thousandths of an inch, and check the results after each modification. While increasing IFR size is merely a matter of drilling with the appropriate drill bit, decreasing the orifice is considerably more difficult. Some enthusiasts obtain a similar metering block but with smaller IFRs and carefully work up in minute steps. Others have the IFRs removed and smaller restrictions installed. Since removal is a delicate operation requiring special tools, contact a professional shop for this operation.

Whether you're increasing or decreasing IFR size, patience is absolutely essential. Take your time and only make small changes. As you approach optimum IFR size, the idle will begin to take on the following characteristics: 1) the mixture screws will produce a good, stable idle when adjusted between 1/2- and 1-1/2-turns from a lightly seated position, and 2) off-idle hesitation and stumble, that occurs when the throttle is opened slowly, will disappear.

Race Cam Idling Problems

A long-duration camshaft with extended overlap produces a much lower vacuum signal at idle than stock camshafts. While competition double-pumper and Dominator carburetors are designed to compensate for these "big" cams, most street carburetors are calibrated with idle and power-valve circuits for high manifold vacuum conditions. If street carburetors are used with racing cams they almost always require modifications.

The first step is to perform a vacuum check. Make sure to use a non-damped, quality vacuum gauge connected to a non-ported, direct manifold vacuum source. The power valve opening point, indicated by the number stamped on the valve, must be lower than the measured vacuum. If vacuum at idle is lower than the power valve number, there will be main-metering irregularities as the power valve opens and closes during light acceleration. Note: Although the power valve may be open at idle, this does not cause a rich idle mixture; idle mixture is controlled by the much smaller idle-feed restriction orifices. Idle mixture is almost completely independent of main-jet, PVCR, and power valve sizes.

With the correct power valve in place, idle and off-idle problems may still exist with racing cams. Engines with long-duration camshafts often will not idle at speeds under 1200 rpm. These high idle speeds require screwing in the idle-stop screw substantially, which may expose more than the recommended .040- to .060-inch of the idle-transfer slots. This adversely affects the start-up of the main circuit, and is evident in off-idle flat spots and a lack of response to the idle-mixture screws. If these symptoms are present, leave the idle stop adjustment where it is, and remove the carburetor from the engine. Measure the transfer-slot exposure with a small ruler (see photos). If it is beyond .060-inch, there are two common methods used to reduce this exposure. First, try adjusting the secondary idle-stop screw 1/2- to 1-turn further open. Additional fuel from the secondary idle circuits often allows the primary idle stop to be returned to within the normal range. If, after this adjustment, excessive transition-slot exposure is still required to maintain sufficient idle speed, drill a 1/16-inch hole in each primary throttle blade on the same side as the idle transfer slot. It may be necessary to increase these "bleed" orifices to 1/8-inch for very radical camshafts.

If a light-throttle, "tip-in" bog remains after idle transfer-slot exposure is optimized, refer to the sidebar *Accelerator Pump Basics,* on page 92 for more tuning tips. Finally, if the problem is reduced but not eliminated after pump-circuit adjustments, the idle-feed restrictions may need to be enlarged, see the section IDLE FEED RESTRICTIONS on page 118).

Holley offers this convenient 0-30-in(HG) vacuum gauge, *part 26-501.* It is very helpful for tuning and troubleshooting; a check of manifold vacuum can determine if the power valve is opening too early.

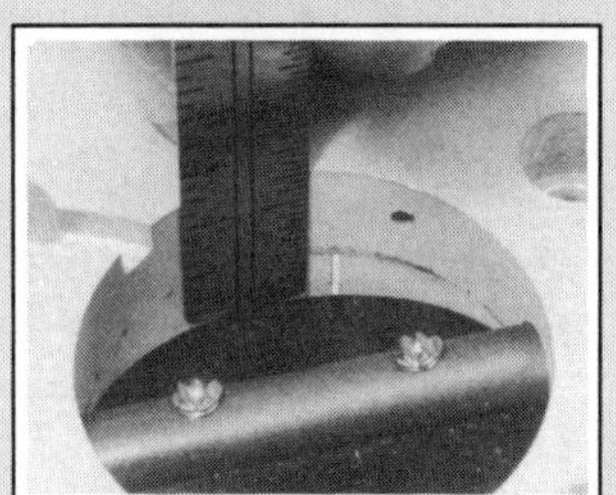

No more than .040- to .060-inch of the idle-transfer slots should be exposed. More than this can cause an off-idle flat spots and a lack of response to the idle-mixture screws.

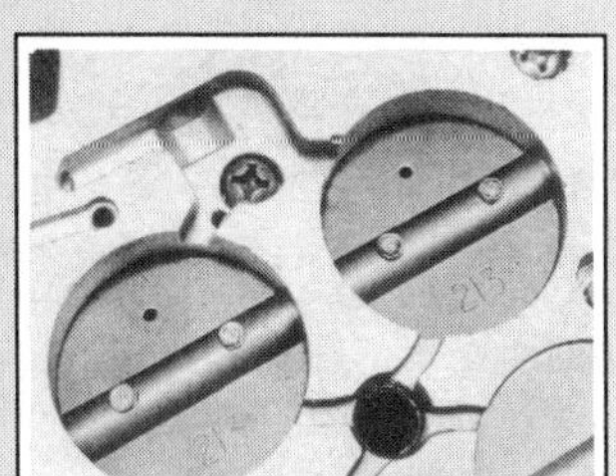

If opening the secondary idle stop does not correct the problem, you may have to drill 1/16- or 1/8-inch holes in the primary butterflies.

Holley Numerical Listing

Carburetor Part No.	Carburetor Model Number	CFM	Renew Kit	Trick Kit	Primary Main Jet	Secondary Main Jet or Plate	Power Valves	Primary Discharge Nozzle	Secondary Nozzle Or Spring Color	Primary Throttle Bore Diameter	Secondary Throttle Bore Diameter
0-1848-1	4160	465	37-119	37-912	122-57	134-3	125-85	.025	Green	1-1/2	1-1/2
0-1849	4160	550	37-119	37-912	122-62	134-5	125-85	.025	Plain	1-1/2	1-1/2
0-1850-2	4160	600	37-119	37-912	122-66	134-9	125-65	.025	Plain	1-9/16	1-9/16
0-1850-3	4160	600	37-119	37-912	122-66	134-9	125-65	.025	Plain	1-9/16	1-9/16
0-2818-1	4150	600	37-333	37-905	122-65	122-76	125-65	.025	Purple	1-9/16	1-9/16
0-3310-1*	4150	780	37-1084	37-915	122-72	122-76	125-105(P),125-85(S)	.025	Plain	1-11/16	1-11/16
0-3310-2*	4160	750	37-754	37-921	122-72	134-21	125-65	.025	Plain	1-11/16	1-11/16
0-3310-3*	4160	750	37-754	37-921	122-72	134-21	125-65	.025	Plain	1-11/16	1-11/16
0-3418-1	4150	855	37-1273	N/A	122-78T/122-82C	122-80T/122-82C	125-65 (Both)	.028	Yellow	1-3/4	1-3/4
0-4118*	4150	725	37-1184	37-906	122-68	122-78	125-85	.025	Yellow	1-11/16	1-11/16
0-4224	4160	660	37-424	N/A	122-76	134-12	N/R	.025	.025	1-11/16	1-11/16
0-4412	2300	500	37-474	37-901	122-73	N/R	125-50	.028	N/R	1-11/16	N/R
0-4412-1	2300	500	37-474	37-901	122-73	N/R	125-50	.028	N/R	1-11/16	N/R
0-4452-1	4160	600	37-119	37-912	122-63	134-39	125-85	.031	Purple	1-9/16	1-9/16
0-4548	4160	450	37-119	N/A	122-57	134-30	25R-219-85	.031	Brown	1-1/2	1-1/2
0-4575**	4500	1050	37-487	37-920	122-84	122-84	125-65 (Both)	.035	.035	2	2
0-4742	4150	600	37-433	N/A	122-63	122-72	25R-237-85	.031	Purple	1-9/16	1-9/16
0-4776**	4150	600	37-485	37-910	122-69	122-71	125-65	.025	.032	1-9/16	1-9/16
0-4776-1**	4150	600	37-485	37-910	122-66	122-76	125-65	.028	.032	1-9/16	1-9/16
0-4776-2**	4150	600	37-485	37-910	122-66	122-76	125-65	.028	.032	1-9/16	1-9/16
0-4776-3**	4150	600	37-485	37-910	122-66	122-73	125-65	.028	.032	1-9/16	1-9/16
0-4777**	4150	650	37-485	37-910	122-71	122-76	125-65	.025	.025	1-11/16	1-11/16
0-4777-1**	4150	650	37-485	37-910	122-67	122-76	125-65	.028	.028	1-11/16	1-11/16
0-4777-2**	4150	650	37-485	37-910	122-67	122-76	125-65	.028	.028	1-11/16	1-11/16
0-4777-3**	4150	650	37-485	37-910	122-67	122-73	125-65	.028	.028	1-11/16	1-11/16
0-4778**	4150	700	37-485	37-910	122-66	122-71	125-65	.028	.032	1-11/16	1-11/16
0-4778-1**	4150	700	37-485	37-910	122-66	122-76	125-65	.028	.031	1-11/16	1-11/16
0-4778-2**	4150	700	37-485	37-910	122-66	122-76	125-65	.028	.031	1-11/16	1-11/16
0-4778-3**	4150	700	37-485	37-910	122-69	122-78	125-65	.028	.031	1-11/16	1-11/16
0-4779**	4150	750	37-485	37-910	122-75	122-76	125-85	.025	.032	1-11/16	1-11/16
0-4779-1**	4150	750	37-485	37-910	122-70	122-80	125-85	.028	.031	1-11/16	1-11/16
0-4779-2**	4150	750	37-485	37-910	122-70	122-80	125-65	.028	.031	1-11/16	1-11/16
0-4779-3**	4150	750	37-485	37-910	122-70	122-73	125-65	.028	.031	1-11/16	1-11/16
0-4779-4*	4150	750	37-485	37-910	122-70	122-80	125-65	.028	.031	1-11/16	1-11/16
0-4780**	4150	800	37-485	37-910	122-72	122-76	125-65(P),125-85(S)	.031	.031	1-11/16	1-11/16
0-4780-1	4150	800	37-485	37-910	122-70	122-76	125-65(P),125-85(S)	.031	.031	1-11/16	1-11/16
0-4780-2**	4150	800	37-485	37-910	122-70	122-85	125-65	.031	.031	1-11/16	1-11/16
0-4780-3**	4150	800	37-485	37-910	122-71	122-85	125-65	.031	.031	1-11/16	1-11/16
0-4781**	4150	850	37-485	37-916	122-80	122-80	125-65 (Both)	.035	.025	1-3/4	1-3/4
0-4781-1**	4150	850	37-485	37-916	122-80	122-80	125-65 (Both)	.031	.031	1-3/4	1-3/4
0-4781-2**	4150	850	37-485	37-916	122-80	122-80	125-65 (Both)	.031	.031	1-3/4	1-3/4
0-4781-3**	4150	850	37-485	37-916	122-80	122-78	125-65 (Both)	.031	.031	1-3/4	1-3/4
0-4782	2300	355	37-396	N/A	122-64	N/R	125-65	.031	N/R	1-1/2	N/R
0-4783	2300	500	37-396	N/A	122-82	N/R	N/R	.028	N/R	1-3/4	N/R
0-4788**	4150	830	37-485	37-916	122-80	122-80	125-65 (Both)	.031	.031	1-11/16	1-11/16
0-4788-1**	4150	830	37-485	37-916	122-80	122-80	125-65 (Both)	.031	.031	1-11/16	1-11/16
0-6109	4150	750	37-485	37-910	122-75	122-76	125-85	.025	.032	1-11/16	1-11/16
0-6210-1	4165	650	37-605	37-918	122-602	122-632	125-85(P),125-65(S)	.025	.037	1-3/8	2
0-6210-2	4165	650	37-605	37-918	122-602	122-83	125-85	.025	.037	1-3/8	2
0-6210-3	4165	650	37-605	37-918	122-602	122-83	125-85	.025	.037	1-3/8	2
0-6211	4165	800	37-605	37-918	122-62	122-85	125-85(P),125-65(S)	.025	.037	1-3/8	2
0-6211-1	4165	800	37-605	37-918	122-602	122-85	125-85(P),125-65(S)	.025	.037	1-3/8	2
0-6212**	4165	800	37-606	37-919	122-63	122-86	125-85(P),125-65(S)	.025	.037	1-3/8	2
0-6213**	4165	800	37-606	37-919	122-62	122-85	125-85(P),125-65(S)	.025	.037	1-3/8	2
0-6214**	4500	1150	37-608	37-914	122-95	122-95	N/R	.026	.026	2	2
0-6262	4165	800	37-605	37-918	122-62	122-85	125-85(P),125-65(S)	.025	.037	1-3/8	2
0-6270-1	4160	600	37-397	N/A	122-64	34R-6153-39	25R-237-85	.032	Orange	1-9/16	1-9/16
0-6291	4160	600	37-119	N/A	122-62	134-39	125-85	.031	Purple	1-5/16	1-9/16
0-6299-1	4160	390	37-654	37-911	122-50	134-34	N/A	.025	Plain	1-7/16	1-7/16
0-6425	2300	650	37-656	37-903	122-82	N/R	125-65	.031	N/R	1-3/4	N/R
0-6464**	4500	1050	37-487	37-920	122-88	122-88	N/R	.035	.035	2	2
0-6468-1	4165	650	37-605	37-918	122-60	122-83	125-85	.025	.037	1-3/8	2
0-6468-2	4165	650	37-605	37-918	122-602	122-83	125-85	.025	.037	1-3/8	2
0-6497	4165	650	37-605	37-918	122-582	122-602	125-85(P),125-65(S)	.025	.037	1-3/8	1-3/8
0-6498	4165	650	37-605	37-918	122-592	122-602	125-85(P),125-65(S)	.025	.037	1-3/8	2

* Indicates Center Inlet Dual Feed Bowls; ** Indicates Dual Accel. Pump & Dual Feed Bowls; N/YA Indicates Not Yet Available; N/A Indicates Not Available; N/R Indicates Not Required

Holley Numerical Listing

Carburetor Part No.	Carburetor Model Number	CFM	Renew Kit	Trick Kit	Primary Main Jet	Secondary Main Jet or Plate	Power Valves	Primary Discharge Nozzle	Secondary Nozzle Or Spring Color	Primary Throttle Bore Diameter	Secondary Throttle Bore Diameter
0-6499**	4165	650	37-606	37-919	122-60	122-63	125-85(P),125-65(S)	.025	.037	1-3/8	2
0-6512	4165	650	37-605	37-918	122-60	122-60	125-85(P),125-65(S)	.025	.037	1-3/8	2
0-6520	4160	600	37-119	N/A	122-62	134-39	125-85	.031	Purple	1-9/16	1-9/16
0-6528	4165	650	37-605	37-918	122-61	122-60	125-85(P),125-65(S)	.025	.037	1-3/8	2
0-6619-1	4160	600	37-720	37-912	122-642	134-39	125-65	.031	Black	1-9/16	1-9/16
0-6708**	4150	650	37-1272	32-925	122-552	122-752	125-65(P),125-35(S)	.025	.037	1-1/2	1-3/4
0-6708-1**	4150	650	37-1272	32-925	122-542	122-85	125-65	.025	.037	1-1/2	1-3/4
0-6709**	4150	750	37-1272	32-925	122-652	122-76	125-65(P),125-35(S)	.025	.037	1-1/2	1-3/4
0-6710**	4165	800	37-606	37-919	122-63	122-86	125-65(P),125-35(S)	.025	.037	1-3/8	2
0-6711	4165	650	37-605	37-918	122-602	122-632	125-65(P),125-35(S)	.025	.028	1-3/8	2
0-6772	4165	650	37-605	37-918	122-592	122-602	125-85(P),125-65(S)	.025	.040	1-3/8	2
0-6773	4165	650	37-605	37-918	122-592	122-602	125-85(P),125-65(S)	.025	.040	1-3/8	2
0-6774	4165	650	37-605	37-918	122-572	122-602	125-85(P),125-65(S)	.025	.037	1-3/8	2
0-6853	4165	650	37-605	37-918	122-62	122-60	125-85(P),125-65(S)	.025	.037	1-3/8	2
0-6895	4150	390	37-739	N/A	122-50	122-62	125-85	.025	.025	1-7/16	1-7/16
0-6909	4160	600	37-119	37-912	122-622	134-39	125-65	.031	Black	1-9/16	1-9/16
0-6910**	4165	800	37-606	37-919	122-612	122-86	125-85(P),125-65(S)	.025	.037	1-3/8	2
0-6919	4160	600	37-1415	37-912	122-622	134-39	125-206	.031	Black	1-9/16	1-9/16
0-6946-1	4160	600	3-1016	N/A	122-612	134-41	25R-609-13A	.025	Plain	1-1/2	1-1/2
0-6947	4160	600	3-1011	N/A	122-612	134-41	25R-475-18A	.025	Plain	1-1/2	1-1/2
0-6979	4160	600	37-747	37-912	122-642	134-39	125-85	.031	Black	1-9/16	1-9/16
0-6979-1	4160	600	37-747	37-912	122-642	134-39	125-208	.031	Black	1-9/16	1-9/16
0-6989	4160	600	37-1415	37-912	122-622	134-39	125-206	.031	Black	1-9/16	1-9/16
0-7001	4165	650	37-743	37-918	122-582	122-602	25R-475A-13,125-65	.025	.037	1-3/8	2
0-7002-1	4175	650	37-732	37-922	122-582	134-21	125-85	.025	Black	1-3/8	2
0-7004-1	4175	650	37-741	37-922	122-562	134-45	125-212	.025	Plain	1-3/8	2
0-7004-2	4175	650	37-741	37-922	122-542	134-50	125-211	.025	Plain	1-3/8	2
0-7005-1	4175	650	37-741	37-922	122-562	134-45	125-212	.025	Plain	1-3/8	2
0-7005-2	4175	650	37-741	37-922	122-542	134-50	25R-619A-16	.025	Plain	1-3/8	2
0-7706-1	4175	650	37-741	37-922	122-562	134-45	125-212	.025	Plain	1-3/8	2
0-7006-2	4175	650	37-741	37-922	122-542	134-50	125-211	.025	Plain	1-3/8	2
0-7009-1	4160	600	37-1415	37-912	122-622	134-39	125-206	.031	Black	1-9/16	1-9/16
0-7010*	4160	780	37-740	37-925	122-662	134-42	125-65	.025	Black	1-1/2	1-3/4
0-7053-1	4160	600	37-119	N/A	122-632	134-39	125-85	.031	Purple	1-9/16	1-9/16
0-7054	4165	650	37-605	37-918	122-592	122-602	125-85(P),125-65(S)	.025	.037	1-3/8	2
0-7154	4160	600	37-119	N/A	122-62	134-43	125-85	.031	Purple	1-9/16	1-9/16
0-7320**	4500	1150	37-487	37-920	122-95	122-95	N/A	.031	.035	2	2
0-7351	4175	650	37-741	37-922	122-592	134-21	25R-475-13	.037	Black	1-3/8	2
0-7397	4175	650	37-741	37-922	122-582	134-21	25R-475-13	.037	Black	1-3/8	2
0-7410	4150	340	37-739	N/A	122-50	122-62	125-85	.025	.025	1-7/16	1-7/16
0-7411	4150	370	37-739	N/A	122-50	122-62	125-85	.025	.025	1-7/16	1-7/16
0-7413	4160	600	37-119	N/A	122-632	134-39	125-85	.031	Purple	1-9/16	1-9/16
0-7448	2300	350	37-749	37-901	122-61	N/A	125-85	.031	N/R	1-1/2	N/R
0-7850	4160	600	37-830	N/A	122-622	134-39	125-85	.031	Plain	1-9/16	1-9/16
0-7855	4175	650	37-741	N/A	122-562	134-45	125-212	.028	Plain	1-3/8	2
0-7985	4160	600	37-840	37-912	122-632	134-39	125-208	.031	Black	1-9/16	1-9/16
0-7986	4160	600	37-840	37-912	125-652	134-39	125-208	.031	Black	1-9/16	1-9/16
0-7987	4160	600	37-840	37-912	122-612	134-39	125-208	.031	Black	1-9/16	1-9/16
0-8004	4160	600	37-840	37-912	122-632	134-39	125-208	.031	Black	1-9/16	1-9/16
0-8005	4160	600	37-840	37-912	122-622	134-39	125-208	.031	Black	1-9/16	1-9/16
0-8006	4160	600	37-840	37-912	122-622	134-39	125-208	.031	Black	1-9/16	1-9/16
0-8007	4160	390	37-720	37-912	122-51	134-34	125-65	.025	Plain	1-7/16	1-7/16
0-8059	4175	650	37-741	N/A	122-582	134-21	25R-475A-13	.037	Black	1-3/8	2
0-8059-1	4175	650	37-741	N/A	122-582	134-49	125-211	.025	Black	1-3/8	2
0-8060	4175	650	37-741	N/A	122-582	134-21	25R-475A-15	.037	Black	1-3/8	2
0-8060-1	4175	650	37-741	N/A	122-582	134-49	125-24	.025	Black	1-3/8	2
0-8082**	4500	1050	37-487	37-920	122-84	122-84	125-65	.035	.035	2	2
0-8082-1**	4500	1050	37-487	37-920	122-88	122-88	125-65 (Both)	.035	.035	2	2
0-8156**	4150	750	37-485	37-910	122-70	122-83	125-65	.028	.031	1-11/16	1-11/16
0-8162	4150	850	37-485	37-916	122-80	122-80	125-65	.031	.031	1-3/4	1-3/4
0-8181	4160	600	37-1415	N/A	122-80	122-80	125-65 (Both)	.031	.031	1-3/4	1-3/4
0-8207	4160	600	37-830	N/A	122-622	134-39	125-85	.031	Plain	1-9/16	1-9/16
0-8276	4175	650	37-732	N/A	122-572	134-21	125-85	.025	Black	1-3/8	2
0-8302	4175	650	37-732	N/A	122-582	134-21	125-85	.025	Black	1-3/8	2
0-8546	4175	650	37-732	N/A	122-582	134-21	125-85	.025	Black	1-3/8	2

* Indicates Center Inlet Dual Feed Bowls; ** Indicates Dual Accel. Pump & Dual Feed Bowls; N/YA Indicates Not Yet Available; N/A Indicates Not Available; N/R Indicates Not Required

Holley Numerical Listing

Carburetor Part No.	Carburetor Model Number	CFM	Renew Kit	Trick Kit	Primary Main Jet	Secondary Main Jet or Plate	Power Valves	Primary Discharge Nozzle	Secondary Nozzle Or Spring Color	Primary Throttle Bore Diameter	Secondary Throttle Bore Diameter
0-8679	4175	650	37-732	37-922	122-592	134-27	125-85	.025	Plain	1-3/8	2
0-8700	4175	650	37-732	37-922	122-582	134-21	125-85	.025	Black	1-3/8	2
0-8804**	4150	830	37-485	37-916	122-80	122-80	125-65 (Both)	.028	.028	1-11/16	1-11/16
0-8879	4175	650	37-732	37-922	122-592	134-21	125-65	.025	Black	1-3/8	2
0-8896**	4500	1050	37-487	37-920	122-88	122-88	N/R	.035	.035	2	2
0-9002	4160	600	37-1176	N/A	122-632	134-37	125-208	.031	Black	1-9/16	1-9/16
0-9040	4160	600	37-1177	N/A	122-661	134-23	125-211	.031	Plain	1-9/16	1-9/16
0-9188	4150	780	37-1184	37-915	122-72	122-76	125-65(P),125-85(S)	.025	Plain	1-11/16	1-11/16
0-9210	4160	600	37-1176	N/A	122-612	134-39	125-208	.031	Black	1-9/16	1-9/16
0-9219	4160	600	37-1176	N/A	122-632	134-39	125-208	.031	Black	1-9/16	1-9/16
0-9254	4160	600	37-1275	N/A	122-622	134-39	125-211	.031	Black	1-9/16	1-9/16
0-9375**	4500	1050	37-487	37-920	122-92	122-92	N/R	.035	.035	2	2
0-9377**	4500	1150	37-487	37-920	122-94	122-94	N/R	.035	.035	2	2
0-9379**	4150	750	37-485	37-910	122-68	122-81	125-65	.028	.031	1-11/16	1-11/16
0-9380**	4150	850	37-485	37-916	122-78	122-78	125-65 (Both)	.031	.031	1-3/4	1-3/4
0-9381**	4150	830	37-485	37-916	122-78	122-78	125-65 (Both)	.028	.028	1-11/16	1-11/16
0-9626	4160	600	37-1274	N/A	122-612	134-39	125-206	.031	Black	1-9/16	1-9/16
0-9645	4150	750	37-1312	37-923	122-80	122-80	125-165	.045	.045	1-11/16	1-11/16
0-9646	4150	850	37-1312	37-923	122-92	122-92	125-165	.045	.045	1-3/4	1-3/4
0-9647	2300	500	37-1311	37-924	122-81	N/R	125-45	.040	N/R	1-11/16	N/R
0-9776	4160	450	37-1321	N/A	122-582	134-6	125-85	121-31	N/R	1-1/2	1-1/2
0-9834	4160	600	37-720	37-912	122-642	134-39	125-65	.031	Black	1-9/16	1-9/16
0-9834-1	4160	600	37-720	37-912	122-661	134-39	125-65	.031	Black	1-9/16	1-9/16
0-9834-2	4160	600	37-720	37-912	122-68	134-39	125-65	.031	Black	1-9/16	1-9/16
0-9895	4175	650	37-741	N/A	122-592	134-21	25R-475-13	.037	Black	1-3/8	2
0-9923	4175	650	37-741	N/A	122-542	134-50	125-211	.025	Black	1-3/8	2
0-9948	4175	650	37-741	N/A	122-563	134-50	125-211	.025	Black	1-3/8	2
0-9976	4175	650	37-741	N/A	122-582	134-49	125-211	.025	Black	1-3/8	2
0-80073	4175	650	37-1421	N/A	122-642	134-52	125-213	.037	Black	1-3/8	2
0-80095	2305	500	37-749	N/A	122-55	122-73	125-85	.035	.028	1-11/16	1-11/16
0-80098	4180	600	37-1346	N/A	122-612	34R-6153-61	125-215	.028	Purple	1-9/16	1-9/16
0-80099	4180	600	37-1346	N/A	122-622	34R-6153-61	125-218	.028	Purple	1-9/16	1-9/16
0-80111	4180	600	37-1346	N/A	122-612	34R-6153-30	125-216	.028	Orange/Red	1-9/16	1-9/16
0-80112	4180	600	37-1346	N/A	122-622	34R-6153-30	125-217	.028	Orange/Red	1-9/16	1-9/16
0-80120	2305	350	37-749	N/A	122-52	122-65	125-85	.035	.028	1-11/16	1-11/16
0-80128	4175	650	37-741	N/A	122-582	134-50	125-211	.031	White	1-3/8	2
0-80133	4180	600	37-1346	N/A	122-611	34R-6153-30	125-216	.028	Orange/Red	1-9/16	1-9/16
0-80134	4180	600	37-1346	N/A	122-612	34R-6153-30	25R-609-8A	.028	Brown	1-9/16	1-9/16
0-80135	4180	600	37-1346	N/A	122-612	34R-6153-30	25R-609-8A	.028	Brown	1-9/16	1-9/16
0-80136	4180	600	37-1346	N/A	122-612	34R-6153-30	25R-609-8A	.028	Brown	1-9/16	1-9/16
0-80137	4180	600	37-1346	N/A	122-612	34R-6153-30	25R-609-8A	.028	Brown	1-9/16	1-9/16
0-80139	4175	600	37-741	N/A	122-592	134-21	25R-475-13A	.037	Black	1-3/8	2
0-80140	4175	650	37-1421	N/A	122-642	134-52	125-213	.037	Black	1-3/8	2
0-80145	4150	600	37-1184	N/A	122-68	122-70	125-65	.031	Plain	1-9/16	1-9/16
0-80155	4175	650	37-741	N/A	122-632	134-21	25R-475-13A	.037	Black	1-3/8	2
0-80163	4180	600	37-1346	N/A	122-622	34R-6153-30	25R-609-8A	.028	Purple	1-9/16	1-9/16
0-80164	4180	600	37-1422	N/A	122-612	34R-6153-30	25R-609-8A	.028	Brown	1-9/16	1-9/16
0-80165	4180	600	37-1346	N/A	122-612	34R-6153-30	25R-609-8A	.028	Pink	1-9/16	1-9/16
0-80166	4180	600	37-1346	N/A	122-612	34R-6153-30	25R-609-8A	.028	Pink	1-9/16	1-9/16
0-80169	4175	650	37-741	N/A	122-543	134-53	125-211	.025	Black	1-3/8	2
0-80186	4500	750	37-487	37-920	122-70	122-70	125-65	.035	.035	2	2
0-80431	4160	550	37-119	37-912	N/YA	N/YA	N/YA	N/A	N/A	1-1/2	1-1/2
0-80432	4160	550	37-119	37-912	N/YA	N/YA	N/YA	N/A	N/A	1-1/2	1-1/2
0-80436	4150	850	N/YA	N/YA	N/YA	N/YA	N/YA	N/A	N/R	1-3/4	1-3/4
0-81850	4160	600	37-119	37-912	122-66	134-9	125-65	.025	Plain	1-9/16	1-9/16
0-83310	4160	750	37-754	37-921	122-72	134-21	125-65	.025	Plain	1-11/16	1-11/16
0-83311	4160	750	37-754	37-921	122-72	134-21	125-65	.025	Plain	1-3/4	1-3/4
0-83312	4160	750	37-754	37-921	122-72	134-21	125-65	.025	Plain	1-3/4	1-3/4
0-84412	2300	500	37-474	37-901	122-73	N/R	125-50	.028	N/R	1-11/16	N/R
0-84778	4150	700	37-485	37-910	122-69	122-78	125-65	.028	.031	1-11/16	1-11/16
0-84779	4150	750	37-485	37-910	122-70	122-73	125-65	.028	.031	1-11/16	1-11/16
0-84780	4150	800	37-485	37-910	122-71	122-85	125-65	.031	.031	1-11/16	1-11/16
0-84781	4150	850	37-485	37-916	122-80	122-78	125-65	.031	.031	1-3/4	1-3/4
0-87448	2300	350	37-749	37-901	122-61	n/a	125-85	.031	N/R	1-1/2	N/R
0-89834	4160	600	37-720	37-912	122-68	134-39	125-65	.031	Black	1-9/16	1-9/16

* Indicates Center Inlet Dual Feed Bowls; ** Indicates Dual Accel. Pump & Dual Feed Bowls; N/YA Indicates Not Yet Available; N/A Indicates Not Available; N/R Indicates Not Required

Carburetor Removal

Location Of Fuel Stains Or Other Removal Notes:

____

____

____

____

____

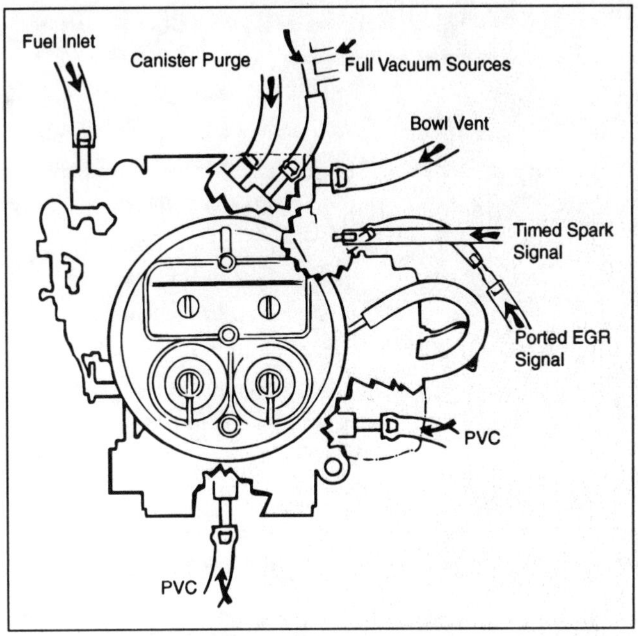

Holley List Number ____________

Enter Information From Holley Numerical Listing Below (see pages 120-122):

Model________ CFM________ *Renew Kit*________

Trick Kit ________ Primary Main Jet ________

Secondary Main Jet or Jet Plate Number________

Power Valves ________ Primary Discharge Nozzle________ Secondary Nozzle or Spring Color________

Primary Throttle Bore Diameter________ Secondary Throttle Bore Diameter________

Carburetor Disassembly

Primary Jet Size: Left (throttle side)________ Right (choke side)________

Secondary Jet Size: Left (throttle side)________ Right (choke side)________

Sec. Metering Plate: Secondary Metering Plate Number________ Main Body Surface Had Cavity: ☐ Yes ☐ No

Support Plate And Thin Gasket Used: ☐ Yes ☐ No

Manual Choke Cable Bracket Or Auto Choke Ground Wire Location (circle screws used to attach these accessories):

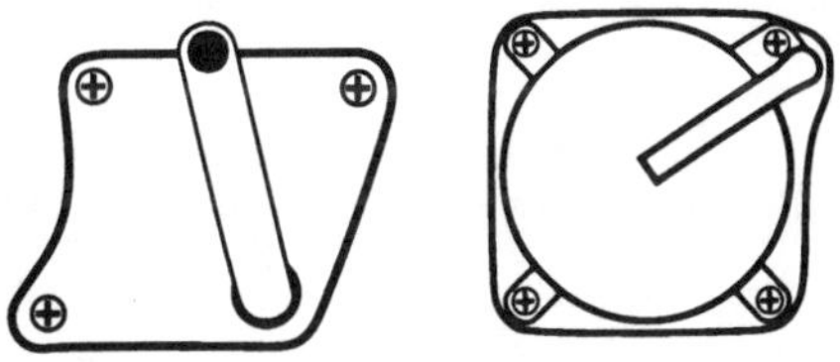

Choke Removal Notes:

____

____

____

Auto Choke Cover Setting (circle choke setting found):

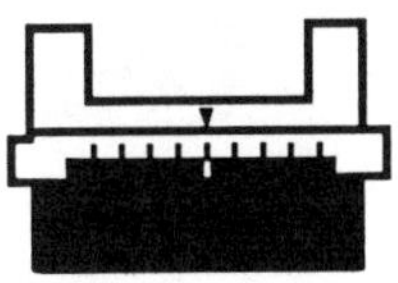

Vacuum Passage Used For Choke (circle passage):

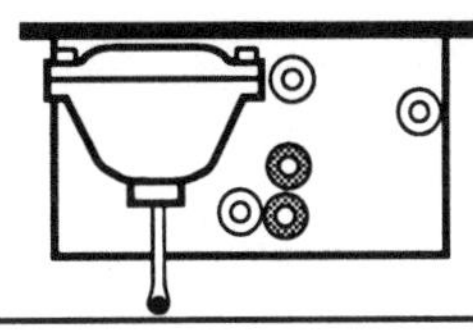

Was Gasket In Place: ☐ Yes ☐ No

Throttle Plate Screws (circle those installed):

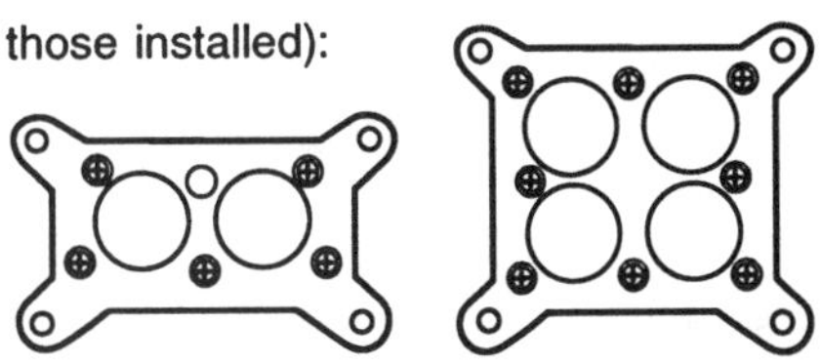

Accelerator Pump Circuit

<u>Pump Cams</u>

Primary: ☐ 30cc ☐ 50cc No.________ Screw Location: ☐ 1 ☐ 2 ☐ 3

Secondary: ☐ 30cc ☐ 50cc No.________ Screw Location: ☐ 1 ☐ 2 ☐ 3

<u>Discharge Nozzles</u>

Primary: Number________ Type: ☐ Straight ☐ Tube ☐ Anti-pullover

Check Valve Under Nozzle: ☐ Yes ☐ No

Secondary: Number________ Type: ☐ Straight ☐ Tube ☐ Anti-pullover

Check Valve Under Nozzle: ☐ Yes ☐ No

Power Valves

Pri.: Number________

☐ Window ☐ Holes ☐ 2-Stage

Sec.: Number________

☐ Window ☐ Holes ☐ 2-Stage

Secondary Diaphragm

Spring Color________

☐ Restriction ☐ Check Ball ☐ Neither

Idle Screw Turns To Bottoming

Primary: Left (throttle side) Turns______ Right (choke side) Turns______

Secondary: Left (throttle side) Turns______ Right (choke side) Turns______

Carburetor Disassembly (Continued)

Fuel Bowl Specifications

❑ Center Pivot Type: ❑ Single Inlet Per Bowl ❑ Dual Inlets Per Bowl; Fuel Plugs on: ❑ Throttle side ❑ Choke Side

❑ Side Pivot Type: ❑ Single Fuel Inlet For Carburetor ❑ Dual Inlets For Carburetor

Accelerator Pump: Primary: ❑ 30cc Pump ❑ 50cc Pump Secondary: ❑ 30cc Pump ❑ 50cc Pump

Pump Check Valve: Primary: ❑ Red "Umbrella" ❑ White "Cup" ❑ Steel Check Ball
Secondary: ❑ Red "Umbrella" ❑ White "Cup" ❑ Steel Check Ball

Inlet Filter Type: Primary: ❑ Wire Screen ❑ Sintered Bronze Element ❑ Paper Element ❑ None
Secondary: ❑ Wire Screen ❑ Sintered Bronze Element ❑ Paper Element ❑ None

Needles-and-Seats: Orifice Diameter________________ ❑ Holley Type ❑ W/Damper Spring ❑ Non-Holley Type
❑ Internally Adjustable ❑ Externally Adjustable

Float Type: ❑ Brass ❑ "Duracon" White Plastic ❑ Nitrophyl

Float Weight: Primary__________grams Secondary__________grams

Fuel Bowl Notes:____

Carburetor Reassembly Or Tuning

Note: Enter values and specifications here if different from those recorded during disassembly.

Carburetor Kit Specifications

Kit Manufacturer__________________________________ Part Number________________

Fuel Bowl Specifications

❑ Center Pivot Type: ❑ Single Inlet Per Bowl ❑ Dual Inlets Per Bowl—Fuel Plugs on: ❑ Throttle side ❑ Choke Side

❑ Side Pivot Type: ❑ Single Fuel Inlet For Carburetor ❑ Dual Fuel Inlets For Carburetor

Accelerator Pump: Primary Diaphragm/Cover: ❑ 30cc ❑ 50cc Secondary Diaphragm/Cover: ❑ 30cc ❑ 50cc

Pump Check Valve: Primary: ❑ Red "Umbrella" ❑ White "Cup" ❑ Steel Check Ball
Secondary: ❑ Red "Umbrella" ❑ White "Cup" ❑ Steel Check Ball

Inlet Filter Type: Primary: ❑ Wire Screen ❑ Sintered Bronze Element ❑ Paper Element ❑ None
Secondary: ❑ Wire Screen ❑ Sintered Bronze Element ❑ Paper Element ❑ None

Needles-and-Seats: Orifice Diameter________________ ❑ Holley Type ❑ W/Damper Spring ❑ Non-Holley Type
❑ Internally Adjustable ❑ Externally Adjustable

Float Type/Weight: ❑ Brass ❑ "Duracon" White Plastic ❑ Nitrophyl Primary____grams Secondary____grams

Float Height: Primary__________________ Measured From: ❑ Heal ❑ Toe ❑ Center
Secondary__________________ Measured From: ❑ Heal ❑ Toe ❑ Center

Secondary Diaphragm

Spring Color____________________

❑ Restriction ❑ Check Ball ❑ Neither

Power Valves

Primary: Number____________ Secondary: Number____________

❑ Window ❑ Holes ❑ 2-Stage ❑ Window ❑ Holes ❑ 2-Stage

Accelerator Pump Circuit

Pump Cams

Primary: ❑ 30cc ❑ 50cc No.________ Screw Location: ❑ 1 ❑ 2 ❑ 3

Secondary: ❑ 30cc ❑ 50cc No.________ Screw Location: ❑ 1 ❑ 2 ❑ 3

Discharge Nozzles

Primary: Number_________ Type: ❑ Straight ❑ Tube ❑ Anti-pullover
Check Valve Under Nozzle: ❑ Yes ❑ No

Secondary: Number_________ Type: ❑ Straight ❑ Tube ❑ Anti-pullover
Check Valve Under Nozzle: ❑ Yes ❑ No

Primary Jet Size:

Left (throttle side)____________

Right (choke side)____________

Secondary Jet Size:

Left (throttle side)____________

Right (choke side)____________

Sec. Metering Plate:

Plate Number____________

Miscellaneous Notes:____

Recommended Torque Range For For Specific Holley Carburetor Fasteners (Dry Torque)		
Steel Fasteners		
Fastener Size—Threads Per Inch	**Application**	**Torque Range Minimum—Maximum Inch-Pounds**
12—24	Bowl Screws	25—30
1/4—32	Main Jets	30—40
1/4—32	Adjustable Needle-And-Seat Lock Screws	50—60
5/16—24	Carburetor To Manifold Attaching Screws	60—80
5/16—24	Internally Adjustable Needles-And-Seats	60—80
7/16—24	Internally Adjustable Needles-And-Seats	70—90
1/2—28	Power Valves With Multiple Drilled Holes	40—50
1/2—28	Power Vavles With Large Window Openings	100
9/16—24	Fuel Bowl Inlet Fitting	175—200
7/8—20	Fuel Bowl Inlet Fitting	200—250
Plugs		
1/16—NPT	1/16-Inch Pipe Plug	50—60
1/8—NPT	1/8-Inch Pipe Plug	100—150
1/4—NPT	1/4-Inch Pipe Plug	160—240

Thread Sizes And Tap Drills				
Thread Size	**Major Diameter (In.)**	**Root Diameter (In.)**	**Tap Drill To Produce 75% Of Full Thread**	**Tap Drill Decimal Equivalent (In.)**
6-32	.1380	.0974	36	.1065
6-40	.1380	.1055	33	.1130
8-32	.1640	.1234	29	.1360
8-36	.1640	1279	29	.1360
10-24	.1900	.1359	25	.1495
10-32	.1900	.1494	21	.1590
12-24	.2160	.1619	16	.1770
12-28	.2160	.1696	14	.1820
1/4-20	.2500	.1850	7	.2010
1/4-28	.2500	.2036	3	.2130
5/16-18	.3125	.2403	F	.2570
5/16-24	.3125	.2584	I	.2720
3/8-16	.3750	.2938	5/16	.3125
3/8-24	.3750	.3209	Q	.3320
7/16-14	.4735	.3447	U	.3680
7/16-20	.4375	.3726	25/64	.3906
1/2-13	.5000	.4001	27/64	.4219
1/2-20	.5000	.4351	29/64	.4531
9/16-12	.5625	.4542	31/64	.4844
9/16-18	.5625	.4903	33/64	.5156
5/8-11	.6250	.5069	17/32	.5312
5/8-18	.6250	.5528	37/64	.5781
3/4-10	.7500	.6201	21/32	.6562
3/4-16	.7500	.6688	11/16	.6875

RECOMMENDED MAXIMUM TORQUE IN FOOT-POUNDS FOR GENERAL AUTOMOTIVE APPLICATIONS SEE ABOVE FOR SPECIFIC HOLLEY FASTENERS (* Indicates torque readings in Inch Pounds)													
U.S. Standards For Bolt Dimensions			**Grade: SAE 2 Tensile: 74,000**		**Grade: SAE 5 Tensile: 120,000**		**Grade: SAE 7 Tensile: 133,000**		**Grade: SAE 8 Tensile: 150,000**		**Special Alloys 180,000+**	**U.S. Regular Wrench Sizes**	
Bolt Diameter (inch)	Threads per inch	Decimal Equivalent	Dry	Lubed	Dry	Lubed	Lubed	Dry	Dry	Lubed	Lubed /w Washers	Bolt Head	Nut
1/4	20	.250	*66 **	*50 **	8	*75 **	10	8	10	8	12	3/8	7/16
1/4	28	.250	*76 **	*56 **	10	*86 **	12	9	14	10	15	3/8	7/16
5/16	18	.3125	11	8	17	13	21	16	25	18	26	1/2	9/16
5/16	24	.3125	12	9	19	14	24	18	25	20	26	1/2	9/16
3/8	16	.375	20	15	30	23	40	30	45	35	36	9/16	5/8
3/8	24	.375	23	17	35	25	45	30	50	35	38	9/16	5/8
7/16	14	.4375	32	24	50	35	60	45	70	55	61	5/8	3/4
7/16	20	.4375	36	27	55	40	70	50	80	60	67	5/8	3/4
1/2	13	.500	50	35	75	55	95	70	110	80	105	3/4	13/16
1/2	20	.500	55	40	90	65	100	80	120	90	110	3/4	13/16
9/16	12	.5625	70	55	110	80	135	100	150	110	155	7/8	7/8
9/16	18	.5625	80	60	120	90	150	110	170	130	175	7/8	7/8
5/8	11	.625	100	75	150	110	190	140	220	170	215	15/16	1
5/8	18	.625	110	85	180	130	210	160	240	180	250	15/16	1
3/4	10	.750	175	130	260	200	320	240	380	280	340	1 1/8	1 1/8
3/4	16	.750	200	150	300	220	360	280	420	320	380	1 1/8	1 1/8
7/8	9	.875	170	125	430	320	520	400	600	460	540	1 15/16	1 1/8
7/8	14	.875	180	140	470	360	580	440	660	500	600	1 15/16	1 1/8
1	8	1.00	250	190	640	480	800	600	900	680	820	1 1/2	1 1/2
1	12	1.00	270	210	710	530	860	660	1000	740	-	1 1/2	1 1/2

Unless otherwise indicated, the torque specifications given on this page are for dry (non-lubricated) fasteners. However, you may find that lubrication allows fasteners to screw in easier and prevents thread seizure. If you decide to use lubrication, use Vaseline or petrolatum. To compensate for lubrication, torque value are often reduced by about 10 to 25%. This reduction may not be required for the small (under 1/4-inch), high-quality fasteners used on Holley carburetors. Refer to the sidebar *Fastener/Bolt Torquing Tips* on page 83 for more information about bolt torquing techniques before applying the torque specification listed on this page or elsewhere in this book.

Common Small O-Rings, Gaskets, And Balls Used On Holley Carburetors

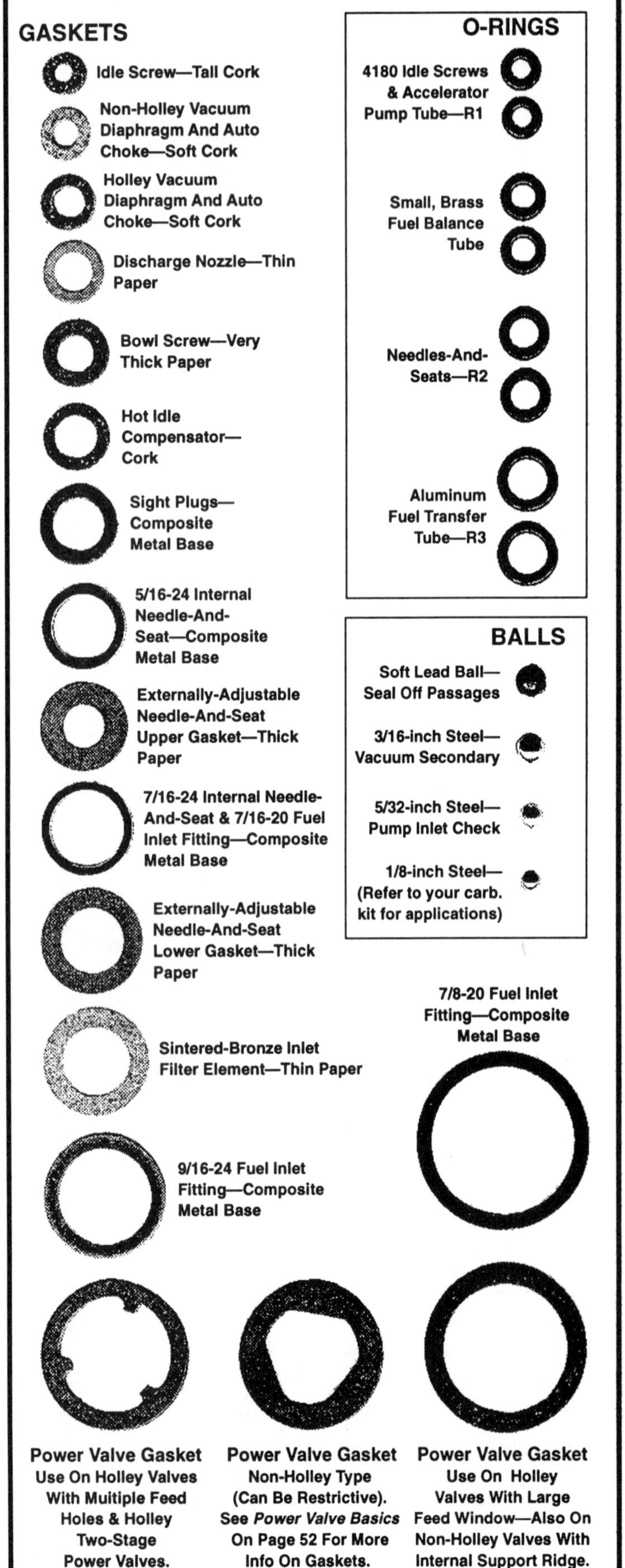

HOLLEY SECONDARY METERING PLATES LISTED BY MAIN JET SIZE			
Stamped Part #	Main Jet Orifice	Idle Orifice	Holley Part Number
7	.052	.026	134-7
34	.052	.029	134-34
3	.055	.026	134-3
4	.059	.026	134-4
32	.059	.029	134-32
40	.059	.035	134-40
5	.063	.026	134-5
18	.064	.028	134-18
30	.064	.029	134-30
13	.064	.031	134-13
33	.064	.043	134-33
8	.067	.026	134-8
23	.067	.028	134-23
16	.067	.029	134-16
9	.067	.031	134-9
36	.067	.035	134-36
6	.070	.026	134-6
19	.070	.028	134-19
20	.070	.031	134-20
41	.070	.033	134-41
35	.071	.029	134-35
39	.073	.029	134-39
37	.073	.031	134-37
17	.073	.040	134-17
10	.076	.026	134-10
22	.076	.028	134-22
43	.076	.029	134-43
12	.076	.031	134-12
53	.076	.035	134-53
28	.076	.040	134-28
38	.078	.029	134-38
52	.078	.040	134-52
11	.079	.031	134-11
24	.079	.035	134-24
44	.081	.029	134-44
49	.081	.033	134-49
21	.081	.040	134-21
31	.081	.052	134-31
29	.081	.063	134-29
46	.082	.031	134-46
25	.086	.043	134-25
47	.089	.031	134-47
5	.089	.037	134-55
27	.089	.040	134-27
26	.089	.043	134-26
4	.093	.040	134-54
15	.094	.070	134-15
50	.096	.031	134-50
45	.096	.040	134-45
14	.098	.070	134-14
42	.133	.026	134-42

Number, Decimal, & Fractional Drill Size Conversions		
Drill Number	Decimal Size	Fractional Size
80	.0135	
79	.0145	
	.0156	1/64
78	.0160	
77	.0180	
76	.0200	
75	.0210	
74	.0225	
73	.0240	
72	.0250	
71	.0260	
70	.0280	
69	.0292	
68	.0310	
	.0312	1/32
67	.0320	
66	.0330	
65	.0350	
64	.0360	
63	.0370	
62	.0380	
61	.0390	
60	.0400	
59	.0410	
58	.0420	
57	.0430	
56	.0465	
	.0469	3/64
55	.0520	
54	.0550	
53	.0595	
	.0625	1/16
52	.0635	
51	.0670	
50	.0700	
49	.0730	
48	.0760	
	.0781	5/64
47	.0785	
46	.0810	
45	.0820	
44	.0860	
43	.0890	
42	.0935	
	.0938	3/32
41	.0960	
40	.0980	
39	.0995	
38	.1015	

Vacuum Secondary Diaphragm Spring Color And Tension	
Spring Color	Spring Tension
White	Lightest
Yellow	Lighter
Yellow	Light
Purple	Medium Light
Plain (No Color)	Medium
Brown	Medium Heavy
Black	Heavy

Holley Jet Numbers And Orifice Sizes			
Jet Number	Orifice Size	Jet Number	Orifice Size
40	.040	71	.076
41	.041	72	.079
42	.042	73	.079
43	.043	74	.081
44	.044	75	.082
45	.045	76	.084
47	.047	77	.086
48	.048	78	.089
49	.048	79	.091
50	.049	80	.093
51	.050	81	.093
52	.052	82	.093
53	.052	83	.094
54	.053	84	.099
55	.054	85	.100
56	.055	86	.101
57	.056	87	.103
58	.057	88	.104
59	.058	89	.104
60	.060	90	.104
61	.060	91	.105
62	.061	92	.105
63	.062	93	.105
64	.064	94	.108
65	.065	95	.118
66	.066	96	.118
67	.068	97	.125
68	.069	98	.125
69	.070	99	.125
70	.073	100	.128

Single Stage Holley Power Valves	
Part Number	Opening Vacuum In(Hg)
Standard Flow	
125-25	2.5
125-35	3.5
125-45	4.5
125-50	5.0
125-65	6.5
125-75	7.0
125-85	8.5
125-95	9.5
125-105	10.5
High Flow	
125-165	6.5
125-1005	10.5

AUTOTRONIC CONTROLS CORP.
MSD IGNITION
1490 Henry Brennan Dr.
El Paso TX 79936
915-857-5200-Order & Tech
8:00am to 5:00pm
Retail Catalog: $2.00
An industry leader that developed its reputation for quality in multiple-spark-discharge ignition controls. Autotronic also manufacturers coils, distributors, ignition wire, ignition timing controls (automatic and manual), rev limiters, retard controls, crank-trigger systems, and many ignition accessories.

B&B PERFORMANCE
29752 Banderas
Rancho Santa Margarita CA 92688
714-589-5956--Order & Tech
8:00am to 5:00pm
Retail Catalog: $3.00
Manufactures high-performance parts for street, off-road, drag and circle-track racing. Offers cast-aluminum carburetor stands.

CARBURETOR SHOP, THE
1457 E. Philadelphia, Unit #24
Ontario CA 91761
714-947-9722-Order & Tech
8:00am to 5:00pm
Retail Catalog: Free
The Carburetor Shop specializes in performance modifications for Holley carburetors. They also offer complete rebuilding services and perform a full test (on a custom engine stand) of all completed carburetors before shipment. Also offer tune-ups, distributor service, and other custom services.

EASTWOOD COMPANY, THE
580 Lancaster Avenue
Box 296
Malvern, PA 19355-0296
800-345-1178-Order & Tech
8:00am to 9:00pm
Retail Catalog: Free
Eastwood offers a unique line of refinishing, painting, and metalworking tools and supplies. They also offer a sandblast cabinet in kit form for the home shop.

EDELBROCK CORPORATION
2700 California Street
Torrance CA 90503
213-781-2222-Order & Tech
8:00am to 5:00pm
Retail Catalog: $2.00
Edelbrock is one of the oldest names in performance. They manufacture a wide range of performance equipment for most American-made vehicles.

FORD POWER PARTS
14504-C South Carmenita
Norwalk CA 90650
213-921-5300-Order & Tech
12:00pm to 7:00pm-Tuesday—Friday
10:00am to 4:00pm-Saturday
Retail Catalog: Free
A supplier of both exotic and stock components for all Ford engines, including rare Holley carburetors and components.

GALE BANKS ENGINEERING
546 Duggan Avenue
Azusa CA 91702-5136
818-969-9605
Retail Catalog: Call
Manufacturer of high performance and turbocharged engines, systems and performance components for automotive, light truck and marine applications.

HOLLEY REPLACEMENT PARTS
11955 E. Nine Mile Road
Warren, MI 48089-2003
For technical information, ordering parts and literature call:
615-859-4924 from 8:00am to 4:30pm
Or write for a performance catalog to:
Holley Parts Division
601 Spacepark North
Goodlettsville, TN 37072
Retail Catalog: $3.00
Holley manufactures an extensive line of replacement and performance carburetors and accessories, including intake manifolds and air cleaners.

K&N ENGINEERING
561 Iowa Avenue
PO Box 1329
Riverside CA 92502
800-858-3333 &
714-684-9762-Order & Tech
8:00am to 5:00pm
Retail Catalog: $3.00
Manufactures a line of high-flow, washable air cleaners and elements for hundreds of high-performance and custom applications. Also manufactures "SubStack" high-velocity airhorns for Holley carburetors.

MALLORY IGNITION
550 Mallory Way
Carson City NV 89701
702-882-6600-Order & Tech
8:00am to 5:00pm
Retail Catalog: $4.00
Mallory manufactures a wide selection of distributors for domestic to foreign engines. Also produces coils, tachometers, gauges, ignition control boxes, spark-plug wires, and OEM ignition components.

MIKE JONES CARBURETION
7602 Talbert Avenue, #6
Huntington Beach CA 92647
714-848-5505-Order
714-848-5500-Tech
8:00am to 5:00pm
Retail Catalog: Brochure Free
Mike Jones Carburetion offers parts and services for Rochester Q-Jet, Holley, and Carter carburetors. Also offers custom designed kits to improve economy and/or performance.

MR. GASKET CO.
8700 Brookpark Road
Cleveland, OH 44121
216-398-8300-Ext. 333 or 335-Order & Tech
8:30am to 5:00pm
Retail Catalog: Free

OFFENHAUSER SALES CORP.
5300 Alhambra Ave.
PO Box 32218
Los Angeles CA 90032
213-225-1307
9:00am to 4:30pm-Order
9:00am to 12noon-Tech
Retail Catalog: $3.00
Offenhauser manufactures street and competition intake manifolds for 4- and 6-cylinder, and V8 engines from flatheads to present.

SNAP-ON TOOLS CORPORATION
28th Avenue
Kenosha, WI 53241-1410
414-656-5372-Tech
8:00am to 5:00pm
Retail Catalog: $3.00
Snap-on Tools Corporation is the largest independent manufacturer of quality hand, power, and electronic tools and test equipment.

VERICOM
Minnetonka Supply Warehouse
6008 Culligan Way
Minnetonka, MN 55345
612-933-3968-Tech
8:00am to 5:00pm
Retail Catalog: None
Manufactures electronic testing devices that easily mount to the inside of the windshield or side glass. Very useful for carburetion and engine tuning.

WEIAND AUTOMOTIVE INDUSTRIES
2316 San Fernando Road
PO Box 65977
Los Angeles CA 90065
213-225-4138-Order & Tech
8:00am to 4:30pm
Retail Catalog: $4.00
One of the original manufacturers of performance equipment, Weiand sells a long list of high-performance automotive parts and equipment, including an extensive line of intake manifolds and GMC blowers and drive kits.

About The Author

Jeff Williams was born in 1950; the beginning of Detroit's golden age of performance. In those early years, there was a palpable excitement in the air; an expectation that unlimited power and speed was to come. And virtually unrestricted by emissions standards and the cost of fuel (over-100 octane gas was less than 30 cents a gallon), everyone knew that Detroit would deliver on this promise.

Jeff was an early "victim" to the passion for power, and by age twelve he successfully replaced the 2-1/2hp engine in his mini-bike with a 10hp McClough chainsaw motor. Two-wheel performance continued to fascinate Jeff through high school where he became proficient in tuning and modifying dirt and street bikes. His obvious wrenching ability soon found an outlet in helping his friends get their hot rods to run, often remarkably well. By the time he entered college, the late 60's had arrived, and it was during this time that the most exciting "iron" Detroit ever produced was rolling off the assembly line, fueling the horsepower wars to a fever pitch. Despite the typically tight budget of a first-year college student, Jeff was able to participate in this unique era by replacing the mild-mannered slant six in his 1963 Dodge Dart with a Stage-III 426cid "Ramcharger" wedge engine, easily capable of producing one horsepower per cubic inch. Many of his friends can still vividly remember the acceleration that innocuous black Dart could generate in street trim. Subsequent blasts at local southern California drag strips, including the legendary Lion's Dragstrip, produced elapsed times in the low eleven-second bracket. For four years the fastest car in the college parking lot was a black '63 Dart.

After graduating from college, Jeff continued to pursue his love for "things mechanical." When not consulting for a racing parts development company or building engines for other enthusiasts, he campaigned a 1968 Plymouth Barracuda in the now defunct AHRA stock eliminator class. During several seasons of racing, he set many national records, saw thousands of miles of America from the front seat of his truck, and made many friends. His racing successes increased the demand for his talents as an engine builder, and he eventually opened his own business where high-performance and stock engine building and machine work have been offered to thousands of satisfied customers.

One of Jeff's many talents is his mastery of explaining how complex things work or why a component must be assembled precisely in a certain way. Able now to fully exploit these skills as a *Workbench*® book editor for S-A Design, Jeff says that he can finally put his college-ripened technical writing skills to good use. But it's the lessons he learned from building that fondly-remembered black Dodge Dart, and the many race cars and engines since, that bring a conspicuous "hands-on" feel of experience to his well-received written efforts.